ELEMENTARY MATHEMATICAL MODELING

Functions and Graphs

ELEMENTARY MATHEMATICAL MODELING

Functions and Graphs

SECOND EDITION

Mary Ellen Davis
Georgia Perimeter College

C. Henry Edwards
University of Georgia

Upper Saddle River, New Jersey 07458

Library of Congress Cataloging-in-Publication Data
Davis, Mary Ellen.
  Elementary mathematical modeling : functions and graphs / Mary Ellen Davis, C. Henry Edwards. — 2nd ed.
    p. cm.
  Includes index.
  ISBN 0-13-145035-2
  1. Functions. 2. Algebra—Graphic methods. I. Edwards, C. H. (Charles Henry) II. Title.
  QA331.3.D395 2007
  511'.8—dc22   2006036631

Acquisitions Editor: Chuck Synovec
Project Manager: Michael Bell
Production Management/Composition: Techbooks
Senior Managing Editor: Linda Mihatov Behrens
Assistant Managing Editor: Bayani Mendoza de Leon
Executive Managing Editor: Kathleen Schiaparelli
Manufacturing Manager: Alexis Heydt-Long
Manufacturing Buyer: Maura Zaldivar
Director of Marketing: Patrice Jones
Marketing Manager: Wayne Parkins
Marketing Assistant: Jennifer Leeuwerk
Director of Creative Services: Paul Belfanti
Creative Director: Juan R. López
Art Director/Cover Designer: Heather Scott
Interior Designer: Maureen Eide
Art Editor: Thomas Benfatti
Editorial Assistant/Supplements Editor: Joanne Wendelken
Art Studio: LaserWords
Cover Image: ©Kennan Ward–Corbis. All rights reserved.

© 2007 by Prentice Hall, Inc.
Pearson Prentice Hall
Pearson Education, Inc.
Upper Saddle River, NJ 07458.

All rights reserved. No part of this book may be reproduced, in any form or by any means, without permission in writing from the publisher.

Pearson Prentice Hall™ is a trademark of Pearson Education, Inc.

Printed in the United States of America

10 9 8 7 6 5 4 3 2 1

ISBN 0-13-145035-2

Pearson Education, Ltd., London
Pearson Education Australia PTY. Limited, Sydney
Pearson Education Singapore, Pte., Ltd
Pearson Education North Asia Ltd, Hong Kong
Pearson Education Canada, Ltd., Toronto
Pearson Education de Mexico, S.A. de C.V.
Pearson Education – Japan, Tokyo
Pearson Education Malaysia, Pte. Ltd

Contents

PREFACE vii

1 FUNCTIONS AND MATHEMATICAL MODELS 2

1.1 Functions Defined by Tables 3
1.2 Functions Defined by Graphs 9
1.3 Functions Defined by Formulas 18
1.4 Average Rate of Change 26
Chapter 1 Review 34

2 LINEAR FUNCTIONS AND MODELS 38

2.1 Constant Change and Linear Growth 39
2.2 Linear Functions and Graphs 56
2.3 Piecewise-Linear Functions 66
2.4 Fitting Linear Models to Data 77
Chapter 2 Review 94

3 NATURAL GROWTH MODELS 98

3.1 Percentage Growth and Interest 99
3.2 Percentage Decrease and Half-Life 113
3.3 Natural Growth and Decline in the World 122
3.4 Fitting Natural Growth Models to Data 135
Chapter 3 Review 147

4 CONTINUOUS GROWTH AND LOGARITHMIC MODELS 150

4.1 Compound Interest and Continuous Growth 151
4.2 Exponential and Logarithmic Functions 165
4.3 Exponential and Logarithmic Data Modeling 181
Chapter 4 Review 194

5 QUADRATIC FUNCTIONS AND MODELS 198

5.1 Quadratic Functions and Graphs 199
5.2 Quadratic Highs and Lows 213
5.3 Fitting Quadratic Models to Data 223

Chapter 5 Review 236

6 POLYNOMIAL MODELS AND LINEAR SYSTEMS 240

6.1 Solving Polynomial Equations 241
6.2 Solving Pairs of Linear Equations—Lots of Ways! 255
6.3 Linear Systems of Equations 270
6.4 Polynomial Data Modeling 285

Chapter 6 Review 300

7 BOUNDED GROWTH MODELS 304

7.1 Limited Populations 305
7.2 Fitting Logistic Models to Data 317
7.3 Discrete Models and Chaos 330

Chapter 7 Review 344

8 TRIGONOMETRIC MODELS 348

8.1 Periodic Phenomena and Trigonometric Functions 349
8.2 Trigonometric Models and Periodic Data 366

Chapter 8 Review 375

ANSWERS TO SELECTED PROBLEMS 378

INDEX 391

Preface

This textbook is designed for an entry-level college mathematics course at the same academic level as college algebra, but is intended for students who are not necessarily preparing for subsequent courses in calculus. Our approach is based on the exploitation of graphing-calculator technology to engage students in concrete modeling applications of mathematics. The mathematical ideas of this course center on functions and their graphs ranging from linear functions and polynomials to exponential and trigonometric functions which we hope will become familiar and accesible to students who complete the course.

BRIEF DESCRIPTION

This book presents an introduction to mathematical modeling based on the use of elementary functions to describe and explore real-world data and phenomena. It demonstrates graphical, numerical, symbolic, and verbal approaches to the investigation of data, functions, equations, and models. We emphasize interesting applications of elementary mathematics together with the ability to construct useful mathematical models and analyze them critically and to communicate quantitative concepts effectively. In short, this is a textbook for

- a graphing technology intensive course that is
- an alternative to the standard college algebra course, and is
- solidly based on functions, graphs, and data modeling.

RATIONALE FOR "REFORMED" COURSE

The content of the traditional college algebra course is defined largely by the paper-and-pencil skills (mainly symbolic manipulation) that are needed by students whose curricula point them towards a subsequent calculus course. However, many of the students in a typical college algebra course are not really headed for calculus or never make it there. For too many of these students, college algebra consists in revisiting the skills and concepts, either mastered or not, that were "covered" in several previous mathematics courses. This experience leaves students with little enhancement of the quantitative skills they most need for their subsequent studies. It is a missed opportunity for them to begin college with a useful mathematics course that is interesting both to them and to instructors and that offers a solid chance for progress and success.

There is wide agreement on the need for an alternative approach to fill this void. Both the National Council of Teachers of Mathematics' *Principles and Standards for School Mathematics* and the American Mathematical Association of Two Year Colleges' *Crossroads in Mathematics: Standards for Introductory College Mathematics Before Calculus* recommend that mathematics courses teach students to reason mathematically, to model real-world situations, and to make use of appropriate technologies. *Beyond Crossroads–Implementing Mathematics Standards in the First Two Years of College*, the AMATYC's latest report on mathematics education, states that "Becoming an efficient, independent problem solver should be a goal of every mathematics student. But for many students, mathematics is viewed as a 'string of procedures to be memorized, where right answers count more than right thinking.'" We offer this as a textbook that values right thinking as the best path to right answers.

The evolution of these materials began with a Web site that was originally developed (starting in 1996) to support University of Georgia students taking pilot sections of a new entry-level course centering on mathematical modeling. About 2000 students used preliminary versions of the textbook, and thousands more across the United States and abroad have used the first edition. We are gratified that students have reacted with an enthusiasm belying their typical lack of success in prior mathematical experiences.

PURPOSE AND OBJECTIVES

Our primary objective is the development of the quantitative literacy and savvy that college graduates need in order to function effectively in society and the workplace. This text exploits technology and real-world applications to motivate necessary skill development and the ability to reason and communicate mathematically, to use elementary mathematics to solve applied problems, and to make connections between mathematics and the real world.

Combining functions and graphs with data modeling, the text is based largely on the use of graphing calculator methods in lieu of traditional symbolic manipulations to solve both familiar and nonstandard problems. The focus is "mathematical modeling" and the use of elementary mathematics— numbers and measurement, algebra, geometry, and data exploration—to investigate real-world problems and questions.

As an alternative to the standard college algebra course, though at the same academic level, this course is intended for students who are not necessarily headed for calculus-based curricula, but still need a solid quantitative foundation both for subsequent studies and for life as educated citizens and workers. Graphing technology enables these students to experience the power of mathematics and to enjoy success in solving interesting and significant problems, an experience that they all too rarely enjoy in traditional college algebra courses.

WHAT'S NEW IN THE SECOND EDITION

Each chapter begins with a "real-world" example that sets the stage for the type of function discussed in the chapter. The example is revisited in the chapter review exercises, in which the student applies the skills and concepts learned in the chapter to the opening example. The examples and exercises contain a wider variety of applications, which are based on current data sources. Exercises for each section are presented at two levels, the more routine ones "Building Your Skills" and subsequent ones "Applying Your Skills."

Chapter 1 provides a detailed introduction to functions from numerical, graphical and symbolic perspectives, with a section devoted to average rate of change and its graphical interpretation. Chapter 2 presents linear functions, and includes a new section on piecewise-linear functions. In order to distinguish clearly between change at a constant rate and change at a constant percentage rate, the chapters on exponential functions now follow Chapter 2. Quadratic functions are presented in Chapter 5, which includes a section on optimization.

These changes are based on the experiences of instructors using the first edition. We believe that this revision makes the text more student-friendly, with more concrete and accessible language and examples throughout.

CONTENT AND ORGANIZATION

The book consists of the following chapters:

1. Functions and Mathematical Models
2. Linear Functions and Models
3. Natural Growth Models
4. Exponential and Logarithmic Models
5. Quadratic Functions and Models
6. Polynomial Models and Linear Systems
7. Bounded Growth Models
8. Trigonometric Models

Most of these chapters fit a single pattern:

- The first section is a low-key introduction to the type of function to be used as a mathematical model throughout the chapter.
- The next section or two illustrate real-world applications of this new function.
- The final section of the chapter is devoted to data modeling using this type of function.

For example, in Chapter 3:

- Section 3.1 begins with the concept of constant-percentage growth and exponential functions of the form

$$A(t) = A_0(1 + r)^t \tag{1}$$

  which models the amount at time t in an account with annual growth rate r and initial investment A_0 (e.g., an investment of $A_0 = 1000$ with $r = 0.12$ for 12% annual growth).
- Section 3.2 focuses on constant-percentage decline, and Section 3.3 illustrates a wide variety of real-world growth and decay problems. These sections deal with such prototypical questions as the half-life of a declining quantity and how long is required for an initial investment to double. Determining when the aforementioned investment would double calls for solving the equation

$$1000 \times 1.12^t = 2000. \tag{2}$$

In a traditional course one might use logarithms, but here we can simply use a graphing calculator's intersection-finding capability to locate the intersection of the two graphs $y = 1000 \times 1.12^x$ and $y = 2000$, and thereby see that this takes about 6.12 years:

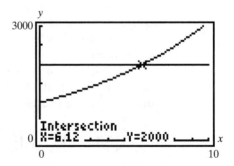

The conceptual core of this problem is the realization that Equation (2) must be solved, rather than the particular method used to do this. Essentially all students can use graphing technology successfully for this purpose, including many who might fail to use logarithms correctly. Moreover, graphical solution is a much more widely applicable approach than the use of logarithms.

- Section 3.4, the final section of Chapter 3 is devoted to the problem of choosing the values of the parameters A_0 and r in Equation (1) so that the resulting natural growth function best fits given data. Again, this is done using graphing calculator facilities.

As one chapter follows another, each developing the same theme with a new class of functions, the story of mathematical modeling comes to be a familiar one. Contrast this with the traditional algebra course, which many students perceive as a sequence of unrelated topics. In pilot sections, students selected mainly because of low placement test scores exhibited success rates not frequently seen in traditional college algebra courses.

APPLICATIONS

The use of real-world applications is vital for making mathematics more lively and interesting to students and in helping them to see connections with the world around them. This text begins with a simple but very real example of a function that is taken from the menu of the popular Waffle House Restaurant chain. Throughout the book, combinations of simulated data and real (sourced) data are used in examples and problems. Contrived data are often more effective in providing clear and straightforward explanations of new concepts. Once concepts have been introduced, more robust data-based applications help to make these concepts more concrete and to underscore their connections to real life.

Each chapter Review has an extended exercise revisiting the chapter-opening discussion, and each chapter concludes with a data-based Investigation that may be assigned as an individual or group project.

TECHNOLOGY

Graphing Calculators

This book assumes no technology other than student use of graphing calculators. Indeed, TI-83/84 syntax and calculator screens are shown throughout the text.

However, it is entirely possible to mix different graphing calculators in the same class—and we have done so—if the instructor is willing and prepared to discuss all of them when necessary.

Spreadsheets

In each historical era, real-world practitioners (if not teachers and academics) have always assimilated rapidly the best available technology to assist with their mathematical computations, whether it be a sandbox, an abacus, a slide rule, or a desktop calculating machine. The principal computational instrument used in today's workplace is the spreadsheet (rather than the graphing calculator). We have therefore explored the use of spreadsheets to augment and reinforce graphing techniques initially introduced with calculators.

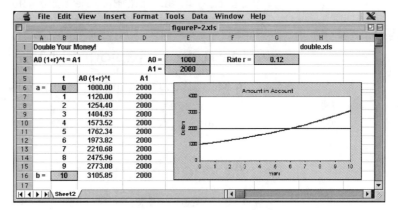

For instance, the foregoing spreadsheet image illustrates the solution of Equation (2). Each of the five shaded cells is a "live cell" whose numerical content can be changed by the student. The accompanying graph and table then change dynamically. We can therefore "zoom in" by table and by graph, vividly and simultaneously. Thus if we enter the new endpoint values $a = 6.1$ and $b = 6.2$, then the chart instantly changes as indicated in the following spreadsheet image in which we see the approximate solution $t \approx 6.12$.

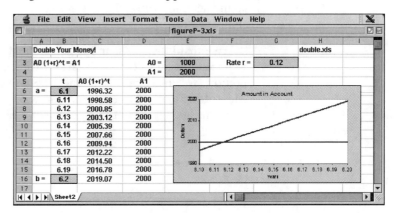

Where feasible, student use of spreadsheets not only can develop valuable familiarity with the modern world's predominant calculating technology, but it can also reinforce understanding of "solution by zooming" through comparison of several variants (ranging from tables to graphs). In particular, solution of the same problems using both graphing calculators and spreadsheets emphasizes that it is the general mathematical approach that we study, rather than the specific technology that is used to implement this approach.

The website www.prenhall.com will provide a suite of spreadsheets that can be downloaded by students and instructors for use as described.

SUPPLEMENTS

Student Solutions Manual (0-13-145036-0)

by Mary Ellen Davis and Henry Edwards

The Student Solutions Manual provides detailed solutions for the odd-numbered exercises in the text. Because these solutions were prepared by the textbook authors themselves, consistency in language and method of solution is guaranteed.

Instructor's Resource Manual (0-13-614210-9)

by Mary Ellen Davis and Henry Edwards

Solutions to the even-numbered exercises appear in the Instructor's Solutions Manual. Sample exams and other resource materials are also included.

Acknowledgments

Experienced and knowledgeable reviewers with extensive classroom teaching experience are crucial to the success of any textbook. We profited greatly from the advice, assistance, criticism, and enthusiasm of the following very fine reviewers:

Scott Adamson, *Chandler-Gilbert Community College*
Dennis C. Ebersole, *Northampton Community College*
Demetria N. Gibbs, *Gwinnett Technical College*
Marko Kranjc, *Western Illinois University*
John Gosselin, *University of Georgia*
Philip W. Gwanyama, *Northeastern Illionois University Chicago*

We thank Sally Yagan for her interest and enthusiasm for this project since its inception and Petra Recter and Chuck Synovec for overseeing the second edition. We also appreciate the fine work of our production editor, Sarvesh Mehrotra. Finally, we must mention with gratitude the understanding, patience, and constant support of George Davis and Alice Edwards.

Mary Ellen Davis
Georgia Perimeter College
mdavis@gpc.edu

C. Henry Edwards
University of Georgia
h.edwards@mindspring.com

About the Authors

MARY ELLEN DAVIS Associate Professor of Mathematics, Georgia Perimeter College, Davis received her Master of Arts degree in Mathematics from the University of Missouri-Columbia. She has taught mathematics at the secondary level in Missouri and New Mexico and at Georgia State University and the University of Birmingham (England). She joined the mathematics department at Georgia Perimeter College (then DeKalb College) in 1991, and has taught a wide range of courses from developmental mathematics through calculus and statistics. She was instrumental in the development of the college's Introduction to Mathematical Modeling course, and has been a leader in the creation of GPC's online statistics course. She was selected as a Georgia Governor's Teaching Fellow in 1996, received GPC's Cole Fellow award for outstanding teaching in 2005, and is a member of the college's inaugural class of Vice President's Teaching Scholars.

C. HENRY EDWARDS (Ph.D. University of Tennessee) Emeritus professor of mathematics at the University of Georgia, Edwards recently retired after 40 years of undergraduate classroom teaching at the universities of Tennessee, Wisconsin, and Georgia. Although respected for his diverse research interests, Edwards' first love has always remained teaching. Throughout his teaching career he has received numerous college- and university-wide teaching awards, including the University of Georgia's honoratus medal in 1983 and its Josiah Meigs award in 1991. In 1997, Edwards was the first university-level faculty recipient of the Georgia Board of Regents newly instituted state-wide award for teaching excellence. A prolific author, Edwards is co-author of well-known calculus and differential equations textbooks and has written a book on the history of mathematics, in addition to several instructional computer manuals. During the 1990s, Edwards worked on three NSF-supported projects that fostered a better integration of technology into the mathematics curriculum. The last three years of his long teaching career were devoted principally to the development of a new technology-intensive entry-level mathematics course on which this new textbook is based. Additional information is provided on his web page: www.math.uga.edu/~hedwards.

CHAPTER 1

FUNCTIONS AND MATHEMATICAL MODELS

1.1 FUNCTIONS DEFINED BY TABLES

1.2 FUNCTIONS DEFINED BY GRAPHS

1.3 FUNCTIONS DEFINED BY FORMULAS

1.4 AVERAGE RATE OF CHANGE

How hot is it *really*? Anyone who has ever visited Phoenix surely has heard the adage that the dry heat of the desert doesn't feel as hot as the temperature would suggest. Indeed, when it is 100°F in Phoenix with a relative humidity of 10%, it "feels like" it is only 95°F—a full five degrees cooler than the thermometer says!

The "heat index" is a number calculated by meteorologists to measure the effect of humidity on the apparent temperature as felt by the human body. It is designed to report how hot you actually feel at a particular combination of air temperature and relative humidity. The chart in Fig. 1.0.1 shows the heat index for various levels of relative humidity at a fixed air temperature of 85°F.

The chart indicates that when the humidity is 10% the apparent temperature is only 80°F. However, if the humidity is 90%, then a temperature of 85°F seems like 102°F. Thus the humidity has a big effect on how hot we actually feel.

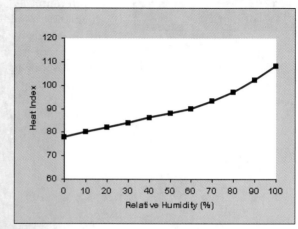

FIGURE 1.0.1

Source: National Weather Service, Buffalo, New York.

We can see from the chart that as the relative humidity increases from 0% to 60%, the apparent temperature increases uniformly from 78°F to 90°F. Notice that this portion of the chart appears to be a straight line. After we reach a relative humidity of 60%, however, the apparent temperature increases more rapidly as the relative humidity increases. That is, this portion of the chart appears to curve upward.

All around us we see quantities, like apparent temperature and relative humidity, that seem to be related to each other in some systematic way. We often describe this dependence of one quantity on another by using the word *function*. Thus, we might say that your weight is (ideally) a function of your height; that your grade on a history test is a function of how long you studied; or that your income is a function of your education level. In each case we are noting a dependence of the first quantity on the second.

In this chapter, we introduce the mathematical concept of a function, we study functions defined by tables, graphs, and rules, and we explore increasing and decreasing functions.

1.1 FUNCTIONS DEFINED BY TABLES

The menu at a local Waffle House restaurant gives the price of a breakfast of eggs, toast, jelly, and grits based on the number of eggs the customer orders, as illustrated in the following table of 2005 prices:

NUMBER OF EGGS	PRICE OF BREAKFAST
3	$2.60
2	$2.25
1	$1.80

Most applications of mathematics involve the use of numbers or *variables* to describe real-world quantities. In this situation, suppose we let n represent the number of eggs ordered and p the price of the breakfast. Then the table describes a relationship between n and p. This relationship is an example of a mathematical *function* because for each number n of eggs ordered there is a corresponding price p charged for the breakfast.

The key concept here is that there is only **one price** associated with each number of eggs. The menu, in effect, provides a "rule" for determining price: If you know how many eggs were ordered, you know the price of the breakfast. If you and your friend each ordered a 1-egg breakfast, then you would expect to be charged the same amount. Indeed, if one of you were charged more than $1.80 for your breakfast, then you surely would complain!

DEFINITION: Function
A **function** f defined on a collection D of numbers is a rule that assigns to each number x in D a specific number $f(x)$.

We will refer to the numbers in the set D as "inputs" and the corresponding $f(x)$ numbers as "outputs." In the Waffle House example, we can say that the price of the breakfast is a function of the number of eggs ordered. This means that for each input (number of eggs), there is only one output (price).

We can also look at the table "backwards," and say that the number of eggs ordered is a function of the price of the breakfast. Then each input (price) has only one output (number of eggs).

EXAMPLE 1 Is It a Function?

The following table gives the average price of a 30-second commercial airing during the Super Bowl in the indicated year.

a. Is the price of the commercial a function of the year?
b. Is the year a function of the price of the commercial?

YEAR	PRICE OF COMMERCIAL (IN MILLIONS)
1998	$1.3
2000	$2.1
2001	$2.3
2003	$2.1
2005	$2.4

Source. www.superbowl-ads.com.

SOLUTION

a. When we ask "Is the price a function of the year?" we are considering the year as the input and the price as the output. Thus, the price of the commercial *is* a function of the year because for each input (year) there is only one output (price).

b. When we ask "Is the year a function of the price?" we are considering the price as the input and the year as the output. Thus, the year *is not* a function of the price because for the input $2.1 million, there are two different outputs (2000 and 2003).

In the definition of function, the number $f(x)$—we say "f of x"—is called the *value* of the function f at the number x. The "rule" mentioned in the definition can be specified by a table, by a formula, or by a graph, or even by a verbal description that tells how the value $f(x)$ is found when the number x is given. While we frequently use x to denote the variable and f to denote the function, we can use any other letters that we like or that seem more natural in a particular situation.

When we consider the Waffle House data as defining price as a function of the number of eggs ordered, the set D is the collection of all possible numbers of eggs—the set of numbers 1, 2, and 3. Since the inputs are numbers of eggs, we will denote them by n (rather than x). Since the outputs are prices, we will denote them by $p(n)$.

n	$p(n)$
1	$1.80
2	$2.25
3	$2.60

Given a number n in the first column of the table, we simply look in the second column to find the corresponding price $p(n)$. For instance, $p(3) = 2.60$ because $2.60 is the price "assigned" to a 3-egg breakfast.

So the mathematical question "Find $p(n)$" is equivalent to the English question "If the input is n, what is the output?" It is convenient for us to think of this as the "frontward" question "Here is the input, what is the output?"

We can also ask the "backward" question "Here is the output, what is the input?" This is equivalent to the mathematical question "For what value(s) of n is $p(n) = 2.25$?" From the table, the answer is $n = 2$.

Notice that when we use this notation, the *input* value appears *inside* the parentheses. In the "frontward" question "Find $p(3)$," we are given the input 3 and asked to find the output. In the "backward" question "For what value(s) of n is $p(n) = 2.25$," we are looking for the input n.

EXAMPLE 2 | Finding Input and Output Values

Rather than keeping a "running total," the Bureau of the Public Debt uses a daily accounting method to calculate the public debt of the United States. At the end of each day, approximately 50 different agencies (such as Federal Reserve Banks) report certain financial information to the Bureau. At around 11:30 AM EST the next morning, the accounting system produces a figure for the public debt, correct to the nearest penny, for the previous day. Thus, to each date is assigned an official public debt amount, and we therefore can say that the debt depends on the day chosen. In other words, the debt is a function of the date. Here is a portion of a table reporting the public debt in February 2005.

DATE x	PUBLIC DEBT $g(x)$
02/18/2005	$7,689,935,780,269.27
02/17/2005	$7,689,847,469,266.70
02/16/2005	$7,671,700,332,790.18
02/15/2005	$7,674,137,053,033.57
02/14/2005	$7,630,849,109,540.36

a. Find $g(02/15/2005)$.
b. Find x if $g(x) = \$7,689,847,469,266.70$.

SOLUTION

From the table:

    a. $g(02/15/2005) = \$7{,}674{,}137{,}053{,}033.57$.
    b. When $g(x) = \$7{,}689{,}847{,}469{,}266.70$, $x = 02/17/2005$.

Domain and Range of a Function

When we use functions to describe relationships between variables, it is important that we know what numbers are sensible values to substitute for those variables.

> **DEFINITION: Domain**
>
> The collection (or set) of all numbers for which the number $f(x)$ is defined is called the ***domain*** (of definition) of the function f. These numbers are the input values of the function.

We saw in Example 1 that the average price of a 30-second Super Bowl ad is a function of the year it airs. Based on the table, the domain of this function is $\{1998, 2000, 2001, 2003, 2005\}$.

> **DEFINITION: Range**
>
> The set of all possible values $y = f(x)$ is called the **range** of the function.

Based on the table in Example 1, the range of the Super Bowl ad function is $\{1.3, 2.1, 2.3, 2.4\}$, where the values are in millions of dollars. Notice that it is not necessary to list 2.1 twice, because the range indicates what numbers occur as output values, regardless of how many times each one appears.

EXAMPLE 3 | Finding Domain and Range

Consider the following table giving calories from fat and milligrams of cholesterol for several sandwiches from the Subway restaurant:

SUBWAY SANDWICH	CALORIES FROM FAT x	CHOLESTEROL (mg) $C(x)$
Honey Mustard Ham	45	25
Cheese Steak	90	35
Buffalo Chicken	130	50
Italian BMT	190	55
Double Meat Cold Cut Combo	250	105

Source. www.subway.com.

This table defines milligrams of cholesterol as a function of calories from fat.

    a. What is the domain of this function?
    b. What is the range of this function?

SECTION 1.1 Functions Defined by Tables

SOLUTION

a. The domain of the function is {45, 90, 130, 190, 250}.
b. The range of the function is {25, 35, 50, 55, 105}.

In this section, we have applied the definitions of function, function value, domain, and range to data given in table form. We will continue to apply these same important definitions when we encounter functions defined by graphs or by rules.

1.1 Exercises | Building Your Skills

In Exercises 1–4, determine whether B is a function of A and then, if B is a function of A, give the domain and the range of the function.

1.

A	10	10.5	13	15	16	20
B	3	4	9	13	15	23

2.

A	10	10.5	13	15	16	20
B	3	4	4	9	13	23

3.

A	2	4	5	5	8	11
B	3	4	9	13	15	23

4.

A	2	4	6	8	10	12
B	3	3	3	3	3	3

In Exercises 5–8, find **(a)** $f(2)$ *and* **(b)** *the value(s) of x for which* $f(x) = 2$.

5.

x	0	2	4	6	8	10
$f(x)$	3	4	9	2	7	3

6.

x	−1	0	1	2	3	4
$f(x)$	−4	2	2	9	−13	23

7.

x	2	4	5	7	8	11
$f(x)$	6	4	7	13	15	23

8.

x	−4	−1	0	2	10	17
$f(x)$	2	2	2	2	2	2

Applying Your Skills

9. The total number of fat grams and calories in various chicken sandwiches are given in the following table:

Fat Grams	20	29	5	33	9	33	26	43
Calories	430	550	320	680	300	750	530	710

Source: Chick-Fil-A Gram Comparison.

a. Based on this table, are calories a function of fat grams?
b. Based on this table, are fat grams a function of calories?

10. In 2005, Kaiser Permanente advertised the following monthly rates for the Personal Advantage health plan for male subscribers in the state of Georgia:

Age (years)	1	4	18	27	35	40	44	45
Cost ($)	45	45	45	64	68	90	90	125

a. Based on this table, is cost a function of age?
b. Based on this table, is age a function of cost?

11. The College Board reports the average verbal and math SAT scores for college-bound seniors by state. The accompanying table shows averages for the year 2004:

STATE	AVERAGE VERBAL SCORE	AVERAGE MATH SCORE
Arkansas	569	555
Connecticut	515	515
Illinois	585	597
Hawaii	487	514
Texas	493	499

a. Based on this table, is average math score a function of average verbal score?
b. Based on this table, is average verbal score a function of average math score?

12. For a person who is 5 feet, 8 inches tall, the following table gives weight in pounds and corresponding body mass index (BMI). A person whose BMI is less than 18.5 is considered underweight.

Weight (pounds)	121	128	130	133	148	155	166	170	173
BMI	18.4	19.5	19.8	20.2	22.5	23.6	25.2	25.8	26.3

Source: Centers for Disease Control and Prevention.

a. Explain why, based on this table, BMI is a function of weight in pounds.
b. Explain why, based on this table, weight in pounds is a function of BMI.

c. What is the BMI for a person 5 feet, 8 inches tall who weighs 133 pounds?
d. A person whose BMI is 25.0–29.9 is considered overweight. Based on this table, for which weights would a person 5 feet, 8 inches tall be considered overweight?

13. The following table shows the year that the Dow Jones Average first reached certain milestone levels:

Dow Jones Average	500	1000	1500	3000	4000	5000	8000	10,000
Year First Attained	1956	1972	1985	1991	1995	1995	1997	1999

Source: Dow Jones Industrial Average.

a. Explain why, based on this table, year is a function of Dow Jones Average.
b. Explain why, based on this table, the Dow Jones Average is not a function of year.
c. When did the Dow Jones average first reach 3000?
d. What value(s) did the Dow Jones first attain in 1995?

14. Use a newspaper, a magazine, or the Internet to find
a. an example of a table of data that represents a function;
b. an example of a table of data that does not represent a function.

1.2 FUNCTIONS DEFINED BY GRAPHS

In Example 3 of Section 1.1, we saw that the table showing calories from fat and cholesterol for Subway sandwiches represented cholesterol $c(x)$ as a function of calories x:

Calories from Fat	45	90	130	190	250
Cholesterol (mg)	25	35	50	55	105

To graphically represent this data, we create a scatter plot, a set of points in the coordinate plane with the input values on the horizontal axis (x-axis) and the output values on the vertical axis (y-axis). Figure 1.2.1 shows a calculator scatter plot for these data with calories from fat on the horizontal axis and milligrams of cholesterol on the vertical axis.

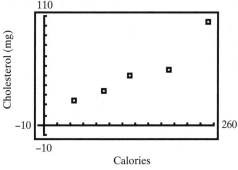

FIGURE 1.2.1 Scatter plot of cholesterol versus calories data.

Recall that our definition of function tells us that each input can have only one output. For the Subway sandwiches, there is only one point in the scatter plot for each of the input values, corresponding to the unique output value for that input.

On the other hand, the table showing fat grams and calories for chicken sandwiches (in Exercise 9 of Section 1.1) did not represent calories as a function of fat grams because the input 33 had two different outputs, 680 and 750:

Fat Grams	20	29	5	33	9	33	26	43
Calories	430	550	320	680	300	750	530	710

Figure 1.2.2 shows a scatter plot for these data with fat grams (input) on the horizontal axis and calories (output) on the vertical axis.

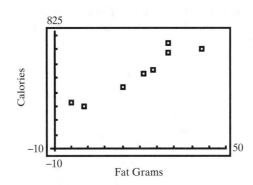

FIGURE 1.2.2 Scatter plot of calories versus fat grams data.

In this scatter plot, we see the two outputs (680 and 750) for the input 33, yielding two different points "stacked" vertically above the input 33.

This understanding that more than one output for a given input shows up on a graph as two points lying on the same vertical line is the source of the "vertical line test" for a function. In order for a graph to represent a function, any vertical line must cross the graph only once.

EXAMPLE 1 Is It a Function?

The scatter plot in Fig. 1.2.3 illustrates the population (in thousands) of St. Louis, Missouri, for the census years 1950–2000. Does this scatter plot represent a function?

FIGURE 1.2.3 Population of St. Louis.
Source: www.census.gov.

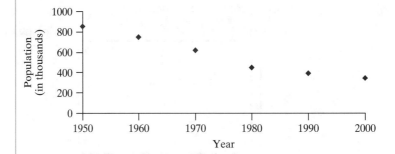

SOLUTION

This plot describes a function $P(x)$ that is defined for each of the years 1950, 1960, 1970, 1980, 1990, and 2000 because for each input (year x), the plot shows *exactly one* dot, indicating only one output (population P).

Throughout this book we will see how quantities measured over time represent functions with time as the input and the quantity measured (such as population) as the output. Using functions to make predictions of how values will change with time is an important use of mathematics. Planning in business and government for future revenue, expenditure, and resource allocation frequently depends on identifying functions to model the behavior of such quantities.

A scatter plot is a representation of a set of individual points. Frequently mathematical relationships are displayed as connected lines or curves. In these situations, we can also apply the vertical line test to determine whether a particular graph represents a function.

EXAMPLE 2 | Is It a Function?

The graph in Fig. 1.2.4 shows the percentage chance per month that a woman 15–45 years of age will become pregnant. Does this graph represent chance of conception as a function of mother's age?

FIGURE 1.2.4 Fertility odds.
Source: Atlanta Journal-Constitution.

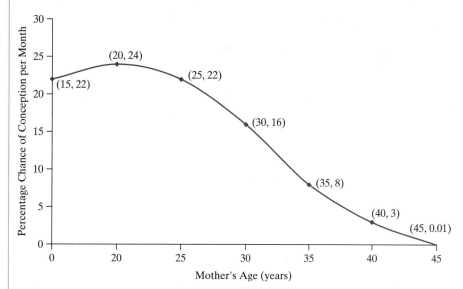

SOLUTION

We see that any vertical line would cross this graph only once. This indicates that this plot represents a function because each input (age) is associated with only one output (percentage chance of conception).

Just as we did with functions defined by tables, we can find input and output values if functions are given graphically. However, if points on the graph are not labeled or accompanied by a table of values, we can only estimate the values in question.

If we consider the fertility odds graph in Fig. 1.2.4, we can ask the "frontward" question "What is the output when the input is 20?" This question asks us to find the percentage chance of conception when the mother is 20 years old. Since age is represented on the horizontal axis, we find 20 on the horizontal axis and then find the point on the graph directly above it. This appears to be the highest point on the graph, located a bit lower than 25, perhaps 24. So we estimate that the output for an input of 20 is about 24; that is, if a woman is 20 years old, she has about a 24% chance of conceiving per month.

Similarly, suppose we are asked the "backward" question "What is the input when the output is 10?" This question asks us to find the mother's age when chance of conception is 10%. Since chance of conception is represented on the vertical axis, we find 10 on the vertical axis and find the point (or points) on the graph that lie directly to its right. This point lies a little to the left of 35, so we estimate that the input for an output of 10 is about 34. That is, if the chance of conception is 10%, the mother's age is about 34 years.

EXAMPLE 3 | Finding Input and Output Values

Figure 1.2.5 shows the number of hospitals and the percentage of beds occupied in the hospitals for several U.S. states.

a. Find the percentage of beds occupied if the number of hospitals in the state is 83.

b. Find the number of hospitals in the state if 74% of beds are occupied.

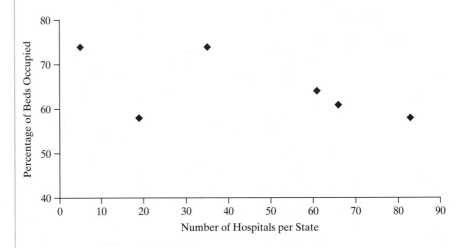

FIGURE 1.2.5 Hospital occupancy.
Source: The World Almanac and Book of Facts 2005.

SOLUTION

a. This question asks us to find the output if the input is 83. If we find 83 on the horizontal axis and look for the point directly above it, we see that the point is located slightly below 60 on the vertical axis. Thus we estimate that about 58% of hospital beds are occupied.

b. This question asks us to find the input if the output is 74. There seem to be two points lying at a height of 74. One seems to have a first coordinate of about 5; the second seems to have a first coordinate of about 35. Therefore, we say that the states with 5 or 35 hospitals have 74% of their beds occupied.

If a function is represented graphically, we can find (at least approximately) its domain and range. Recalling that the domain is the set of input values, we look at the hospital occupancy function in Fig. 1.2.5. Estimating these values from the graph, we see that the domain of this function is {5, 19, 35, 61, 66, 83}.

Similarly, since the range is the set of output values, we estimate those values from the graph and find that the range of the function is {58, 61, 64, 74}. Notice that although there are six values in the domain, the range only contains four different values. This occurs because the inputs 5 and 35 have the same output, 74, and the inputs 19 and 83 both have the output 58.

EXAMPLE 4 Finding Domain and Range

Figure 1.2.6 shows the typical distance (in feet) that a car travels after the brakes have been applied for various speeds (in miles per hour).

a. Find the domain of this function.
b. Find the range of this function.

FIGURE 1.2.6 Typical stopping distances.
Source: www.highwaycode.gov.uk.

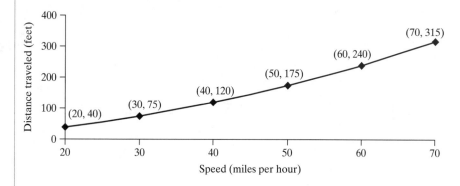

SOLUTION

a. Because the graph is given as a connected curve, the domain consists of all values that lie between the smallest input value and the largest input value, inclusive. That is, the domain contains all real numbers between 20 and 70, including both 20 and 70. Using interval notation, we write the domain as [20, 70].

b. Similarly, the range of the function consists of all values that lie between the smallest output value and the largest output value, inclusive. So the domain contains all real numbers between 40 and 315, inclusive. In interval notation, the range is [40, 315].

1.2 Exercises | Building Your Skills

In Exercises 1 and 2, make a scatter plot of the data, using A as input (on the horizontal axis) and B as output (on the vertical axis). Then explain why the scatter plot does or does not represent B as a function of A.

1.

A	6	8	9	11	13	14
B	3	6	9	6	14	0

2.

A	2	4	5	5	8	11
B	2	8	9	13	15	23

14 CHAPTER 1 Functions and Mathematical Models

In Exercises 3–6, first determine whether the graph represents a function (with input on the horizontal axis and output on the vertical axis) and then, if the graph represents a function, give its domain and range.

3.   **4.**

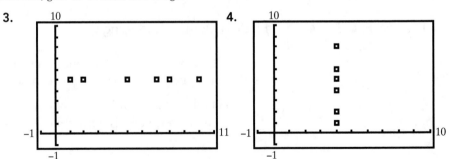

5.

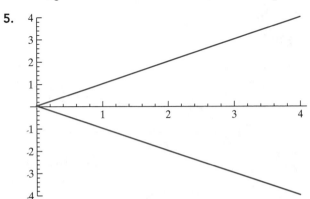

6.

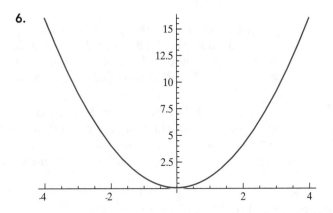

*In Exercises 7–10, let $y = f(x)$. Find **(a)** $f(2)$ and **(b)** the value(s) of x for which $f(x) = 2$. Each tick mark on the axes represents one unit.*

7.   **8.**

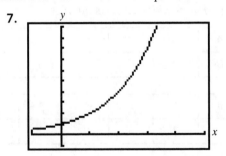 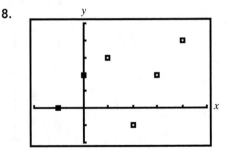

9.

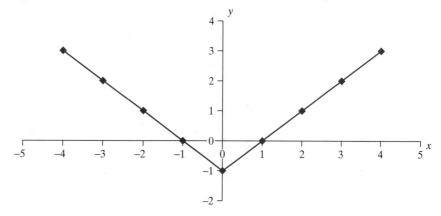

10.

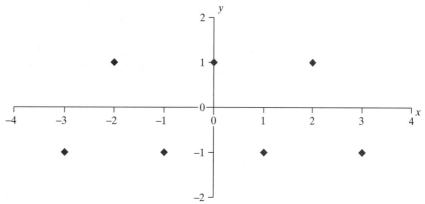

Applying Your Skills

11. The scatter plot in Fig. 1.2.15 shows the height and weight for 12 pitchers on the Texas Rangers active roster in June 2005. Based on this scatter plot, is weight a function of height? Why or why not?

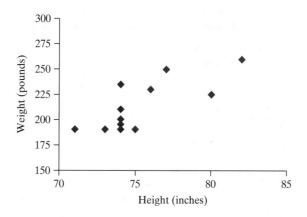

FIGURE 1.2.15 Texas Rangers pitchers.

Source: www.texasrangers.mlb.com.

12. The scatter plot in Fig. 1.2.16 shows the latitude and length of growing season for several U.S. cities. Based on this scatter plot, is length of growing season a function of latitude? Why or why not?

16 CHAPTER 1 Functions and Mathematical Models

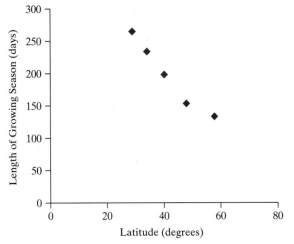

FIGURE 1.2.16 Growing season versus latitude.

Source: www.census.gov and *The Old Farmer's Almanac.*

13. The scatter plot in Fig. 1.2.17 gives the death toll for California earthquakes of various magnitudes that occurred in the twentieth century.

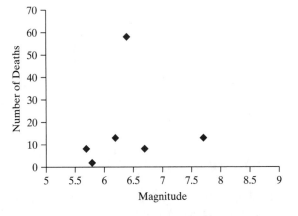

FIGURE 1.2.17 California earthquakes.

Source: National Geophysical Data Center (NGDC).

   a. Explain why, based on the scatter plot, the number of deaths is a function of the magnitude of the earthquake.
   b. Give the magnitudes for two earthquakes that have the same number of deaths.
   c. How many deaths occurred in the earthquake with magnitude 6.4?

14. Fig. 1.2.18 shows data collected by a mathematical modeling student in 2005, giving the price of a used Corvette as a function of its age.

FIGURE 1.2.18 Price of a used Corvette.

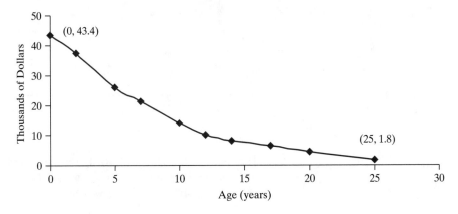

a. What is the domain of this function?
b. What is the range of this function?
c. What is the price of a 10-year-old Corvette?
d. How old is a Corvette that costs 10 thousand dollars?

15. In an August 2004 article concerning the political power of the United Auto Workers union, *The Detroit News* included Fig. 1.2.19, which gives the union's membership as a function of the year.

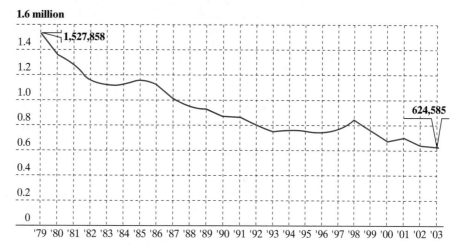

FIGURE 1.2.19 Membership in the United Auto Workers union.
Source: The Detroit News.

a. What is the domain of this function?
b. What is the range of this function?
c. How many UAW members were there in 1987?
d. In what year were there 900,000 UAW members?

16. Fig. 1.2.20 shows the number of cases (in millions) of 7-Up sold for as a function of the year.

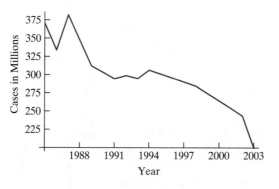

FIGURE 1.2.20 Sales of 7-Up.
Source: Atlanta Journal-Constitution.

a. What is the domain of this function?
b. What is the range of this function?
c. In what year were 250 million cases of 7-Up sold?
d. How many cases were sold in 1994?

17. Use a newspaper, a magazine, or the Internet to find:

a. An example of a graph that represents a function.
b. An example of a graph that does not represent a function.

1.3 FUNCTIONS DEFINED BY FORMULAS

In Sections 1.1 and 1.2, we examined functions that were defined using either a table of values or a graph. In this section, we look at functions defined by formulas, either symbolic or verbal, and consider the relationship among a formula, a table, and a graph.

The Body Mass Index, or BMI, is a measure that assesses the amount of fat in the body. It is commonly used to determine whether individuals are underweight or overweight and is used with both adults and children.

When weight is measured in pounds, the BMI for a child who is 36 inches tall is calculated by the following formula:

$$\text{BMI} = 703\left(\frac{\text{weight}}{36^2}\right).$$

From this formula, we can see that BMI is a function of weight because for each input value (weight in pounds), we will get exactly one BMI value. For example, if a child's weight is 40 pounds, then his or her BMI is $703\left(\frac{40}{36^2}\right) = 21.7$, rounded to the nearest tenth.

Furthermore, we can use the formula to determine BMI for any given weight. By entering this function rule into our calculator's **Y=** menu as **Y1=703(X/36²)**, we can use the table feature to generate a table of function values, as shown in Figs. 1.3.1 and 1.3.2. The values in the **X** column are our inputs

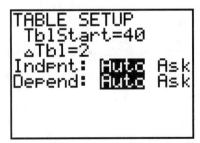

FIGURE 1.3.1 Table setup for BMI function.

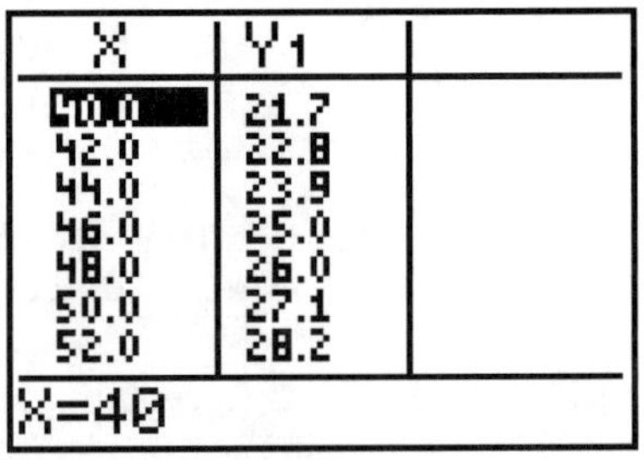

FIGURE 1.3.2 Table of BMI function values.

(weights) and the values in the **Y1** column are the outputs (BMI). We see the BMI values displayed correct to one decimal place, and notice that the value obtained for a weight of 40 pounds corresponds to the value we found previously.

To examine this relationship graphically, we can create a scatter plot of BMI versus weight using the data in Fig. 1.3.2 or we can create a graph of the function directly from the function rule. Since the function is already entered into the **Y=** menu of the calculator, all we need do to display the graph is to choose an appropriate viewing window. In order to see all the ordered pairs displayed in Fig. 1.3.2, we must set the window so that our minimum x-value is smaller than 40, our maximum x-value is larger than 52, our minimum y-value is smaller than 21, and our maximum y-value is larger than 28. Using the settings displayed in Fig. 1.3.3, we obtain the graph in Fig. 1.3.4.

This graph shows BMI not only for the integer weights given in Fig. 1.3.2, but also for any weight between 35 pounds and 55 pounds. For example, to find the

SECTION 1.3 Functions Defined by Formulas 19

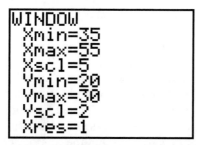

FIGURE 1.3.3 Setting the window for the BMI function.

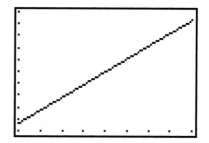

FIGURE 1.3.4 Graph of BMI function.

BMI for a weight of 42.5 pounds, you can use the **value** command in the **CALC** menu. Figure 1.3.5 shows that the BMI for a weight of 42.5 is 23.1.

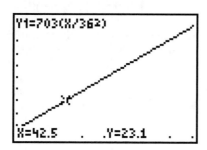

FIGURE 1.3.5 Finding the output value for an input of 42.5.

EXAMPLE 1

Is It a Function?

In June 2005, T-Mobile advertised a "Get More" Plan for cell phone service that included 600 "whenever minutes" plus unlimited weeknight and weekend minutes each month for $39.99, with additional minutes charged at 40¢ each. Let's assume that only weekday minutes count as "whenever minutes" against the 600-minute total under the $39.99 basic monthly charge. Does this formula describe the monthly cost of cell phone service as a function of the number of weekday minutes used?

SOLUTION

This plan describes a function $C(n)$ that is defined for nonnegative values of n, the number of weekday minutes used, because for each input (number of weekday minutes), there is only one output (cost in dollars and cents).

Suppose that we want a symbolic, rather than verbal, formula for this function. If we look at the process we use to find specific output values, we are led to a symbolic rule.

Let's call weekday minutes simply "minutes" for the time being, since weeknight and weekend minutes are free. Our "frontward" question "Here's the input, what's the output?" then becomes "How much would a person be charged for cell phone service if she used (for example) 123 minutes?" The formula tells us that her cost would be $39.99, or symbolically, $C(123) = \$39.99$.

What if she used 130 minutes, 218 minutes, 467 minutes, or 600 minutes? The answer to each of these questions is the same—$39.99. So we can say that $C(n) = \$39.99$ if $0 \leq n \leq 600$.

But that is only part of the story. Suppose we were asked the cost if a person used 601 minutes. We can easily see that there is only one "extra" minute, costing 40¢, which must be added to the basic charge of $39.99. So $C(601) = \$39.99 + \$0.40 = \$40.39$. Similarly, if a person used 738 minutes, there would be $738 - 600 = 138$ additional minutes, each charged at 40¢, making $C(738) = \$39.99 + 138(\$0.40) = \$95.19$. If a person used 919 minutes, there would be $919 - 600 = 319$ additional minutes, each charged at 40¢, making $C(919) = \$39.99 + 319(\$0.40) = \$167.59$. In each case, to calculate the cost, we needed to subtract 600 from the minutes used to find the number of additional minutes charged at 40¢ each. In general, $C(n) = \$39.99 + (n - 600)(\$0.40)$ if $n > 600$.

Thus, in order to give a symbolic rule for this function, we must write it in two parts:

$$C(n) = \begin{cases} 39.99 \text{ if } 0 \leq n \leq 600, \\ 39.99 + (n - 600)(0.40) \text{ if } n > 600, \end{cases}$$

where n represents the number of weekday minutes used and $C(n)$ is cost in dollars and cents. The graph of this function is shown in Fig. 1.3.6.

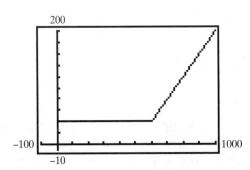

FIGURE 1.3.6 Graph of the "Get More" Plan function.

EXAMPLE 2 Finding Input and Output Values

If a certain savings account earns 4% simple interest per year, then the interest I earned each year is given in terms of the amount A in the account by the function $I(A) = 0.04A$.

a. Find the amount in the account if the yearly interest earned is $18.00.
b. Find the interest earned each year if the amount in the account is $360.00.

SOLUTION

a. This question asks us to find the input if the output is $18.00. Since $I(A)$ represents the output for an input of A dollars, we need to solve the equation

$$18.00 = 0.04A$$

for A. Thus, $A = \dfrac{18.00}{0.04} = 450.00$, and there is $450.00 in an account earning $18.00 interest yearly.

We can also find this input graphically, using the calculator. To the calculator, inputs are x's and outputs are y's, so we enter $0.04x$ into **Y1** and 18.00 into **Y2**. Then we set the window so we can see the graphs of these functions. Since the input x represents the dollars invested, we use the window **Xmin = −10, Xmax = 800, Xscl = 100**. The y's represent the dollars of interest earned, so we let **Ymin = −1, Ymax = 25, Xscl = 5,** and then graph the functions. The point where these two graphs intersect has the input value that gives an output of 18.00. Using the **intersect** command in the **CALC** menu of our calculator, we hit **ENTER** to select the "first curve" **Y1, ENTER** again to select the "second curve" **Y2,** and **ENTER** once again to select a "guess". (Sometimes you need to move the cursor to get close to the point of intersection, but usually wherever the cursor "lands" is fine.) We can see from Fig. 1.3.7, that $450 is the amount invested if the interest earned is $18.

b. This question asks us to find the output if the input is $360.00. Here we need to find $I(360.00) = 0.04(360.00) = 14.40$. So, an account of $360.00 earns $14.40 each year.

If we use the **value** command in the **CALC** menu, as we did with the BMI function previously, the calculator returns the output value of 14.4, as shown in Fig. 1.3.8.

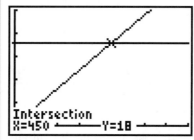

FIGURE 1.3.7 Finding the amount invested for $18.00 interest earned.

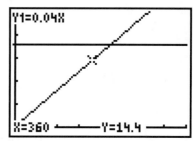

FIGURE 1.3.8 Finding the interest earned for $360.00 invested.

Since we are dealing with interest earned, we express the answer in dollars and
● cents as $14.40.

Examples 1 and 2 illustrate functions that describe relationships between real-world variables. The key to using mathematics to analyze a real-world situation often is recognizing relationships among the variables that describe the situation. The following example illustrates functions defined by formulas that may be familiar from your previous studies in mathematics and science.

EXAMPLE 3 | Finding Input and Output Values

a. The area A of a circle of radius r (Fig. 1.3.9) is given by

$$A(r) = \pi r^2 \quad \text{(where } \pi \approx 3.1416\text{)},$$

using function notation to indicate that the area depends on the radius. In order to make a circular garden with area 30 square feet, how large (to the nearest hundredth of a foot) should the radius be?

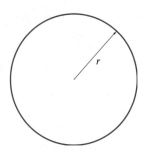

FIGURE 1.3.9 The area A of a circle is a function of its radius r.

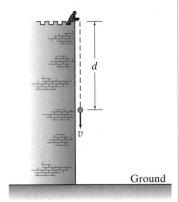

FIGURE 1.3.10 The distance d the rock has fallen is a function of time t.

b. If a rock is dropped from atop a high tower (Fig. 1.3.10) and the acceleration of gravity is 32 ft/sec^2, then its velocity v after t seconds and the distance d it has fallen are given by, respectively,

$$v(t) = 32t \quad \text{and} \quad d(t) = 16t^2.$$

How fast is the rock moving after 2.5 seconds, and how far has it fallen?

c. If the temperature of a 3-gram sample of carbon dioxide is 27°C, then its volume V in liters is given by

$$V(p) = \frac{168}{p},$$

where p is the pressure of the gas in atmospheres. What is the pressure of the gas if its volume is 14 liters?

SOLUTION

a. This question asks us to find the input if the output is 30 square feet. Since $A(r)$ represents the output for an input of radius r, we need to solve the equation

$$30 = 3.1416r^2$$

for r. Thus, $r^2 \approx \dfrac{30}{3.1416}$, and $r \approx \sqrt{\dfrac{30}{3.1416}} \approx 3.09$ feet (rounded to the nearest hundredth).

b. This question asks us to find the outputs for both the velocity and distance functions if the input is 2.5 seconds. So we need to find $v(2.5) = 32(2.5) = 80$ feet per second and $d(2.5) = 16(2.5^2) = 100$ feet.

c. Here we are again given output and asked for input. We need to solve the equation

$$14 = \frac{168}{p}.$$

Therefore, $p = \dfrac{168}{14} = 12$ atmospheres.

Of course, we could also use the calculator techniques from Example 2 to find ● these same input and output values.

EXAMPLE 4 | Finding Domain and Range

Suppose the total cost C of manufacturing n copies of a regional cookbook is $500 to set up the printing press plus $6 for each book actually printed. Then C is given as a function of n by the formula

$$C(n) = 500 + 6n.$$

Find the domain and the range of this function.

SOLUTION

The domain of the cost function C is the set $\{1, 2, 3, \ldots\}$ of all *positive integers*, because it would be meaningless to speak of printing a negative or fractional number of books. The range of C is the set of all numbers described by the list $\{506, 512, 518, \ldots\}$ because each book adds an additional \$6 to the total cost of printing.

EXAMPLE 5 Finding Domain and Range

The *squaring function* defined by

$$f(x) = x^2$$

assigns to each number x its square x^2. What are the domain and the range of f?

SOLUTION

Because every number can be squared, the domain of f is the set—often denoted by $\boldsymbol{R}$—of all real numbers. Since the output of squaring a real number is never negative, the range of f is the set of all nonnegative numbers, $[0, \infty)$.

Figure 1.3.11 shows a calculator graph of $y = x^2$. While the calculator shows only that portion of the graph visible in a standard viewing window (with x and y lying between -10 and 10), we can see that it confirms our belief that no negative numbers are in the range of f.

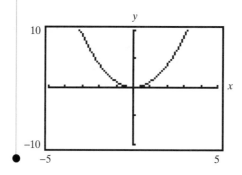

FIGURE 1.3.11 Calculator graph of $f(x) = x^2$.

Examples 4 and 5 illustrate the process of finding the range of a function. Frequently, however, we are more concerned with the domain of a function than with its range because it is crucial for us to know what numbers we can use for the input to get "sensible" values for the output. When we are given a function specified by a formula, we will assume that the domain of the function is the set of real numbers for which corresponding output values "make sense."

When we describe the function f by writing a formula $y = f(x)$, we call x the *independent variable* and y the *dependent variable* because the value of y depends (through the rule or formula of f) on the choice of x. As the independent variable x changes or varies, then so does the dependent variable y. For instance, as x changes from -2 to 0 to 3, the value $y = x^2$ of Example 5 changes from 4 to 0 to 9.

You may find it useful to visualize the dependence of the value $y = f(x)$ on x by thinking of the function f as a kind of machine that accepts as *input* a number x and then produces as *output* the number $f(x)$, perhaps printed or displayed on a monitor (Fig. 1.3.12).

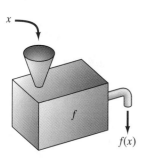

FIGURE 1.3.12 A "function machine" with input x and output $f(x)$.

One such machine is (in effect) the square root key of a simple pocket calculator. When a number x is entered and this key is pressed, the calculator displays (a decimal approximation to) the number $\sqrt{x}$. This square root function $f(x) = \sqrt{x}$ has domain and range both equal to the set of all nonnegative numbers. A calculator like the TI-84 illustrates its knowledge of the domain by displaying an error message if we ask it to calculate the square root of a negative number.

1.3 Exercises

Building Your Skills

In Exercises 1–4 the given formula defines a function. Identify the independent variable and the dependent variable, and then rewrite each formula in function notation.

1. $R = 8\sqrt{s}$
2. $T = 4u - 7$
3. $C = 2\pi r$
4. $V = e^3$

In Exercises 5–10, explain why the formula does or does not represent y as a function of x.

5. The record high temperature y in Denver for each month x of the year.
6. The total cost y of purchasing x hamburgers at your favorite fast-food restaurant.
7. The price y of a Christmas tree based on its height x if
   - a 6-foot tree costs between \$20 and \$50;
   - an 8-foot tree costs between \$25 and \$75;
   - an 11-foot tree costs between \$100 and \$200;
   - a 13-foot tree costs \$300 or more.
8. $y = x^2$
9. $x = y^2$
10. $4x - 3y = 8$

In Exercises 11–14, find and simplify each of the following values: $f(-1)$, $f(0.5)$, and $f(\sqrt{2})$.

11. $f(x) = 2x + 3$
12. $f(x) = x^2 + 1$
13. $f(x) = \dfrac{1}{2x + 1}$
14. $f(x) = \sqrt{x^2 + 2}$

In Exercises 15–17 find all values of a such that $g(a) = 13$.

15. $g(x) = 3x + 4$
16. $g(x) = \sqrt{5x + 4}$
17. $g(x) = x^2 - 36$

In Exercises 18–20 find the domain and the range of the function h.

18. $h(x) = -x + 3$

19. $h(x) = \sqrt{2x - 3}$

20. $h(x) = x^2 + 7$

Applying Your Skills

21. The BMI for a person who weighs 150 pounds is given by

$$\text{BMI} = 703\left(\frac{150}{h^2}\right),$$

where h is the person's height in inches.

   a. Explain why this formula represents BMI as a function of height in inches.
   b. How tall (to the nearest inch) is a 150-pound person whose BMI is 24.2?
   c. What is the BMI for a 150-pound person who is 72 inches tall?

22. On August 2, 2005, the website XE.com gave the rate for exchanging U.S. dollars and Euros as 1.00 United States dollar = 0.820150 Euro.

   a. Write a symbolic function $E(d) = \cdots$ giving the number of Euros received when d dollars are exchanged.
   b. How many Euros would you get for $475?
   c. How many dollars did you exchange if you received 492.09 Euros?
   d. What are the domain and the range of this function?

23. Straight line depreciation is a method for computing an asset's loss of value over time that assumes that the asset will lose an equal amount of value over each year of its useful life. If you buy a $3000 computer today for your home business and assume that it will be worth only $200 at the end of its useful life 5 years from now, it loses $560 of value each year.

   a. Write a symbolic function $V(n) = \cdots$ giving the value of the computer as a function of its age in years.
   b. How much is the computer worth after 3 years?
   c. How old is the computer if it is worth $1880?
   d. What are the domain and the range of V?

24. According to the U.S. Postal Service, "First-Class Mail is used for personal and business correspondence. Any mailable item may be sent as First-Class Mail. It includes postcards, letters, large envelopes, and small packages. The maximum weight is 13 ounces." The Postal Service gives its First-Class Mail rates as follows:

First ounce	$0.37
Each additional ounce	$0.23

By "each additional ounce," the Postal Service means each additional ounce or fractional part of an ounce, if the letter does not weigh a whole number of ounces. Thus a letter weighing 4.5 ounces costs the same amount as a letter weighing 5 ounces.

a. Explain why the Postal Service rule describes the postage rate as a function of weight.
b. What is the cost to send a letter weighing 6.9 ounces? 7 ounces? 7.1 ounces?
c. If a letter costs $1.06, how much does it weigh?
d. Write a multipart rule for the First-Class Mail rate function $R(w)$, giving the rate in dollars and cents as a function of the weight in ounces.
e. What are the domain and the range of the function R?

25. Use a newspaper, a magazine, or the Internet to find
    a. an example of a verbal or symbolic formula that represents a function;
    b. an example of a verbal or symbolic formula that does not represent a function.

1.4 AVERAGE RATE OF CHANGE

According to a July 2005 College Board report, "There's no escaping the fact that college costs are rising." This fact probably comes as no surprise to any college student paying for tuition, housing, and books. The following table shows the average cost of tuition at a public four-year college (for an academic year beginning in the fall of each indicated year):

YEAR	TUITION	YEAR	TUITION
1993	$2535	1999	$3362
1994	$2705	2000	$3487
1995	$2811	2001	$3725
1996	$2975	2002	$4115
1997	$3111	2003	$4694
1998	$3247		

Based on the table, the cost of tuition is a function of the year. As we examine the table, we see that as the year increases, so does the tuition. We therefore say that the tuition cost is an *increasing* function of the year.

In general, a function $y = f(x)$ is increasing over an interval of x-values if, for any two different values x_1 and x_2 in the interval, if $x_1 < x_2$, then $f(x) < f(x_2)$. That is, as the x-values increase, so do the y-values.

What does this mean in terms of the graph of the function? If we consider the graph of the increasing tuition cost function shown in Fig. 1.4.1, we see that, as we look from left to right, the graph "goes uphill." This provides visual confirmation of the fact that as the x-values increase, so do the y-values.

Similarly, a function $y = f(x)$ is *decreasing* over an interval of x-values if, for any two different values x_1 and x_2 in the interval, if $x_1 < x_2$, then $f(x_1) > f(x_2)$. That is, as the x-values increase, the y-values *decrease*. The graph of a decreasing function "goes downhill" as we look from left to right.

Finally, a function $y = f(x)$ is constant over an interval of x-values if, for any two different values x_1 and x_2 in the interval, $f(x_1) = f(x_2)$. That is, no matter what the

FIGURE 1.4.1 Graph of an increasing cost function.

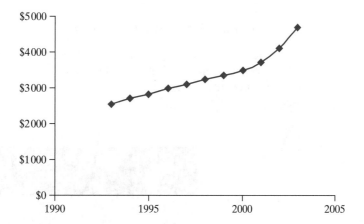

x-value is, the y-value remains the same. Thus, if $f(x) = c$ for every x, then the graph of the function is the line $y = c$. So the graph of a constant function is a horizontal line.

EXAMPLE 1 Increasing, Decreasing, or Constant?

Recall the cell phone cost function

$$C(n) = \begin{cases} 39.99 & \text{if } 0 \leq n \leq 600, \\ 39.99 + (n - 600)(0.40) & \text{if } n > 600, \end{cases}$$

where n represents the number of weekday minutes used and $C(n)$ is cost in dollars and cents. Use the graph of this function, shown in Fig. 1.4.2, to determine the intervals for which C is increasing, decreasing, or constant.

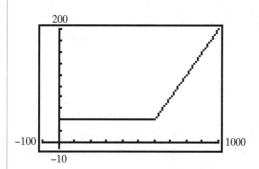

FIGURE 1.4.2 Graph of the "Get More" Plan function.

SOLUTION

For all n between 0 and 600, inclusive, $C(n)$ is $39.99, so C is constant on the interval $[0, 600]$. Once n reaches 600, as n increases, so does $C(n)$, so C is increasing for "sensible" input values greater than or equal to 600. No matter what month we are in, there are no more than 31 days and there are always at least 8 weekend days. Then the largest number of anytime minutes possible is $\frac{60 \text{ minutes}}{\text{hour}} \times \frac{24 \text{ hours}}{\text{day}} \times 23 \text{ days} = 33{,}120$ minutes per month. So C is increasing on the interval $[600, 33120]$. (Most people would probably not argue if you used the interval $[600, \infty)$ to indicate "sensible" values of n greater than or equal to 600.)

• Notice that when we report intervals where the function is increasing, decreasing, or constant, we use intervals of *input* values.

EXAMPLE 2 | Increasing, Decreasing, or Constant?

The accompanying table gives the percentage of the U.S. population that was born outside of the United States as a function of the year. Use the table to determine the intervals where the function is increasing, decreasing, or constant.

YEAR x	PERCENTAGE OF POPULATION BORN OUTSIDE U.S. $f(x)$
1940	8.8
1950	6.9
1960	5.4
1970	4.7
1980	6.2
1990	8.0
2000	10.4

Source: The World Almanac and Book of Facts 2005.

SOLUTION

● $f(x)$ is decreasing on [1940, 1970] and increasing on [1970, 2000].

When we discuss how function values change when input values change, we frequently look at the function's average rate of change. For a function $y = f(x)$, we define the **average rate of change** of the function over the interval $[a, b]$ to be

$$\frac{\Delta y}{\Delta x} = \frac{f(b) - f(a)}{b - a}.$$

Thus the average rate of change is the change in y divided by the change in x. For instance, for the function $f(x)$ in Example 3, the average rate of change over the interval [1940, 1950] is

$$\frac{\Delta y}{\Delta x} = \frac{f(1950) - f(1940)}{1950 - 1940} = \frac{6.9 - 8.8}{10} = \frac{-1.9}{10} = -0.19.$$

EXAMPLE 3 | Average Rate of Change

Complete the accompanying table to find the average rate of change of $f(x)$ over each interval of consecutive x-values. Note that we are recording the average rate of change between two consecutive data points in the same line of the table as the first of the data points. (The last line of the average rate of change column is blank because average rate of change requires two points.)

SECTION 1.4 Average Rate of Change

YEAR x	PERCENTAGE OF POPULATION BORN OUTSIDE U.S. f(x)	AVERAGE RATE OF CHANGE Δy/Δx
1940	8.8	−0.19
1950	6.9	
1960	5.4	
1970	4.7	
1980	6.2	
1990	8.0	
2000	10.4	—

SOLUTION

YEAR	PERCENTAGE OF POPULATION BORN OUTSIDE U.S. f(x)	AVERAGE RATE OF CHANGE Δy/Δx
1940	8.8	−0.19
1950	6.9	−0.15
1960	5.4	−0.07
1970	4.7	0.15
1980	6.2	0.18
1990	8.0	0.24
2000	10.4	—

Recall that we reported that $f(x)$ is decreasing on [1940, 1970] and increasing on [1970, 2000]. Now we notice that the average rate of change is increasing over the entire interval. So we can say that $f(x)$ is decreasing at an increasing rate on [1940, 1970] because the function's output values are getting smaller but average rate of change is getting larger. Similarly, $f(x)$ is increasing at an increasing rate on [1970, 2000] because both the function's output values and its average rate of change are getting larger.

If we consider the graph of this function in Fig. 1.4.3, we can see the intervals of increase and decrease. We also observe that the function bends upward like a bowl sitting right side up. This is a property associated with a function with an increasing average rate of change. We say that the graph of the function is *concave upward*.

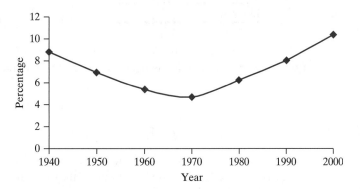

FIGURE 1.4.3 Percentage of U.S. population born outside the United States.

EXAMPLE 4 Describing Average Rate of Change

In Section 1.2, we looked at the graph of a woman's percentage chance of conception each month as a function of her age. The following table shows the function values $f(x)$ for various ages.

a. Complete the table to find the average rate of change of $f(x)$ over each interval of consecutive x-values.

b. Describe the increasing and decreasing behavior of both the function and its average rate of change.

AGE x	ODDS OF CONCEPTION $f(x)$	AVERAGE RATE OF CHANGE $\Delta y/\Delta x$
15	22	0.4
20	24	
25	22	
30	16	
35	8	
40	3	
45	0.01	—

SOLUTION

a. Fortunately, we can use the calculator to complete a table such as this. Store the x-values in **L1** and the function values in **L2**. Then highlighting **L3**, select **2nd**, **STAT, OPS,** and choose **ΔList(**. Type **2nd 2** for **L2**, then**)** and **÷**. Again select **2nd**, **STAT, OPS,** and choose **ΔList(**. Now type **2nd 1** for **L1,** then**)** and **ENTER**. Figure 1.4.4 shows the resulting table.

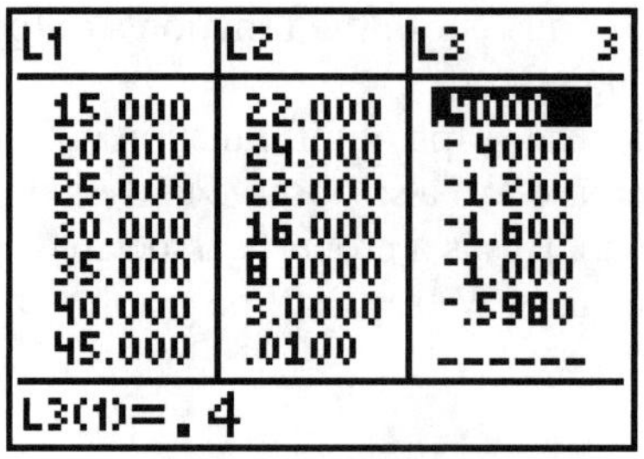

L1	L2	L3 3
15.000	22.000	.4000
20.000	24.000	-.4000
25.000	22.000	-1.200
30.000	16.000	-1.600
35.000	8.0000	-1.000
40.000	3.0000	-.5980
45.000	.0100	------

L3(1)=.4

FIGURE 1.4.4 Finding the average rate of change.

b. $f(x)$ is increasing on [15, 20] and decreasing on [20, 45]. The average rate of change is decreasing on [15, 35] and increasing on [35, 45].

EXAMPLE 5 Describing Average Rate of Change

The function $f(x) = \sqrt{x}$ is increasing on its domain $[0, \infty)$. Complete the following table to describe its average rate of change on [0, 36].

SECTION 1.4 Average Rate of Change 31

x	f(x)	AVERAGE RATE OF CHANGE $\Delta y/\Delta x$
0		
1		
4		
9		
16		
25		
36		

SOLUTION

The calculator table shown in Fig. 1.4.5 shows that the average rate of change of $f(x)$ is decreasing on [0, 36].

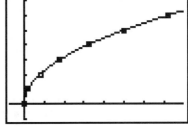

FIGURE 1.4.5 Finding the average rate of change for $f(x) = \sqrt{x}$.

FIGURE 1.4.6 Graph of $f(x) = \sqrt{x}$.

If we examine the graph of the function in Fig. 1.4.6, we see that the graph goes uphill from left to right because the function is increasing. It also curves downward because the average rate of change is decreasing. We say that this graph is *concave downward*.

As we continue through this book we will see the important role that the average rate of change plays in describing a function.

1.4 Exercises | Building Your Skills

In Problems 1–4, determine the intervals over which the indicated function is increasing, decreasing, or constant.

1.

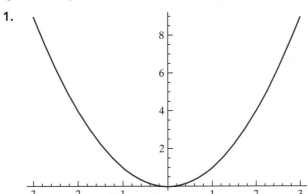

2.

x	−6	−3	0	2	5
$f(x)$	−10	−8	1	3	8

3.

x	0	2	4	6	8	10
$f(x)$	6	7	8	8	8	7

4.
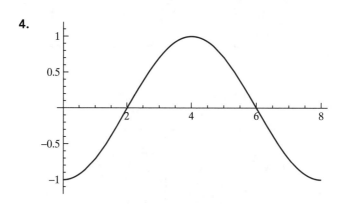

In Exercises 5–8, find the average rate of change for the indicated function, and determine where the average rate of change is increasing, decreasing, or constant.

5.

x	−6	−3	0	2	5
$f(x)$	−128	−62	4	48	114

6.

x	0	5	10	15	20
$f(x)$	25	30	34	33	32

7.

x	1	3	5	7	9
$f(x)$	18	6	−3	−9	−12

8.

x	−1	0	4	11	23
$f(x)$	18	15	9	0	−12

Applying Your Skills

9. The accompanying graph shows the number of farms (in thousands) in Missouri as a function of the years 1996 through 2003. Determine the intervals of years for which this function is increasing, decreasing, or constant.

Farms in Missouri

Source: Missouri Agricultural Statistics Service.

10. The following table gives the median age at first marriage for U.S. women as a function of the year. Use the table to determine the years over which the median age is increasing, decreasing, or constant.

Year	1993	1995	1997	1999	2001	2003
Number	24.5	24.5	25.0	25.1	25.1	25.3

Source: World Almanac and Book of Facts 2005.

11. The following graph shows the percentage y of all music sold that was rap music x years after 1992. The y-intercept of 8.6 indicates that 8.6% of all music sold in 1992 was rap music.

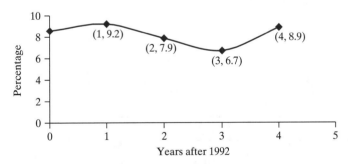

Source: World Almanac and Book of Facts 1998.

a. Determine the intervals of x-values for which the function is increasing.
b. Determine the intervals of x-values for which the function is decreasing.
c. Determine the intervals of x-values for which the average rate of change of the function is increasing.
d. Determine the intervals of x-values for which the average rate of change of the function is decreasing.

12. The accompanying table gives the number (in thousands) of new cases of diagnosed diabetes among adults aged 18–79 years in the United States as a function of the year. Find the average rate of change for this function and use your answer to determine the intervals of years for which,

a. the number of new cases of diabetes is increasing at a decreasing rate;
b. the number of new cases of diabetes is increasing at an increasing rate.

Year	1997	1998	1999	2000	2001	2002	2003	2004
Number	878	921	979	1104	1213	1304	1349	1356

Source: www.cdc.gov.

Chapter 1 | Review

In this chapter, you learned about functions and function models. After completing the chapter, you should be able to

- determine whether a relation described numerically, graphically, or symbolically represents a function;
- find the domain and range of a function;
- find the output value of a function for a given input value;
- find the input value(s) of a function for a given output value;
- determine the intervals over which a function is increasing, decreasing, or constant;
- determine the average rate of change of a function over an interval.

Review | Exercises

In Exercises 1–6, decide whether, based on the table, graph, or formula, y is a function of x.

1.

x	2	4	6	6	8	11
y	3	4	9	13	15	23

2.

x	2	4	5	6	8	11
y	3	4	9	9	15	23

3.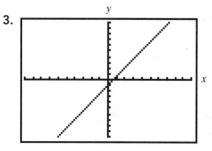

4.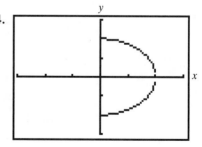

5. $y = x^2 - 4$

6. $x = |y|$

In Exercises 7–9, find the domain and range of each function.

7. The number of fat grams $F(x)$ in a sandwich based on the number of calories x, according to the following table.

Calories	430	550	320	680	300	750	530	710
Fat Grams	20	29	5	33	9	33	26	43

Source: Chick-Fil-A Fat Gram Comparison.

8.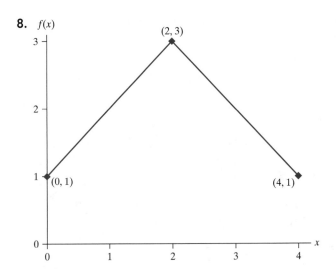

9. $f(x) = |x| - 3$

10. The function graphed as follows is defined on the interval $[-2, 4]$. Determine the intervals over which the function is increasing or decreasing.

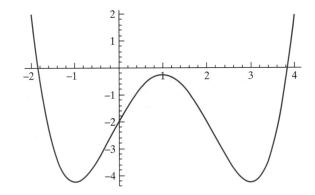

11. The monthly cos't function for a "light" plan for dial-up Internet access is given by

$$C(h) = \begin{cases} 6.95 & \text{if } 0 \le x \le 5, \\ 6.95 + 2(h - 5) & \text{if } x > 5, \end{cases}$$

where C is in dollars and h is the number of hours used. If a subscriber's bill was \$23.65, how many hours of Internet time did she use that month?

12. For many years, insurance companies published guidelines of ideal weights W for various heights h. The most common rule for calculating ideal weight for a woman was 100 pounds for a woman 60 inches tall plus 5 additional pounds for each inch over 60 inches.

   a. Write a function rule $W(h)$ to determine the ideal weight for a woman who is at least 60 inches tall.

   b. Use your rule to determine the ideal weight for a woman who is 66 inches tall.

13. Often in the media, functions are displayed using bar graphs rather than scatter plots, as in the graph below.

   a. Make a table of values for this function, letting x be years after 1998 and $P(x)$ the percentage of young adults who never smoked cigarettes.

   b. Use your calculator to make a scatter plot of this function, using x-window $[-1, 7]$ and y-window $[0, 100]$.

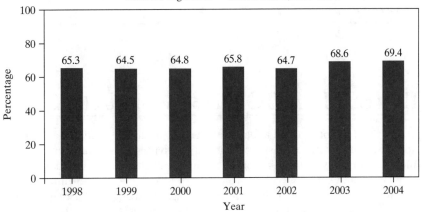

Source: www.cdc.gov/mmwr.

14. The text on the graph showing UAW membership as a function of year states that there has been a steady decline in membership from 1979 to 2003. However, if you look carefully at the graph, you will see that there are brief intervals where the function appears to be increasing or constant. Identify the intervals of years over which this function is increasing or constant.

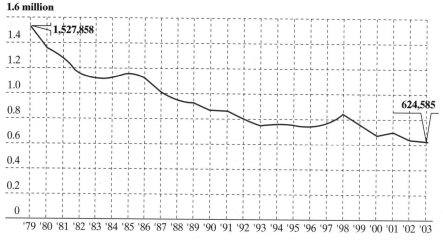

Source: *The Detroit News.*

15. (Chapter Opener Revisited) In the discussion that introduced this chapter, we looked at the graph of the heat index at a fixed air temperature of 85°F as a function of the relative humidity.

   a. Use the given table of values for that function to determine the average rate of change over consecutive input values.
   b. Use your answer from part (a) to determine the intervals of input values where the average rate of change is increasing, decreasing, or constant.
   c. In our discussion, we noted that for relative humidities of 0% to 60%, the graph appeared to be a straight line. Over those input values, was the average rate of change increasing, decreasing, or constant?
   d. In our discussion, we noted that for relative humidities of 60% to 100%, the graph appeared to curve upward. Over those input values, was the average rate of change increasing, decreasing, or constant?

Relative Humidity (%)	0	10	20	30	40	50	60	70	80	90	100
Apparent Temperature (°F)	78	80	82	84	86	88	90	93	97	102	108

INVESTIGATION Exploring Rate of Change

In this chapter, you learned about the behavior of functions and their average rates of change. This activity will allow you to investigate these concepts numerically, graphically, and verbally.

To raise money for their sorority's service project, Kari and Keisha decided to sell cookies outside the student center. After the first week of sales, they examined their sales record and made a chart that includes the number of cookies sold per day along with the change from the previous day. Complete a chart like the one shown for each of the situations described. Then make a plot that shows the day on the *x*-axis and the number of cookies sold on the *y*-axis. Describe in words how each graph looks.

DAY	COOKIES SOLD	AVERAGE RATE OF CHANGE (cookies per day)
1	40	N/A
2		
3		
4		
5		
6		
7		

1. Cookie sales remained constant.
2. Cookie sales increased at a constant rate.
3. Cookie sales increased at an increasing rate.
4. Cookie sales increased at a decreasing rate.
5. Cookie sales decreased at a constant rate.
6. Cookie sales decreased at an increasing rate.
7. Cookie sales decreased at a decreasing rate.

CHAPTER 2: LINEAR FUNCTIONS AND MODELS

2.1 CONSTANT CHANGE AND LINEAR GROWTH

2.2 LINEAR FUNCTIONS AND GRAPHS

2.3 PIECEWISE-LINEAR FUNCTIONS

2.4 FITTING LINEAR MODELS TO DATA

On March 1, 1872, President Ulysses S. Grant signed a law declaring that approximately 2 million acres of land near the headwaters of the Yellowstone River would be "dedicated and set apart as a public park or pleasuring ground for the benefit and enjoyment of the people." This act created the world's first national park, Yellowstone National Park.

Seventy-five percent of all the geysers on Earth are in Yellowstone National Park, and the largest concentration of geysers in the world occurs in the Upper Geyser Basin of the park. This is the area that contains the most famous of the geysers, Old Faithful, which was named for its consistent performance by explorers in 1870. According to the National Park Service, "Old Faithful erupts more frequently than any of the other big geysers, although it is not the largest or most regular geyser in the park. Its average interval between eruptions is about 91 minutes, varying from 65–92 minutes." An eruption lasts from 1.5 to 5 minutes, during which between 3700 and 8400 gallons of boiling water are expelled, with the height of the eruption between 106 and 184 feet.

Visitors to Yellowstone are interested in seeing Old Faithful erupt, and many people mistakenly believe that it erupts every hour. The problem is, as the Park Service indicates, that the geyser doesn't erupt on a fixed schedule, although there do seem to be patterns in its eruptions. Figure 2.0.1 shows a plot of data from selected Old Faithful eruptions in January 2003 as recorded by park rangers and volunteers in the Old Faithful Visitor Center logbook and transcribed for the Internet by Lynn Stephens.

In this graph, the input variable is the duration of an eruption and the output variable is the time to the next eruption. We can make several observations about this graph. First, the graph represents a function since for each input value, there is

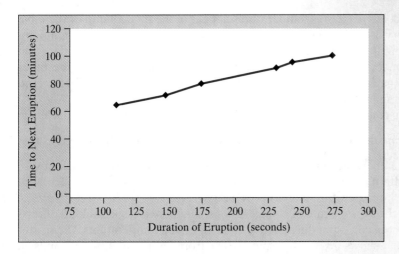

FIGURE 2.0.1 Old Faithful eruptions.
Source: www.geyser-study.org.

only one output value. Second, this function is increasing; that is, as the duration of the eruption increases, so does the time to the next eruption. Finally, the points in this scatter plot lie approximately on a (straight) line.

Since the time to the next eruption is what visitors want to know, it would be useful to determine a function rule that would allow us to make such a prediction. In this chapter, we study functions whose graphs are lines and learn how to determine rules for such functions. These rules will allow us to predict output values for given input values—exactly what we would want to do for the Old Faithful geyser.

2.1 CONSTANT CHANGE AND LINEAR GROWTH

Temperature is commonly measured in the United States in Fahrenheit degrees, but in much of the rest of the world it is measured in Celsius degrees. The following table illustrates the relation between the Celsius temperature C and the Fahrenheit temperature F. (This relation is a function because each Celsius temperature corresponds to exactly one Fahrenheit temperature.)

C (DEGREES)	F (DEGREES)
0	32
10	50
20	68
30	86
40	104
50	122
60	140
70	158
80	176
90	194
100	212

Here, successive Celsius temperatures appear in the first column of the table at equal intervals of 10 degrees. Note also that successive Fahrenheit temperatures appear in the second column at equal intervals of 18 degrees. The fact that equal differences in one column correspond to equal differences in the other indicates a special kind of relationship between corresponding Celsius and Fahrenheit temperatures.

Using the techniques from Section 1.4, we see that the average rate of change $\frac{\Delta F}{\Delta C}$ is *constant*; that is, $\frac{\Delta F}{\Delta C} = \frac{18}{10} = 1.8$. This means that for each 10 degrees that the Celsius temperature increases, the Fahrenheit temperature increases by 18 degrees, or alternatively, for each degree that the Celsius temperature increases, the Fahrenheit temperature increases by 1.8 degrees.

Because the average rate of change is constant, there is a special kind of relationship between corresponding Celsius and Fahrenheit temperatures. The scatter plot in Fig. 2.1.1 shows this relationship. We see from this plot that these points all appear to lie on the same (straight) line. So what is special about the data in our original table of temperatures is this: It is described by a function whose graph is a straight line. That is, because the average rate of change of this function is constant, the graph of the function is a line. We call a function whose graph is a line a *linear* function.

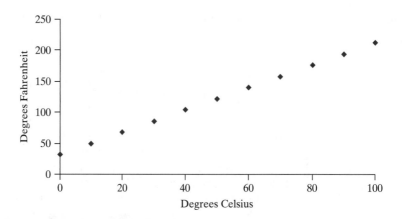

FIGURE 2.1.1 Scatter plot of degrees F versus degrees C.

You can verify that if C and F are corresponding entries in the table, then if we write F as a function of C, we have

$$F(C) = 1.8C + 32. \tag{1}$$

One way to check this is to substitute each value $C = 0, 10, 20, \ldots, 100$ into (1) and verify in each case that the corresponding value of F shown in the table results.

Another way is to rewrite (1) as

$$y = 1.8x + 32 \tag{2}$$

with x in place of C and y in place of $F(C)$. With **Y1 = 1.8X + 32** and using the **TBLSET** menu with **TblStart = 0**, and **ΔTbl = 10**, a calculator yields the table shown in Fig. 2.1.2, the same columns of numerical values as in the Celsius–Fahrenheit table.

If we graph the function in (2) in the calculator window defined by

$$\text{Xmin} = 0, \text{Xmax} = 100, \text{Xscl} = 10,$$
$$\text{Ymin} = 0, \text{Ymax} = 300, \text{Yscl} = 30,$$

we get the picture shown in Fig. 2.1.3. Here we see again that the graph of this function is a line. Furthermore, we see **all** points on the graph (in this particular window), not merely the pairs given in the original table.

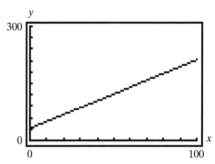

FIGURE 2.1.2 **Y1=1.8X+32** table with **TblStart=0** and **ΔTbl=10**.

FIGURE 2.1.3 The graph $y = 1.8x + 32$.

D E F I N I T I O N : Linear Function

A **linear function** is a function of the form

$$f(x) = ax + b. \tag{3}$$

Note that the right-hand side in (3) contains both a constant term and an x-term (but no higher powers). A particular linear function f is determined when the values of the constant *coefficients* a and b are specified.

Since $f(0) = a \cdot 0 + b = b$, the constant term b is the output when the input is 0. Thus, the graph of the linear function $y = f(x)$ intersects the y-axis at the point $(0, b)$. That is, **the constant b is the y-intercept of the line** (Fig. 2.1.4) and therefore measures its vertical location.

The constant coefficient a is the average rate of change of the linear function and measures how the output value of the function changes for each one-unit change in the input value.

The roles of a and b in a linear function make it easy for us to determine a function's symbolic rule if we are given its average rate of change and the output for an input value of 0.

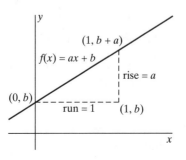

FIGURE 2.1.4 The geometric roles of a and b.

Suppose the following table records the growth in the population of a certain hypothetical city during the 1990s:

YEAR	POPULATION (THOUSANDS)
1990	110
1991	116
1992	122
1993	128
1994	134
1995	140
1996	146
1997	152
1998	158
1999	164
2000	170

We see that each number in the second column is found by adding 6 to the previous one. Thus the population of this city grew each year during the 1990s by the same amount of 6 thousand people. Since its average rate of change is constant, the population of the city is a **linear** function of time t.

In order to find a symbolic rule for this function, let's take $t = 0$ in 1990. (We will call this technique "resetting the clock," and will commonly use it when our input variable is calendar years.) The result of starting with a population of $P(0) = 110$ thousand people and adding 6 thousand people t times in succession—once each year—can be described by saying that t years after 1990 the population is given by

$$P(t) = 110 + 6t \quad \text{or} \quad P(t) = 6t + 110 \quad \text{(thousand)} \tag{4}$$

Note that the constant term in (4) is the city's *initial population* and the coefficient of t is its annual change in population.

More generally, suppose a city has an initial population of

$$P(0) = P_0$$

at time $t = 0$ and that its constant annual change in population is a. Then after t years the population has increased by

$$\Delta P = a \cdot t \tag{5}$$

(using the Δ-notation for differences). Therefore, the city's population after t years is

$$P = \text{initial population} + \text{change} = P_0 + \Delta P = P_0 + at.$$

Thus a population $P(t)$ with **initial population** P_0 and (constant) **annual change** a is described by the linear function

$$P(t) = P_0 + at. \tag{6}$$

Not all populations are described by linear functions, but a population model like the one in (6)—with a constant term and a t-term—is called a *linear population model.* In this section we discuss both linear population models and other linear models for which we know a function's initial value. In Sections 2.2 and 2.3 we discuss linear models for a wider variety of situations.

EXAMPLE 1 — Finding a Linear Population Model

On January 1, 1999, the population of Ajax City was 67,255 and had increased by 2935 people during the preceding year. Suppose this rate of increase continues, with 2935 more people added to the population of Ajax City each subsequent year. Find each of the following:

  a. A linear function $P(t)$ modeling the population of Ajax City.
  b. The predicted population of Ajax City on October 1, 2002, based on the model.
  c. The month and the calendar year in which the population of Ajax City reaches 100 thousand.

SOLUTION

  a. In a given situation, we must decide when to start the clock, and we can count population however we wish—either by the person or by the thousand (for instance). If we reset the clock by letting $t = 0$ on 1/1/1999 and measure population in thousands, it makes it easier to set an appropriate calculator window. Then $P(0) = 67.255$ is our initial population and 2.935 is the function's constant average rate of change. Therefore, the linear function model for this population is

  $$P(t) = 2.935t + 67.225, \text{ for } P \text{ in thousands and } t = 0 \text{ on } 1/1/1999. \quad (7)$$

  b. This is once again our standard question "Here's the input, what's the output?" Since we reset the clock, October 1, 2002, is 3 years and 9 months—that is, 3.75 years—after January 1, 1999. So we substitute $t = 3.75$ in (7) and calculate

  $$P(3.75) = 67.255 + 2.935 \times 3.75 = 78.261$$

  either by hand or by entering our population function into our calculator as **Y1**, and using the **value** command from the **CALC** menu. Thus our linear model predicts an Ajax city population of 78.261 thousand, or 78,261 people on October 1, 2002.

  c. In this case, we need to answer the "backward" question "Here's the output, what's the input?" We can solve this algebraically by setting $P(t) = 100$ and solving for t:

  $$67.255 + 2.935t = 100, \quad (8)$$

  so

  $$2.935t = 100 - 67.255 = 32.745$$

  and

  $$t = \frac{32.745}{2.935} = 11.157 \text{ years.}$$

Starting with January 1, 1999, when $t = 0$, exactly 11 years later is January 1 of the year 1999 + 11 = 2010. So the population of Ajax City hits 100 thousand

$$0.157 \text{ years} = 0.157 \text{ years} \times \frac{12 \text{ months}}{\text{year}} = 1.88 \text{ months}$$

into the year 2010. Note that 1.88 months means that we have completed all of January and are somewhere toward the end of February. So the population of Ajax City is 100 thousand during February 2010.

Note that for the purpose of "calendar month" problems like this, we will consider the year to be divided into 12 equal months. This is slightly inaccurate, but it is standard practice to take the differing lengths of individual months into account only when a question asks for a specific day of a particular month.

In Example 1, we solved part (c) algebraically. You may prefer to find the answer to such questions either graphically or numerically. Examples 2 and 3 illustrate these methods.

EXAMPLE 2   Finding an Input Graphically

Use graphical methods to find the year and month in which the population of Ajax City is 100 thousand.

SOLUTION

In order to determine the year and month in which the population of Ajax City is 100 thousand, we need to determine when the output $P(t)$ has the value 100. Since both $P(t)$ and 100 represent outputs, we enter them into the **Y=** menu of the calculator, plot their graphs, and see where the two graphs intersect. Because a graphing calculator requires that x and y (rather than t and P) be used as the independent and dependent variables, Fig. 2.1.5 shows the **Y=** menu with the appropriate entries

$$y = 67.255 + 2.935x \quad \text{and} \quad y = 100.$$

Figure 2.1.6 shows the resulting plot with viewing window

Xmin = 0	Ymin = 0
Xmax = 20	Ymax = 150

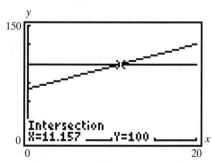

FIGURE 2.1.5 The **Y=** menu for the graphical solution of Example 2.

FIGURE 2.1.6 The graphs **Y1 = 67.225 + 2.935X** and **Y2=100** intersect at (11.157, 100).

We have used the calculator's **CALC intersect** facility to locate the point of intersection.

The indicated result (11.157, 100) agrees with our previous algebraic solution $t = 11.157$ of Equation (8). ●

EXAMPLE 3 Finding an Input Numerically

Use numerical methods to find the year and month in which the population of Ajax City is 100 thousand.

SOLUTION

We now proceed to solve Example 1(c) using the calculator's table-making facility. First we must enter the function we wish to tabulate. This is done in the same way as when we wish to graph a function—using the **Y=** menu. (In this case our function is already stored in **Y1** from Example 2; we can clear **Y2** since we no longer need it.)

To prepare to tabulate values of the function we've defined, we must specify the sort of table we want using the calculator's "table set" menu. We want to start our table at $x = 0$ and proceed by yearly increments of $\Delta x = 1$. Thus our table's *starting point* and its *increment* between successive entries are specified by entering these two values as shown in Fig. 2.1.7.

FIGURE 2.1.7 Initial table setting for the Ajax City population function.

Now we're ready to go! When we execute the calculator's **TABLE** command, we get the table shown in Fig. 2.1.8. This table doesn't go quite far enough to reach a population of 100 thousand, so we use the calculator's down arrow to scroll down. The calculator obligingly fills in additional values, with the result shown in Fig. 2.1.9. Again we see the population is closest to 100 thousand when $x \approx 11$ years (the symbol $\approx$ indicates "approximate equality" as opposed to $=$,

X	Y1
0.000	67.255
1.000	70.190
2.000	73.125
3.000	76.060
4.000	78.995
5.000	81.930
6.000	84.865

X=0

FIGURE 2.1.8 Initial table for the Ajax City population function.

X	Y1
7.000	87.800
8.000	90.735
9.000	93.670
10.000	96.605
11.000	99.540
12.000	102.48
13.000	105.41

X=13

FIGURE 2.1.9 Additional table values for the Ajax City population function.

which means exact equality). But the population is still a bit short of 100 thousand after precisely 11 years, so the exact population is hit sometime during the twelfth year after January 1, 1997.

To see more closely when during the twelfth year the population reaches 100 thousand, we use the calculator's "table set" menu to specify the new starting value $x = 11$ and the new increment $\Delta x = 0.1$ (Fig. 2.1.10). The resulting table (Fig. 2.1.11) indicates that the population is closest to 100 thousand after 11.2 years (rounded off to the nearest *tenth* of a year). This is because 100.13 thousand after 11.2 years is closer to 100 thousand than is 99.834 thousand after 11.1 years.

FIGURE 2.1.10 Second table setting for the Ajax City population function

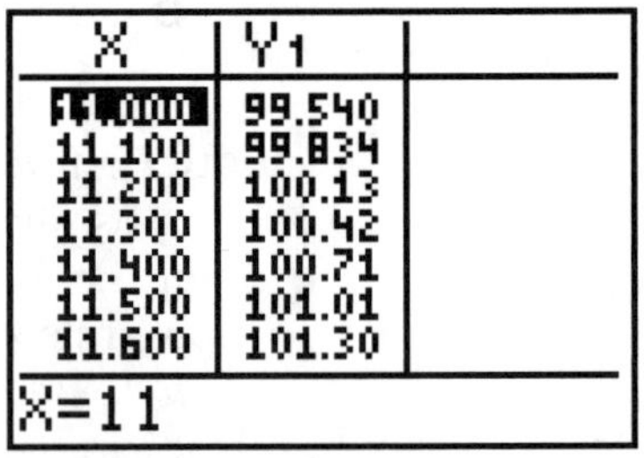

FIGURE 2.1.11 Second table for the Ajax City population function.

To narrow it down still further—in the interval between $x = 11.1$ years and $x = 11.2$ years—we go back to the calculator's table set menu and specify starting point $x = 11.1$ and increment $\Delta x = 0.01$. (Note that we typically divide our desired interval of x-values into 10 "pieces.") Then the **TABLE** command yields the table shown in Fig. 2.1.12. Now we see that the population of Ajax City is closest to 100 thousand after $x \approx 11.16$ years (rounded off accurate to the nearest *hundredth* of a year).

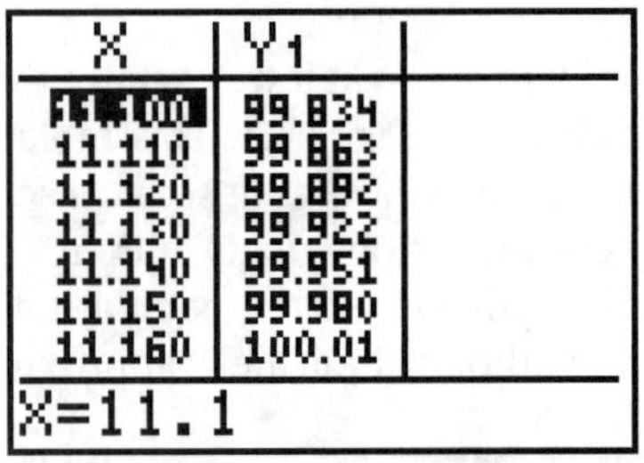

FIGURE 2.1.12 Third table for the Ajax City population function.

To satisfy yourself that we will obtain the same answer here as we did in Example 1(c) and Example 2, you should perform this "table zoom" procedure one more time, with starting point $x = 11.15$ and increment $\Delta x = 0.001$. Do you
● see that the population is closest to 100 thousand when $x = 11.157$ years?

Perhaps it seems to you that this "table zooming" is a great deal more work than solving the linear equation $67.255 + 2.935t = 100$ algebraically or graphically. Most people (including the authors) would agree with you. The method is presented here to illustrate that you will have a choice of algebraic, graphical, and numerical methods for solving the problems you will encounter. In another situation, you may find using the table preferable.

Slope and Rate of Change

You may recall from previous mathematics courses the **slope-intercept equation**

$$y = mx + b \qquad (9)$$

of a straight line in the xy-plane with coefficients

$$m = \text{slope} \quad \text{and} \quad b = y\text{-intercept} \qquad (10)$$

as illustrated in Fig. 2.1.13.

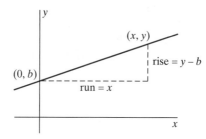

FIGURE 2.1.13 The slope-intercept equation $y = mx + b$ of a straight line.

The population function

$$P(t) = P_0 + at \qquad [\text{or, equivalently, } P(t) = at + P_0] \qquad (11)$$

has the same linear form as the slope-intercept equation (4) but with independent variable t instead of x and with dependent variable P instead of y. The coefficients (or *parameters*) in the linear function (11) are

$$a = \text{average rate of change} \quad \text{and} \quad P_0 = \text{initial population.} \qquad (12)$$

Comparing (9) and (11), we note the correspondence between the **slope** of a straight line and the **rate of change** of a linear function. Indeed, the rate of change a of the linear function $P(t) = P_0 + at$ is simply the slope of its straight line graph in the tP-plane (Fig. 2.1.14).

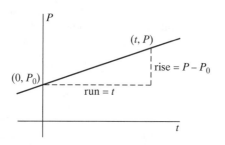

FIGURE 2.1.14 The graph of the linear function $P(t) = P_0 + at$.

48 CHAPTER 2 Linear Functions and Models

EXAMPLE 4 Another Linear Population Model

On January 1, 1992, the population of Yucca City was 46,350 and on July 1, 1994, it was 56,925. Suppose this rate of population increase continues for the foreseeable future.

  a. Write a linear function $P(t) = P_0 + at$ giving the population of Yucca City at time t (where $t = 0$ on January 1, 1992).
  b. Find the predicted population of Yucca City on October 1, 2000.
  c. In what month of what calendar year will the population of Yucca City double?

SOLUTION

  a. The statement that "$t = 0$ on January 1, 1992" means that we are "resetting the clock." Furthermore, we can measure our population in thousands, making our initial population of 46,350 = 46.350 thousand. This procedure makes it easier for us to find a suitable window for viewing the graph of the function.

  July 1, 1994, is 2½ years after January 1, 1992, so the population of Yucca City is

  $$P = 46.350 \quad \text{when} \quad t = 0, \text{ and}$$
  $$P = 56.925 \quad \text{when} \quad t = 2.5.$$

  Since we have our initial population (46.350), all we need in order to write our population function is its average rate of change:

  $$a = \frac{\Delta P}{\Delta t} = \frac{56.925 - 46.350}{2.5 - 0} = \frac{10.575}{2.5} = 4.230 \text{ (thousand people per year)}.$$

  So our population function for Yucca City is

  $$P(t) = 46.350 + 4.230t, \tag{13}$$

  where t is years after 1992 and P is measured in thousands.

  b. On October 1, 2000, we have completed 8 years and 9 months since January 1, 1992, so October 1, 2000 represents $t = 8\tfrac{9}{12}$ or 8.75. Algebraically, to find the population when $t = 8.75$, we use (13) to find

  $$P(8.75) = 46.350 + 4.230 \times 8.75 \approx 83.363.$$

  Thus our linear model predicts 83,363 people in Yucca City on October 1, 2000.

  c. In this context, "double" means reaching the population that is two times the initial population. So we need to determine when $P(t)$ reaches 2×46.350 (thousand) by setting $P(t) = 92.700$ and solving for t:

  $$46.350 + 4.230t = 92.700$$
  $$4.230t = 46.350$$
  $$t = \frac{46.350}{4.230} \approx 10.957$$

So the function $P(t)$ reaches the value 92.700 (thousand) at 10.957 years after January 1, 1992. Ten years after January 1, 1992, is January 1, 2002, and 0.957 year is

$$0.957 \text{ year} \times \frac{12 \text{ months}}{\text{year}} = 11.484 \text{ months}$$

into the year 2002. So 11 months of 2004 have been completed, and we are somewhere in the twelfth month. Therefore, the population of Yucca City should double during the twelfth month of the year 2002, that is, during December 2002.

Figures 2.1.15 and 2.1.16 show graphical solutions for parts (b) and (c), respectively. While you may find the algebraic solutions easier while we are working with linear functions, you should be familiar with graphical techniques as well. You may prefer the graphical techniques when we solve similar problems involving functions that are not linear.

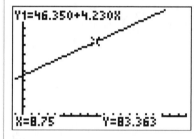

FIGURE 2.1.15 The graphical solution for Example 4(b).

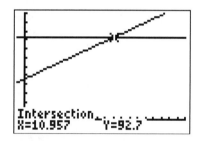

FIGURE 2.1.16 The graphical solution for Example 4(c).

Other Linear Models

The techniques that we used to create linear population models also apply to other linear models. All we need to know is the function's output for an input of 0 (its initial value) and its average rate of change.

EXAMPLE 5 | Finding a Linear Function Model

Suppose that the local garden club wants to raise money for neighborhood beautification and decides to sell a small cookbook containing its members' recipes. A print shop charges $200 to set up the press and $5 for each book produced.

    a. Find a linear function model that gives the total cost C of the book as a function of the number of books produced n.

    b. Find the total cost if 250 books are produced.

    c. How many books can be produced if the garden club's budget for this project is $3000?

    d. If the garden club plans to sell these cookbooks for $12 each, how many cookbooks must the members sell to break even?

SOLUTION

a. In order to write a linear function model, we need to identify both the function's initial value and its constant average rate of change. In this situation, the initial value is \$200 because this is the total cost if we produce 0 books. The constant average rate of change is \$5 because the cost increases by \$5 for each book that is produced. Thus, the total cost function is

$$C(n) = 200 + 5n.$$

b. To find the total cost if 250 books are produced, we need to find the output when the input is 250. So $C(250) = 200 + 5 \cdot 250 = 1450$, and the total cost of the 250 cookbooks is \$1450.

c. Here we need to find the input that produces an output of \$3000. Let's use the calculator's table feature to answer this question. We begin by storing our function in the **Y=** menu, letting **Y1=200+5x**. We know from part (b) that 250 books cost only \$1450, so the input we are looking for is larger than 250. Let's try setting **TblStart=300** and **ΔTbl=50**. Then the table in Fig. 2.1.17 shows that **Y1** is \$2950 when **X** is 550 and \$3200 when **X** is 600. Using the table once again with **TblStart=550** and **ΔTbl =10**, we see in Fig. 2.1.18 that **Y1** is exactly \$3000 when **X** is 560. Therefore, if the garden club's budget is \$3000, then 560 cookbooks can be produced.

X	Y1
300.00	1700.0
350.00	1950.0
400.00	2200.0
450.00	2450.0
500.00	2700.0
550.00	2950.0
600.00	3200.0

X=300

FIGURE 2.1.17 Table 1 for the numerical solution for Example 5(c).

X	Y1
550.00	2950.0
560.00	3000.0
570.00	3050.0
580.00	3100.0
590.00	3150.0
600.00	3200.0
610.00	3250.0

X=550

FIGURE 2.1.18 Table 2 for the numerical solution for Example 5(c).

d. Breaking even means that the garden club sells enough cookbooks to cover the cost of producing the books; that is, the revenue generated by selling the cookbooks must equal the cost of producing them. We already have the linear function $C(n)$ representing the total cost of producing n cookbooks; we must now find a function to represent the revenue earned by the cookbook sales. Since no revenue is earned if no cookbooks are sold, the initial value of our revenue function is \$0. The average rate of change for this function is \$12 per book, so the revenue is a linear function of the number of books n that are sold. If we use $R(n)$ to indicate the revenue function, we then have

$$R(n) = 0 + 12n \qquad \text{or} \qquad R(n) = 12n.$$

Leaving our original cost function $C(n)$ stored in **Y1** and putting the revenue function $R(n)$ in **Y2**, we choose an appropriate window to view the graphs, and then use the **intersect** command from the **CALC** menu as shown in Fig. 2.1.19. Thus, we see that the cost and revenue functions have the same output ($342.86) when the input is 28.57. But recall that the input here is number of cookbooks, and the garden club is not going to sell (nor is anyone likely to buy) a fraction of a cookbook. So to cover the cost of making the books sold, the garden club must both produce and sell 29 books.

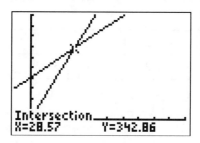

FIGURE 2.1.19 The graphical solution for Example 5(d).

Do you see that if the garden club produces and sells fewer than 29 books, the total cost will be greater than the revenue generated by sales? Since the garden club wants to make money, they need to sell more than 29 books because then the revenue earned is more than the cost of the books. However, it is important for the club to determine what the demand for the cookbooks is likely to be because any unsold books will decrease the overall profit.

Notice that parts (b) and (c) are the same "Here's the input, what's the output?" and "Here's the output, what's the input?" questions that we answered in Example 1, as well as in Chapter 1, when we looked at functions of all different types. Part (d) is a variation on the "Here's the output, what's the input?" question. While we are indeed looking for an input, the output is "hidden" in a sense. The output that we "have" is the output that is the same for our two linear functions. You might wish to practice using algebraic, graphical, and numerical methods to answer such questions so that you can decide which method or methods you prefer.

2.1 Exercises | Building Your Skills

In Exercises 1–4, first determine whether B is a function of A and then, if B is a function of A, decide (based on the average rate of change $\Delta B/\Delta A$) whether B is a linear function of A.

1.

A	10	10.5	13	15	16	20
B	3	4	9	13	15	23

2.

A	10	10.5	13	15	16	20
B	3	4	4	9	13	23

3.

A	2	4	5	5	8	11
B	3	4	9	13	15	23

4.

A	2	4	6	8	10	12
B	3	3	3	3	3	3

In Exercises 5–8 a population function P(t) is given, with t in years and P in thousands. Find the initial population and the function's constant rate of change a.

5. $P(t) = 123 + 6t$

6. $P(t) = 387 - 8t$

7. $P(t) = 487.139 + 20.558t$

8. $P(t) = 666.333 - 42.789t$

In Exercises 9–12, write a population function P(t) with the given initial population P_0 and constant rate of change a.

9. $P_0 = 42$ and $a = 5$

10. $P_0 = 73$ and $a = -6$

11. $P_0 = 324.175$ and $a = 15.383$

12. $P_0 = 786.917$ and $a = -21.452$

In Exercises 13–16 use the "resetting the clock" method to find a linear population function P(t) with the average rate of change a. Be sure to indicate the calendar year in which t = 0.

13. $P(1987) = 375$ and $a = 12$

14. $P(1983) = 685$ and $a = -24$

15. $P(1991) = 227.625$ and $a = 17.234$

16. $P(1993) = 847.719$ and $a = -60.876$

In Exercises 17–20 use the linear model $P(t) = P_0 + at$ to represent each city's population. Be sure to indicate the calendar year in which t = 0.

17. City A had a population of 35,500 on January 1, 1985, and it was growing at the rate of 1700 people per year. Assuming that this annual rate of change in the population of City A continues, find:

  a. Its population on January 1, 2000.
  b. The month of the calendar year in which its population reaches 85 thousand.

SECTION 2.1 Constant Change and Linear Growth

18. City B had a population of 375 thousand on January 1, 1992, and it was growing at the rate of 9250 people per year. Assuming that this annual rate of change in the population of City B continues, find:

   a. Its population on January 1, 2000.
   b. The month of the calendar year in which its population reaches 600 thousand.

19. City C had a population of 45,325 on January 1, 1985, and a population of 50,785 on January 1, 1990. Assuming that this annual rate of change in the population of City C continues, find:

   a. Its population on January 1, 2000.
   b. The month of the calendar year in which its population reaches 75 thousand.

20. City D had a population of 428 thousand on January 1, 1992, and a population of 422 thousand on January 1, 1997. Assuming that this annual rate of change in the population of City D continues, find:

   a. Its population on January 1, 2000.
   b. The month of the calendar year in which its population is 400 thousand.

21. Find the month of the calendar year during which Cities A and C of Problems 17 and 19 have the same population.

22. Find the month of the calendar year during which Cities B and D of Problems 18 and 20 have the same population.

Applying Your Skills

23. Consumer credit experts are concerned that Americans' credit card debt continues to increase. *USA Today* reported that in 1995, U.S. bank credit card loans totaled $358 billion, while in 2004, they totaled $697 billion.

   a. Find a linear function model $C(t) = \cdots$ that gives the U.S. bank credit card loans as a function of t, number of years after 1995. (Round the average rate of change to three decimal places.)
   b. According to your function model, in what month and year will U.S. bank credit card loans reach $1000 billion (1 trillion dollars)?
   c. Use your model to predict the U.S. bank credit card loans in 2006.

24. In 1997, Allergan, the maker of Botox, earned $90 million in sales of Botox, used as injections for both cosmetic and medical purposes. Allergan's revenues from Botox in 2001 were $310 million.

   a. Find a linear function model $R(t) = \cdots$ that gives Botox revenue as a function of t, the number of years after 1997.
   b. Use your model to predict Allergan's Botox revenue in 2010.
   c. According to your function model, in what month and year will Allergan's Botox revenue reach $500 million?

25. According to the U. S. Census, the population of St. Louis was 397 thousand in 1990 and 348 thousand in 2000.

   a. Find a linear function model $P(t) = \cdots$ that gives the population of St. Louis as a function of t, the number of years after 1990.
   b. Use your model to predict the population of St. Louis in 2003.
   c. If the actual population of St. Louis in 2003 was 332 thousand, was your answer in **(b)** a reasonable prediction?

d. Use your model to predict the month of the calendar year in which the population of St. Louis falls to 300 thousand.

26. According to the U.S. Census, the population of Lexington, Kentucky, was 225 thousand in 1990 and 261 thousand in 2000.
    a. Find a linear function model $P(t) = \cdots$ that gives the population of Lexington as a function of t, the number of years after 1990.
    b. Use your model to predict the population of Lexington in 2003.
    c. If the actual population of Lexington in 2003 was 267 thousand, was your answer in **(b)** a reasonable prediction?
    d. Use your model to predict the month of the calendar year in which the population of Lexington grows to 300 thousand.

27. Use your models from Exercises 25 and 26 to predict the month of the calendar year in which the populations of St. Louis and Lexington are the same.

28. One of the services that Statewide Landscape provides is grinding out stumps of trees that have fallen or been otherwise removed. The company charges a $45 "trip fee" and $2.00 per inch of diameter of the stump (or stumps) to be ground.
    a. Find a linear function model $P(i) = \cdots$ that gives the price of stump grinding as a function of the number of inches i of diameter.
    b. If the customer has a budget of $275 for stump grinding, how many inches of diameter of tree stumps can be ground?
    c. If a total diameter of 200 inches of tree stumps needs to be ground, how much will the customer be charged?

29. In 2006, AT&T advertised a "one rate" plan for long distance customers in New Mexico, with a $3.95 monthly fee and long distance calls charged at 7¢ per minute. A competing company, Qwest, offered a "15¢ Single Rate Plan" for long distance, which featured a 99-cent monthly fee and a charge of 15¢ per minute for calls.
    a. Find a linear function model $A(t) = \cdots$ that gives the monthly cost of AT&T long distance service as a function of t, the number of minutes used.
    b. How much is the monthly cost of the AT&T long distance if the customer uses 1 hour of long distance time?
    c. Find a linear function model $Q(t) = \cdots$ that gives the monthly cost of Qwest long distance service as a function of t, the number of minutes used.
    d. How much is the monthly cost of the Qwest long distance if the customer uses 1 hour of long distance time?
    e. When is it cheaper to use the AT&T plan, and when is it cheaper to use the Qwest plan?

30. In 2006, Bell South advertised both a "Dollar Plan" and a "Nickel Plan" to its long distance customers in Charlotte, North Carolina. The Dollar Plan had a service charge of $1.00 a month, and a charge of 10¢ per long distance minute. The Nickel Plan had a $5.95 monthly fee, with long distance calls charged at 5¢ per minute.
    a. Find a linear function model $D(t) = \cdots$ that gives the monthly cost of the Dollar Plan for long distance as a function of t, the number of minutes used.
    b. How much is the monthly cost of the Dollar Plan if the customer uses 1 hour of long distance time?
    c. Find a linear function model $N(t) = \cdots$ that gives the monthly cost of the Nickel Plan as a function of t, the number of minutes used.
    d. What is the monthly cost of the Nickel Plan if the customer uses 1 hour of long distance time?

e. When is it cheaper to use the Dollar Plan, and when is it cheaper to use the Nickel Plan?

31. As a person ages, the number of additional years he or she can expect to live declines. U.S. government statisticians use population data to provide estimates of how much longer a person will live based on gender and current age. A newborn boy (0 years old) in 2001 was expected to live 74.4 years, while a man 30 years old in 2001 was expected to live an additional 46.2 years.

    a. Find a linear function model $L(a) = \cdots$ that gives the years of life expectancy as a function of a, a male's age in 2001.
    b. Use your model to predict how much longer a male is expected to live if he is 18 in 2001.
    c. In 2001 how old is a male who is expected to live 32.1 additional years?

32. The excerpt shown in Fig 2.1.20 is from the Pew Hispanic Center's June 2005 paper "Unauthorized Migrants: Numbers and Characteristics" discusses the

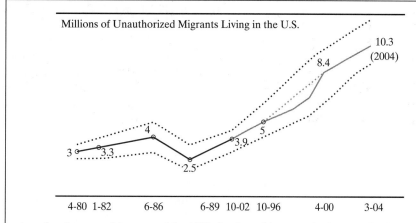

A major demographic story of the 1990s is a broad increase in the unauthorized population. This chart portrays the growth trend in the unauthorized population while illustrating the uncertainty involved with the dotted bands of error and the alternative trend line at the end incorporating the results based on the 2000 Census and subsequent March CPSs through 2004. (Note that smooth lines should not be interpreted to mean that there are not annual fluctuations in growth. The lines merely connect the dates for which stock estimates are available.)

Because of inherent uncertainties in the residual technique, the difference in successive annual estimates of the unauthorized population is not a valid measure of growth. However, it is possible to use differences taken over longer intervals to measure growth. Thus, the average annual change over the 2000–2004 period is about 485,000 or 10.3 million minus 8.4 million divided by 4. For the entire decade of the 1990s, growth averaged just about 500,000 per year. However, there are a number of data sources that point to substantially larger growth increments at the very end of the 1990s (and possibly at the end of the 1980s and the very early 1990s).

The apparent slowdown in growth after 1996 in probably not a real decline but is attributable to undercoverage in the data used to estimate unauthorized flows. Similarly, the apparent very rapid growth to 2000 may (or may not) be an accurate depiction of the trend but may reflect data anomalies in the CPSs of the late 1990s.

The decrease in size from 1986 to 1989 is caused by the IRCA legalizations that removed immigrants from the unauthorized population by granting them legal status, not by making them leave the country.

FIGURE 2.1.20 Unauthorized at New Heigh—Details of Trend Uncertain

difficulty of accurately predicting the number of illegal immigrants in the United States.

a. This paper suggests that considering the average rate of change over longer intervals is a reasonable way to measure the growth of the unauthorized migrant population. Use the data from April 1980 (3 million migrants) and from April 2000 (8.4 million migrants) to find a linear model that gives the number of unauthorized migrants as a function of years after April 1980.

b. Use your model to predict the number of unauthorized migrants in April 2004. How does your answer compare to the value given in the chart for March 2004?

c. In what year does the model predict that the number of unauthorized migrants will grow to 15 million?

2.2 LINEAR FUNCTIONS AND GRAPHS

Recall that in the last section we investigated the relationship between temperature measured in Celsius degrees and temperature measured in Fahrenheit degrees. We determined that, because the average rate of change was constant, we could express Fahrenheit temperature as a linear function of Celsius temperature.

The table of values indicated that for a Celsius temperature of 0 degrees, the Fahrenheit temperature was 32 degrees. Furthermore, we found that the average rate of change was 1.8 degrees F per degree C. Using the techniques of Section 2.1, we can write this temperature function as $F(C) = 32 + 1.8C$.

Suppose that, instead of having the function's initial value (its output when input $= 0$), we were given two pairs of (C, F) values, say $(30, 86)$ and $(70, 158)$. As before, we can find the function's average rate of change:

$$\frac{\Delta F}{\Delta C} = \frac{158 - 86}{70 - 30} = \frac{72}{40} = 1.8.$$

But how do we determine a rule for this function? We know that the function has this same average rate of change between any two ordered pairs. So consider $(C, F(C))$ any pair of Celsius and Fahrenheit temperatures. Then

$$\frac{\Delta F}{\Delta C} = \frac{F(C) - 86}{C - 30} = 1.8, \quad \text{and so} \quad F(C) - 86 = 1.8(C - 30).$$

Therefore

$$F(C) = 86 + 1.8(C - 30). \tag{1}$$

To verify that this is the very same linear function obtained earlier (as well as in Section 2.1), we can simplify (1) to get

$$F(C) = 86 + 1.8C - 54, \text{ so}$$

$$F(C) = 32 + 1.8C.$$

In general, suppose that we know a function's average rate of change a and a pair of input and output values $(x_1, f(x_1))$. Then (as before)

$$\frac{\Delta f}{\Delta x} = \frac{f(x) - f(x_1)}{x - x_1} = a.$$

So
$$f(x) - f(x_1) = a(x - x_1),$$
and then
$$f(x) = f(x_1) + a(x - x_1). \tag{2}$$

This is the **point-slope form** of a linear function. In this form, we see visibly displayed both the slope a of the function's graph and a particular point $(x_1, f(x_1))$ lying on the graph.

EXAMPLE 1 — Finding a Linear Function Model

Use the point-slope form to find a linear function $f(x)$ such that $f(2) = 1$ and $f(4) = 15$.

SOLUTION

To find a linear function having these two pairs of input-output values, we first find the average rate of change a:

$$\frac{\Delta f}{\Delta x} = \frac{f(4) - f(2)}{4 - 2} = \frac{15 - 1}{4 - 2} = \frac{14}{2} = 7.$$

Now we substitute appropriate values in (2), the point-slope form of the linear function:

$$f(x) = f(2) + 7(x - 2),$$
$$f(x) = 1 + 7(x - 2), \tag{3}$$

or

$$f(x) = f(4) + 7(x - 4),$$
$$f(x) = 15 + 7(x - 4). \tag{4}$$

If we prefer to have the slope-intercept form of the function (as in Section 2.1), we merely simplify either (3) or (4).
    That is, $f(x) = 1 + 7x - 14 = 7x - 13$ or $f(x) = 15 + 7x - 28 = 7x - 13$. Notice that whichever version of the point-slope form we use, we end up with the same slope-intercept form of the function. ●

We can use the point-slope form to find linear functions in applied settings as well.

EXAMPLE 2 — Finding a Linear Cost Function

AT&T's "one rate" plan for long distance customers in New Mexico charged a monthly fee with long distance calls charged at 7¢ per minute. In June 2006, a customer used 87 long distance minutes, and her bill was $10.04.

   a. Use the point-slope form to find a linear function $C(t)$ giving the monthly cost of long distance service as a function of the number of minutes of long distance calls made.

b. What was the monthly service charge for the "one rate" plan?

c. How many minutes did the customer use in July 2006 if her bill was \$8.92?

SOLUTION

a. Here the average rate of change a is the per-minute charge (\$0.07), and $C(87) = 10.04$. So, the cost of long distance service as a function of minutes is given by

$$C(t) = 10.04 + 0.07(t - 87). \qquad (5)$$

b. The service charge is the fee just for having the calling plan, even if no minutes of long distance time are used. So the service charge is the output for the input of 0, or the function's initial value. This is once again our standard question, "Here's the input, what's the output?" with the given input being 0. So we substitute $t = 0$ in (5) and calculate

$$C(0) = 10.04 + 0.07(0 - 87).$$

Then

$$C(0) = 10.04 + 0.07(-87),$$

and so

$$C(0) = 3.95.$$

Thus, the monthly service charge for the "one rate" plan is \$3.95.

c. In this case, we are given an output and are required to find the appropriate input. As always, we have several choices of methods for determining the input that makes the function output 8.82. We might first notice that now that we have found the function's initial value [in part (b)], we can write an equivalent slope-intercept form for our function:

$$C(t) = 3.95 + 0.07t. \qquad (6)$$

We can store our original function from (5) in **Y1** and the slope-intercept form from (6) in **Y2**. Using the table settings **TblStart=0** and **ΔTbl=10**, we see in Figs. 2.2.1 and 2.2.2 that the function values are the same. So the two different forms do describe the same function, and we are free to use whichever one we prefer.

X	Y1	Y2
0.0000	3.950	3.950
10.000	4.650	4.650
20.000	5.350	5.350
30.000	6.050	6.050
40.000	6.750	6.750
50.000	7.450	7.450
60.000	8.150	8.150

X=0

FIGURE 2.2.1 Table 1 for the numerical solution for Example 2(c).

X	Y1	Y2
70.000	8.850	8.850
80.000	9.550	9.550
90.000	10.250	10.250
100.00	10.950	10.950
110.00	11.650	11.650
120.00	12.350	12.350
130.00	13.050	13.050

X=130

FIGURE 2.2.2 Table 2 for the numerical solution for Example 2(c).

Furthermore, we see that the cost reaches \$8.92 for some t between $t = 70$ and $t = 80$. If we reset our table to start at 70 and use increments of 1 (rather than 10), we obtain the table in Fig. 2.2.3. So, it is an input of 71 that generates an output of 8.92. Therefore, if a customer's bill is \$8.92, she used 71 minutes of long distance.

X	Y1	Y2
70.000	8.850	8.850
71.000	8.920	8.920
72.000	8.990	8.990
73.000	9.060	9.060
74.000	9.130	9.130
75.000	9.200	9.200
76.000	9.270	9.270

X=70

FIGURE 2.2.3 Table 3 for the numerical solution for Example 2(c).

While we solved this problem numerically, we would have obtained the same result if we had used either an algebraic or a graphical method.

The method used in the previous examples can also be used to find linear population models if we choose not to "reset the clock."

EXAMPLE 3 Finding a Linear Population Model

According to the Census Bureau, the population of Providence, Rhode Island, was 160,728 in 1990 and 173,618 in 2000.

a. Use the point-slope form of a linear function to describe the population of Providence as a function of the year.

b. According to your model, in what month and year will the population of Providence be 200,000?

c. According to your model, what will Providence's population be in 2008?

SOLUTION

a. We first need to find the average rate of change a:

$$a = \frac{\Delta P}{\Delta t} = \frac{173618 - 160728}{2000 - 1990} = \frac{12890}{10} = 1289.$$

Then

$$P(t) = 160728 + 1289(t - 1990).$$

b. To find the year in which the population reaches 200,000, we need to find the input for which the function's output is 200000. That is, we need to solve the equation $160728 + 1289(t - 1990) = 200000$. In order to solve this equation graphically, we store our population function in **Y1** and the output value 200000 in **Y2**. Recalling that the input values for our function are actual calendar years, we set the window as indicated in Fig. 2.2.4. Then Fig. 2.2.5 shows the intersection point found

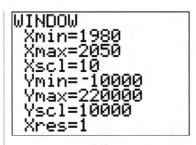 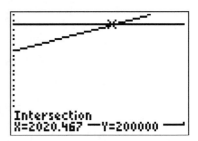

FIGURE 2.2.4 The window setting for Example 3(b).

FIGURE 2.2.5 The intersection point for Example 3(b).

by the **intersect** feature of the calculator. Unless we are told otherwise, we assume that our input years begin on January 1, so 2020.467 would be $0.467 \times 12 = 5.604$ months into the year 2020. Our model predicts that the population of Providence will grow to 200,000 during the sixth month (June) of 2020.

c. Here we are looking for $P(2008)$, which is $160728 + 1289(2008 - 1990) = 183930$. Thus, the population of Providence in 2008 is predicted to be 183,930 people.

Notice that in Example 3, we left our function in point-slope form because we were using actual calendar years there. It makes no sense to simplify this function to put it into slope-intercept form since this intercept would indicate the population of Providence in the year 0 (more than 20 centuries ago). There are several problems with this notion. First, there is no year designated "0" in the Western calendar. We have the year 1 BC, followed by the year 1 AD. Second, even if there were such a year way back then, there was no city of Providence. Finally, if we went backward 2000+ years using this model, we would predict a *negative* value for the population, which is certainly not sensible.

This example demonstrates a problem we may encounter when we create functions to model real-world data. Sometimes our meaningful or "sensible" domain does not include 0, making meaningless any value we might find there for the y-intercept. As we noted in Chapter 1, consideration of the domain of a function is a critical part of the function definition.

Straight Lines and Linear Graphs

The **graph** of the linear function $f(x) = ax + b$ is the straight line consisting of all points (x, y) in the xy-plane that satisfy the equation

$$y = ax + b. \tag{7}$$

Although we often abbreviate and speak of "the straight line $y = ax + b$," it is important to understand the differences among

- the linear *function f*,
- its defining *formula* $f(x) = ax + b$,
- the *equation* $y = ax + b$, and
- the *graph* with equation $y = ax + b$.

In particular, the words *function, equation,* and *graph* should not be used interchangeably. In mathematics (as elsewhere) it's important to "say what you mean and mean what your say."

EXAMPLE 4 Distinguishing Rising and Falling Lines

Define **Y1 = A*X + B** in the **Y=** menu of your graphing calculator. Then enter several different pairs of slope-intercept values as indicated in Fig. 2.2.6, pressing **GRAPH** after each is entered.

```
1→A: -2→B
                    -2
-2→A: 5→B
                    5
3→A: -7→B
                    -7
-0.5→A: 3→B
```

FIGURE 2.2.6 Various slopes a and y-intercepts b.

Describe how changing the parameter a affects the graph of the line.

SOLUTION

As illustrated in Figs. 2.2.7 and 2.2.8, we find that

- the line *rises* (from left to right) if the slope a is *positive*;
- the *larger* is $a > 0$, the more *steeply* the line rises;
- the line *falls* (from left to right) if the slope a is *negative*;
- the *larger* in absolute value is $a < 0$, the more *steeply* the line falls.

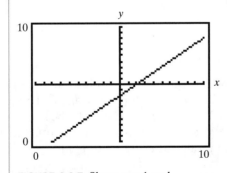

FIGURE 2.2.7 Slope $a = 1$ and y-intercept $b = -2$, so $y = x - 2$.

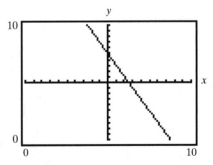

FIGURE 2.2.8 Slope $a = -2$ and y-intercept $b = 5$, so $y = -2x + 5$.

A straight line is a simple example of a graph of an equation.

> **DEFINITION: Graph of an Equation**
> The **graph** of an equation involving two variables x and y consists of all points in the xy-plane whose coordinates (x, y) satisfy the equation.

62 CHAPTER 2 Linear Functions and Models

For instance, there are three types of equations whose graphs are straight lines:
- Equations whose graphs are vertical lines
- Equations whose graphs are horizontal lines
- Equations whose graphs are slanted lines (neither vertical nor horizontal)

EXAMPLE 5 | Recognizing a Vertical Line

Sketch and describe a graph of the equation $x = -3$.

SOLUTION

The graph of the equation $x = -3$, illustrated in Fig. 2.2.9, is a vertical line. Only ordered pairs (x, y) whose first coordinate x equals -3 satisfy this equation. The second coordinate y can be any real number. Notice that in this relationship, y is *not* a function of x because the single value -3 of x is paired with infinitely many different values of y. A graph that *is* a vertical line certainly fails our "vertical line test" for a function.

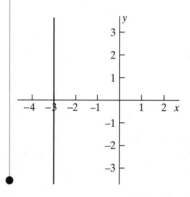

FIGURE 2.2.9 The graph of $x = -3$.

EXAMPLE 6 | Recognizing a Horizontal Line

Sketch and describe a graph of the equation $y = 2$.

SOLUTION

The graph of the equation $y = 2$, illustrated in Fig. 2.2.10, is a horizontal line. In this case, the first coordinate x can be any real number, but the second coordinate y must equal 2. Here y *is* a function of x because each x-value is paired with only one y-value. What makes this function unusual is that every point on the graph has the same y-value as every other point.

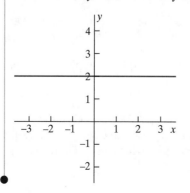

FIGURE 2.2.10 The graph of $y = 2$.

EXAMPLE 7

Recognizing a More General Line

Sketch and describe a graph of the equation $y = -2x + 1$.

SOLUTION

Figure 2.2.11 shows the graph of the equation $y = -2x + 1$. Note that it is a slanted line, falling from left to right, with y-intercept 1. We see that again y is a function of x since each x-value is paired with only one y-value. Unlike the previous example, different points on the graph have different y-values.

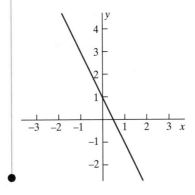

FIGURE 2.2.11 The graph of $y = -2x + 1$.

The graph of a function is a special case of the graph of an equation.

> **DEFINITION: Graph of a Function**
> The **graph** of the function f is the graph of the equation $y = f(x)$.

Thus the graph of the function f consists of all points in the plane whose coordinates have the form $(x, f(x))$ with x being in the domain of f. Observe that the second coordinate of any such point is the value of f at its first coordinate (Fig. 2.2.12).

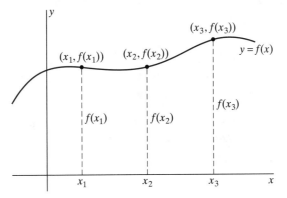

FIGURE 2.2.12 The graph of the function f.

We have been using this definition somewhat intuitively when graphing a function in our calculator. We have typed our function rule into the **Y=** menu in the calculator, using **X** as our input value. Thus the definition confirms our understanding that the notation $f(x)$ is another (somewhat fancy) name for y. This notation is convenient in that it gives us a shorthand for specifying a particular set of input and output values. Rather than having to say (for example) "when $x = 2$, $y = 5$," we can say $f(2) = 5$. Much of the mathematical notation that may appear confusing at first glance actually makes it easier for us to "talk" about mathematical relationships.

2.2 Exercises | Building Your Skills

In Exercises 1–4 use the point-slope form to find a linear function $f(x)$ having the given values.

1. $f(2) = 7$ and $f(5) = 13$
2. $f(3) = 5$ and $f(7) = 17$
3. $f(-2) = 19$ and $f(3) = -16$
4. $f(1) = 8$ and $f(3) = -10$

In Exercises 5–10 write (in slope-intercept form) an equation of the line L described and sketch its graph.

5. L passes through the origin and the point $(2, 3)$.
6. L is vertical and has x-intercept 7.
7. L is horizontal and passes through the point $(3, -5)$.
8. L has x-intercept 2 and y-intercept -3.
9. L passes through $(-1, -4)$ and has slope ½.
10. L passes through $(4, 2)$ and rises (left to right) at a 45-degree angle.

Three points A, B, and C lie on a single straight line if and only if the slope of AB equals the slope of BC. In Exercises 11–14 plot the three given points and then use the slopes AB and BC to determine whether all three points lie on a single line.

11. $A(-1, -2)$, $B(2, 1)$, $C(4, 3)$
12. $A(-2, 5)$, $B(2, 3)$, $C(8, 0)$
13. $A(-1, 6)$, $B(1, 2)$, $C(4, -2)$
14. $A(-3, 2)$, $B(1, 6)$, $C(8, 14)$

In Exercises 15–18 find a linear population function $P(t)$ with the average rate of change a. Do not "reset the clock."

15. $P(1987) = 375$ and $a = 12$
16. $P(1983) = 685$ and $a = -24$
17. $P(1991) = 227.625$ and $a = 17.234$
18. $P(1993) = 847.719$ and $a = -60.876$

Applying Your Skills

19. The Fahrenheit temperature F and the absolute (Kelvin) temperature K are linear functions of each other. Moreover, $K = 273.16$ when $F = 32$, and $K = 373.16$ when $F = 212$.

    a. Find a linear function $K(F)$ that gives the absolute temperature as a function of the Fahrenheit temperature.
    b. Use your function from part (a) to find the Fahrenheit temperature when the Kelvin temperature is 0 ("absolute zero").
    c. What is the Kelvin temperature when the Fahrenheit temperature is 0?

20. The owner of a grocery store finds that she can sell 980 gallons of milk each week at $2.69 per gallon and 1220 gallons of milk each week at $2.49 per gallon.

    a. Find a linear function $G(p)$ that gives the number of gallons sold as a function of the price of the milk.
    b. How many gallons would she then expect to sell at $2.99 per gallon?
    c. According to your function, if the owner sold 800 gallons of milk, what was the price per gallon?

21. The length L in centimeters of a copper rod is a linear function of its Celsius temperature C.

    a. If $L = 124.942$ when $C = 20$ and $L = 125.131$ when $C = 110$, find a linear function $L(C)$ that gives the length of the rod as a function of its Celsius temperature.
    b. What is the length of the rod if the temperature is 50°C?
    c. If the rod is 125.0764 centimeters long, what is its Celsius temperature?

22. Taking a cruise has become an increasingly popular form of vacationing for Americans. In 1995, 4.4 million Americans took a cruise; in 2003 the number had risen to 8.2 million. (Source: *USA Today*)

    a. Assuming that the number of Americans taking cruises was increasing at a constant rate over this period, find a function $A(y)$ that gives the number of Americans taking a cruise as a function of the year.
    b. In what year can the travel industry expect the number of Americans taking a cruise to reach 12 million?

23. According to the U.S. Department of Agriculture, there were 705 farms in Cheyenne County, Nebraska, in 1987, and 615 farms in 2002.

    a. Find a linear function $F(t)$ that gives the number of farms in Cheyenne County as a function of the year.
    b. Use your model to find the year in which the number of farms in Cheyenne County will fall to 580.

24. The number of calories that a person burns depends not only on what type of activity is performed, but also on the person's weight. A person who weighs 130 pounds will burn 590 calories playing racquetball competitively for an hour, while a 155-pound individual will burn 704 calories doing the same activity.

    a. Find a linear function $C(w)$ that gives the number of calories burned per hour of competitive racquetball as a function of weight.

b. According to your model, how many calories will a person weighing 180 pounds burn if he or she plays competitive racquetball for an hour?

25. It is fairly common knowledge that crickets are sensitive to temperature and chirp faster as the temperature rises. There are many different formulas for obtaining the temperature when you know how frequently a cricket chirps. Most of these formulas are simple linear functions.

   a. Suppose that you count the number of times a cricket chirps in 10 seconds and find that it chirped 14 times when the temperature was 61°F. When the temperature was 88°F the cricket chirped 32 times in 10 seconds. Find a linear function $T(c)$ that gives the temperature in degrees Fahrenheit as a function of the number of cricket chirps in a 10-second interval.
   b. What is the temperature if the cricket chirps 24 times in 10 seconds?

26. According to the U.S. Census, the population of Virginia Beach, Virginia, was 393 thousand in 1990 and 425 thousand in 2000.

   a. Find a linear function $P(t)$ that gives the population P as function of the year t.
   b. Use your model to predict the population of Virginia Beach in 2003.
   c. According to your model, what is the yearly rate of change in the population of Virginia Beach?
   d. Use your model to predict the month of the calendar year in which the population of Virginia Beach grows to 500 thousand people.

27. According to the U.S. Census, the population of Milwaukee, Wisconsin was 628 thousand in 1990 and 597 thousand in 2000.

   a. Find a linear function $P(t)$ that gives the population P as function of the year t.
   b. Use your model to predict the population of Milwaukee in 2003.
   c. According to your model, what is the yearly rate of change in the population of Milwaukee?
   d. Use your model to predict the month of the calendar year in which the population Milwaukee is 400 thousand people.

28. Use your models from Exercises 30 and 31 to predict the month of the calendar year in which the populations of Virginia Beach and Milwaukee are the same.

2.3 PIECEWISE-LINEAR FUNCTIONS

Sometimes function relationships cannot be described by a single linear function. There are many situations in which different intervals of input values generate different linear functions. For instance, a population P might grow at one constant rate on one time interval and at a different constant rate on another time interval. We call such functions piecewise-linear functions because each section (or piece) of the graph is a portion of a line.

> **DEFINITION: Piecewise-linear Function**
>
> A *piecewise-linear function* is a function that is defined by different linear functions on different intervals.

EXAMPLE 1　Finding a Piecewise-Linear Population Model

Suppose that the population of Springfield was 150 thousand in 1970 and from 1970 to 1990 the population grew at the rate of 10 thousand per year. However, due to new industry acquired in 1990, additional people started moving in steadily. As a result, after 1990 the population of Springfield grew at the increased rate of 20 thousand people per year. Find a piecewise-linear function $P(t)$ that gives the population of Springfield as a function of the year.

SOLUTION

In 1970 the population was 150 thousand, giving us a (t, P) ordered pair (1970, 150) and an average rate of change $a = 10$. Using the techniques of Section 2.2, we find that the population P is described by the function $P(t) = 150 + 10(t - 1970)$ for t from 1970 to 1990. Noting that the population of Springfield in 1990 was 350 thousand (why?), and the average rate of change thereafter was $a = 20$, we use the same techniques to determine that the city's population is described by $P(t) = 350 + 20(t - 1990)$ for t from 1990 onward. Thus, the piecewise-linear function giving Springfield's population is described for $t \geq 1970$ by

$$p(t) = \begin{cases} 150 + 10(t - 1970) & \text{if } 1970 \leq t \leq 1990, \\ 350 + 20(t - 1990) & \text{if } t > 1990. \end{cases}$$

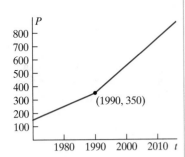

FIGURE 2.3.1 The piecewise-linear population function of Example 1.

Figure 2.3.1 shows the graph of the population function $P(t)$.

In your previous study of mathematics, you've probably encountered the absolute value function, $f(x) = |x|$. The absolute value of a number measures how far the number is away from 0, without regard to its direction from 0. Thus, both $|-2|$ and $|2|$ are equal to 2 because both numbers are two units away from 0; negative 2 is two units to the left of 0, while positive 2 is 2 units to the right of 0.

Figure 2.3.2 shows the calculator graph **Y1 = abs(X)** of the absolute value function defined mathematically by

$$|x| = \begin{cases} -x & \text{if } x < 0, \\ x & \text{if } x \geq 0. \end{cases}$$

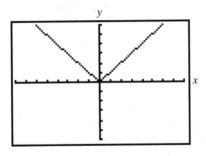

FIGURE 2.3.2　Graph of $y = |x|$.

We see that the function is defined differently for $x < 0$ and for $x \geq 0$. The two "pieces" of the function rule correspond to the two straight line "pieces" of the graph. The graph "turns the corner" at the origin because it consists of

- the left half of the line $y = -x$, which falls from left to right with slope -1, and
- the right half of the line $y = x$, which rises from left to right with slope $+1$.

While such functions definitions may seem odd initially, piecewise functions frequently turn up in applied settings.

EXAMPLE 2 Finding a Piecewise-Linear Cost Function

In 1999, Mindspring Enterprises offered a "light" plan for Internet access in which the monthly cost was $6.95 for up to 5 hours of connection time, with additional hours charged at $2.00 each. Find a piecewise linear function that gives the cost C of Internet access as a function of the number of hours h of connection time.

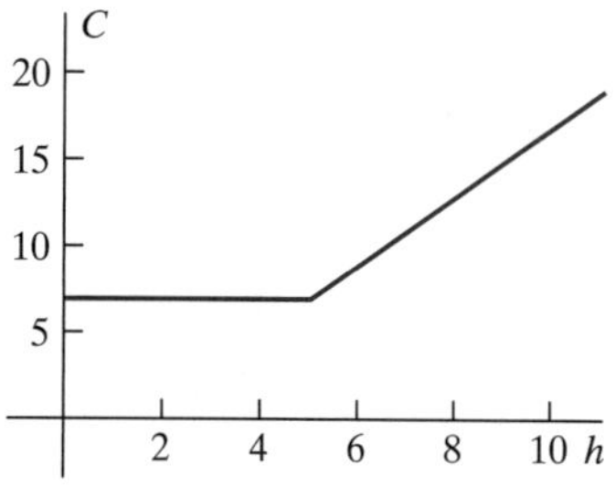

SOLUTION

If a customer uses between 0 and 5 hours of connection time, the monthly cost remains the same. That is, $C(h) = 6.95$ for h between 0 and 5, inclusive. Then, for each additional hour, the customer is charged $2 per hour. So for over 5 hours, $C(h) = 6.95 + 2(h - 5)$. Thus, our two-part rule for this function is

$$C(h) = \begin{cases} 6.95 & \text{if } 0 \leq h \leq 5 \\ 6.95 + 2(h - 5) & \text{if } h > 5. \end{cases}$$

FIGURE 2.3.3 The graph of the cost function $C(h)$ for the Mindspring "light" plan.

Figure 2.3.3 shows the graph of this function.

Note that the graph consists of two straight-line pieces—one horizontal and one slanted—that meet at the point (5, 6.95) and correspond to the two parts of the formula. (You may recognize this symbolic form from Chapter 1, where you answered a "Here's the output, what's the input?" question for this function.)

Graphical, Numerical, and Symbolic Viewpoints

In working examples in Chapter 1 and this chapter we demonstrated graphical, numerical, and symbolic methods of solution, and you have probably used all of these methods at one time or the other as well. Likewise, applications of piecewise functions can involve looking at the same function from different viewpoints.

EXAMPLE 3 Describing Distance As a Function of Time

Suppose that a car begins (at time $t = 0$ hours) in Hartford, Connecticut and travels to Danbury 60 miles away at a constant speed of 60 miles per hour. The car stays in Danbury for exactly 1 hour, and then returns to Hartford, again at a constant speed of 60 miles per hour.

The car's distance from Hartford d is a function of the time t in hours.

a. Describe the car's distance from Hartford graphically.
b. Describe the car's distance from Hartford symbolically.

SOLUTION

a. Here we are asked to create a graph of the function $d(t)$ from its verbal description. For the first hour, the car is traveling a constant rate of speed away from Hartford, so the distance is increasing at a constant rate. This means that for t between 0 and 1, d is an increasing linear function of t. In order to graph $d(t)$ on this interval, we only need two points. We know that at time $t = 0, d = 0$, and at time $t = 1, d = 60$, so our graph is the line segment connecting $(0, 0)$ and $(1, 60)$.

For the next hour, the car remains in Danbury, so for t between 1 and 2, d is a constant function of t. So on this interval, the graph d is the line segment connecting $(1, 60)$ and $(2, 60)$.

During the last hour, the car returns to Hartford, again at a constant rate of speed. Thus, the distance is decreasing at a constant rate, and for t between 2 and 3, d is a decreasing linear function of t. At time $t = 3$, the car has returned to Hartford and $d = 0$. The last "piece" of our graph is the line segment connecting $(2, 60)$ and $(3, 0)$.

The graph of $d(t)$ is shown in Fig. 2.3.4.

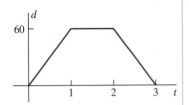

FIGURE 2.3.4 Graph of $d(t)$.

b. Now we will build a symbolic function $d(t)$ based on the verbal description. For the first hour, the constant average rate of change of distance (with respect to time) is 60 miles per hour. The car's initial distance from Hartford is 0. So for t between 0 and 1, $d(t) = 60t$.

During the next hour, the car stays in Danbury, so its distance from Hartford is constant—60 miles. That is, for t between 0 and 1, $d(t) = 60$.

Finally, during the last hour, the car is returning to Hartford, so the constant average rate of change of distance (with respect to time) is -60 miles per hour. Our starting point for the journey back is at $t = 2, d = 60$. Using the point-slope form, we find that $d(t) = 60 - 60(t - 2)$ for t between 2 and 3. [We can simplify this to $d(t) = 180 - 2t$ if we wish, or merely leave it as it is.]

We have used the word "between" in a vague manner in discussing this symbolic form. We can see that the three "pieces" of graph "match up"—there are no breaks or jumps in the graph. This means that our three function rules also "match up" at their endpoints, so we may include the endpoint t-values in whichever "piece" of the domain we prefer. For no particular reason, we will include these two values ($t = 1$ and $t = 2$) with their left-hand rules. Then the distance from Hartford (in miles) as a function of time (in hours) is given by the three-part formula

$$d(t) = \begin{cases} 60t & \text{if } 0 \leq t \leq 1, \\ 60 & \text{if } 1 < t \leq 2, \\ 60 - 60(t - 2) & \text{if } 2 < t \leq 3. \end{cases}$$

The domain of this function is the interval $0 \leq t \leq 3$, and we can see from its graph that its range is the interval $0 \leq d \leq 60$.

70 CHAPTER 2 Linear Functions and Models

In Example 2, we began with a verbal description of the function and then constructed both a graphical description and a symbolic description. In the next example, we begin with a table of values.

EXAMPLE 3 Finding and Graphing a Step Function

Figure 2.3.5 shows a portion of the rate table for UPS Next Day Air delivery from the state of Alaska.

- Any fraction of a pound more than the weight shown in the rate chart requires the next higher rate.

- The Letter rate applies only to document, correspondence and electronic media shipments sent in a UPS Express Envelope. Express Envelopes containing items other than those listed above and Express Envelopes exceeding one pound that are

Zones	22	24
	Alaska Metro	48 Contiguous States, Hawaii Metro and Puerto Rico
Letter	$15.25	$17.25
1 lbs.	21.00	27.25
2	23.75	30.00
3	26.25	32.50
4	28.25	35.50
5	30.25	38.00

FIGURE 2.3.5 Rates for package delivery from Alaska.
Source: www.ups.com.

a. Find a piecewise-linear function $R(w)$ that gives the rate to send a package from Alaska to one of the "lower 48" states as a function of its weight w if the weight is no more than 5 pounds.
b. Graph $R(w)$.

SOLUTION

a. In the table, we see the rates for the whole number of pounds from 1 through 5. Furthermore, the note accompanying the table tells us that any fraction of a pound more than the weight shown requires the next higher rate. That means that any package that weighs up to and including 1 pound costs $27.25; a package weighing more than 1 pound but not more than 2 pounds costs $30.00, and so on. So our rate function is given by

$$R(w) = \begin{cases} 27.75 & \text{if } 0 < w \leq 1, \\ 30.00 & \text{if } 1 < w \leq 2, \\ 32.50 & \text{if } 2 < w \leq 3, \\ 35.50 & \text{if } 3 < w \leq 4, \\ 38.00 & \text{if } 4 < w \leq 5. \end{cases}$$

Notice that $R(w)$ is a special kind of piecewise-linear function because each of its linear pieces is also constant.

b. The easiest way to graph such a function in the calculator is to store each part of the function rule in a separate location in the function editor. In order to get the proper graph, on each section of the domain we

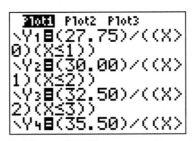

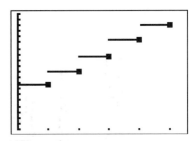

FIGURE 2.3.6 Entering $R(w)$ into the calculator.

FIGURE 2.3.7 The calculator graph of $R(w)$.

put the function rule in parentheses and then use "/" to separate the rule from its domain restriction (also in parentheses). The first three pieces of the rule are shown in Fig. 2.3.6. Then the graph (with a couple of tricks to show the endpoints clearly) is shown in Fig. 2.3.7.

From the graph, you can see why we describe a function like $R(w)$ as a "step" function.

Most people encounter piecewise-linear functions when they pay their income tax on or before April 15 of each year. While the IRS gives us its "rules" in table form, we can use the point-slope form to convert those rules into linear functions.

EXAMPLE 4   Income Tax As a Piecewise-Linear Function

Schedule X of the 2005 Form 1040 instructions specifies the tax paid by a single taxpayer as shown in Fig. 2.3.8.

If your taxable income is:		The tax is:	of the amount over—
Over—	But not over—		
$0	$7,300	---------- 10%	$0
7,300	29,700	$730.00 + 15%	7,300
29,700	71,950	4,090.00 + 25%	29,700
71,950	150,150	14,652.50 + 28%	71,950
150,150	326,450	36,548.50 + 33%	150,150
326,450	----------	94,727.50 + 35%	326,450

FIGURE 2.3.8 Income tax for single taxpayers.

a. Find a piecewise-linear function $T(I)$ that gives the income tax owed for a taxable income of I dollars.

b. Find the income tax owed by a single taxpayer with a taxable income of $50,000.

c. Find the income tax owed by a single taxpayer with a taxable income of $150,000.

SOLUTION

a. The information in the IRS table determines a left-hand endpoint (I, T) and an average rate of change a for each interval. (This average rate of change over each interval is called the *marginal tax rate*. You can see that the marginal tax rate increases as the income increases.) The following table summarizes this information.

INTERVAL	LEFT ENDPOINT (I, T)	AVERAGE RATE OF CHANGE a
$0 < I \leq 7300$	$(0, 0)$	0.10
$7300 < I \leq 29700$	$(7300, 730)$	0.15
$29700 < I \leq 71950$	$(29700, 4090)$	0.25
$71950 < I \leq 150150$	$(71950, 14652.50)$	0.28
$150150 < I \leq 326450$	$(150150, 36548.50)$	0.33
$I > 326450$	$(326450, 94727.50)$	0.35

Using the point-slope form on the first interval, we have $T(I) = 0 + 0.10(I - 0)$, or $T(I) = 0.10I$. Similarly, on the second interval, we have $T(I) = 730 + 0.15(I - 7300)$. If we continue in this fashion for each subsequent interval, we have

$$T(I) = \begin{cases} 0.10I & \text{if } 0 < I \leq 7300, \\ 730 + 0.15(I - 7300) & \text{if } 7300 < I \leq 29700, \\ 4090 + 0.25(I - 29700) & \text{if } 29700 < I \leq 71950, \\ 14652.50 + 0.28(I - 71950) & \text{if } 71950 < I \leq 150150, \\ 36548.50 + 0.33(I - 150150) & \text{if } 150150 < I \leq 326450, \\ 94727.50 + 0.35(I - 326450) & \text{if } I > 326450. \end{cases}$$

b. Because $50,000 falls into the third interval, a single taxpayer with a taxable income of $50,000 owes income tax of

$$T(50000) = 4090 + 0.25(50000 - 29700) = \$9165.$$

c. $150,000 falls into the fourth interval, so a single taxpayer with a taxable income of $150,000 owes

$$T(150000) = 14652.50 + 0.28(I - 71950) = \$36,506.20.$$

This taxpayer has three times the taxable income of the taxpayer in (b) but owes almost four times as much tax. This illustrates the fact that the federal income tax is *progressive*—people with larger incomes pay larger percentages of their income in tax.

Most states also have income tax structures that are progressive and thus are represented by piecewise-linear functions. Several states have a "flat" tax rate (the same percentage regardless of income); such tax structures are simple linear functions. A few states tax only dividends and interest, using a flat tax rate.

2.3 Exercises | Building Your Skills

Sketch the graphs of each function in Exercises 1–4. Label any points where the function rule changes.

1. $f(x) = |x - 1|$
2. $f(x) = |x| - 1$
3. $f(x) = \begin{cases} 2 & \text{if } x < 0 \\ 3 & \text{if } x \geq 0 \end{cases}$
4. $f(x) = \begin{cases} 5 - 2x & \text{if } -1 \leq x < 2 \\ 1 + x & \text{if } 2 < x \leq 6 \end{cases}$

In Exercises 5–8 write a symbolic description of the function whose graph is pictured.

5.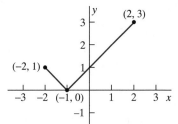

6.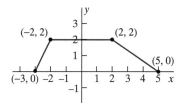

7.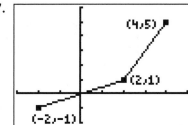

8.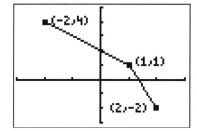

Each of the Exercises 9–12 describes a trip you made along a straight road connecting two cities 120 miles apart. Sketch the graph of the distance d from your starting point (in miles) as a function of the time t elapsed (in hours). Also describe the function d(t) symbolically.

9. You traveled for 1 hour at 45 miles per hour, then realized you were going to be late, and therefore traveled at 75 miles per hour for the next hour.

10. You traveled for 1 hour at 60 miles per hour, were suddenly engulfed in dense fog, and therefore drove back home at 30 miles per hour.

11. You traveled for 1 hour at 60 miles per hour, stopped for ½ hour while a herd of bison crossed the road, and then drove on toward your destination for the next hour at 60 miles per hour.

12. You traveled for ½ hour at 60 miles per hour, suddenly remembered that you'd left your wallet at home, drove back at 60 miles per hour to get it, and finally drove 2 hours at 60 miles per hour to reach your destination.

Applying Your Skills

13. According to the U.S. Census Bureau, the population of Tucson, Arizona, was 45 thousand in 1950, and from 1950 to 1970 it grew at the average rate of 10.9 thousand people per year. After 1970, the population of Tucson grew at the average rate of 7.5 thousand people per year.

   a. Find a piecewise-linear function $P(t)$ that gives the population P of Tucson as a function of the year t.
   b. According to your model, in what year did the population of Tucson reach 100 thousand?
   c. Use your model to predict the population of Tucson in 2010.

14. According to the U.S. Census Bureau, the population of Charlotte, North Carolina, was 18 thousand in 1900, and from 1900 to 1950 it grew at the average rate of 2.32 thousand people per year. After 1950, the population of Charlotte grew at the average rate of 6.26 thousand people per year.

   a. Find a piecewise-linear function $P(t)$ that gives the population of Charlotte as a function of the year.
   b. Use your model to determine the year in which the population of Charlotte grew to 500 thousand.
   c. Based on your model, what was the population of Charlotte in 1980?

15. According to the U.S. Census Bureau, the population of Rochester, New York, was 163 thousand in 1900, and from 1900 to 1950 it grew at the average rate of 3.38 thousand people per year. After 1950, the population of Rochester declined at the average rate of 2.24 thousand people per year.

   a. Find a piecewise-linear function $P(t)$ that gives the population of Rochester as a function of the year.
   b. Based on your model, what was the population of Rochester in 1932?
   c. Use your model to predict the year in which the population of Rochester returns to its 1900 level.

16. According to the U.S. Census Bureau, the population of San Francisco was 775 thousand in 1950, and from 1950 to 1980 it declined at the average rate of 3.2 thousand people per year. After 1980, the population of San Francisco grew at the average rate of 4.9 thousand people per year.

   a. Find a piecewise-linear function $P(t)$ that gives the population of San Francisco as a function of the year.
   b. Based on your model, what was the population of San Francisco in 1975?
   c. Use your model to determine the year in which the population of San Francisco returned to its 1950 level.

17. For an express delivery letter weighing up to 8 ounces sent to a certain destination, the charge C is $8.00 for the first 8 ounces plus 80 cents for each additional ounce or fraction thereof. Sketch the graph of the step function $C(x)$ that gives the charge as a function of the total number x of ounces.

18. In a certain city, the charge C for a taxi trip of at most 10 miles is $5.00 for the first 2 miles (or fraction thereof) plus 75 cents for each half-mile (or fraction thereof) up to a total of 10 miles. Sketch the graph of the step function $C(x)$ that gives the charge as a function of the total number x of miles.

19. In 2006, the first-class mail rate in the United States was 39 cents for the first ounce and 24 cents for each additional ounce or fraction thereof (up to 13 ounces).

a. Find a piecewise linear function $P(x)$ that gives the first-class mail rate for letters not weighing more than 5 ounces.
b. Sketch the graph of $P(x)$.

20. In 2006, the fee for insurance coverage on parcels sent from the United States to Canada was $1.30 for values not over $50, $2.20 for values not over $100, and $1.00 for each additional $100 of value (or fraction thereof), up to a value of $700.

    a. Find a piecewise linear function $F(v)$ that gives the fee for insurance coverage for parcels with value up to $700.
    b. Sketch the graph of $F(v)$.

21. While other soft drinks struggled to retain their share of the market, Mountain Dew saw a dramatic increase in sales beginning in 1993, as indicated in Fig 2.3.13.

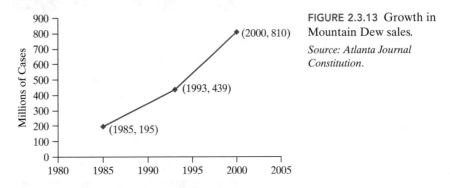

FIGURE 2.3.13 Growth in Mountain Dew sales.

Source: Atlanta Journal Constitution.

a. Find a piecewise linear function $C(t)$ that gives the number of cases sold (in millions) as a function of the year.
b. Use your model to determine when the number of cases of Mountain Dew sold was 300 million.
c. Use your model to determine when the number of cases of Mountain Dew sold was 500 million.
d. According to your model, how many cases of Mountain Dew were sold in 2005?

22. The U.S. Census Bureau maintains a "population clock" that keeps track of the population of the United States. In 1967, as the population clock reached 200,000,000, a Chinese-American baby was born in Atlanta and designated (by *Life Magazine*) as the 200,000,000th American. Since that time, the population has continued to grow (reaching 300,000,000 by the time you read this), and has become increasingly diverse. The percentages of the population that are African-American, Asian, and Hispanic have increased, while the percentage of the population that is white has decreased, as indicated in Fig. 2.3.14.

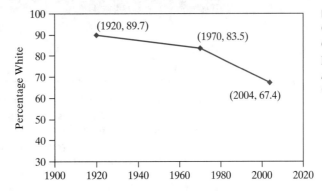

FIGURE 2.3.14 Change in race/ethnicity of U.S. population.

Source: Atlanta Journal-Constitution.

a. Find a piecewise linear function $P(t)$ that gives the percentage of the population that is white as a function of the year.
b. According to your model, what percentage of the population was white in 1949?
c. According to your model, what percentage of the population will be white in 2010?
d. Use your model to determine when the percentage of the population that is white was 75%.

23. The following table gives the 2004 Connecticut state tax rates for persons who are head of household. (Source for tax information in Exercises 23–27 is *The World Almanac and Book of Facts 2005*.)

IF YOUR TAXABLE INCOME IS OVER	BUT NOT OVER	THE TAX YOU OWE IS	OF THE AMOUNT OVER
$0	$16,000	3%	$0
$16,000		$480 + 5%	$16,000

a. Find the tax owed on a taxable income of $5700.
b. Find the tax owed on a taxable income of $16,000.
c. Find the tax owed on a taxable income of $43,000.
d. Write a formula defining a piecewise linear function $T(I) = \cdots$ that gives tax owed in Connecticut as a function of taxable income.

24. The following table gives the 2004 Mississippi state tax rates for single persons.

IF YOUR TAXABLE INCOME IS OVER	BUT NOT OVER	THE TAX YOU OWE IS	OF THE AMOUNT OVER
$0	$5000	3%	$0
$5000	$10,000	$150 + 4%	$5000
$10,000		$350 + 5%	$10,000

a. Find the tax owed on a taxable income of $3350.
b. Find the tax owed on a taxable income of $6000.
c. Find the tax owed on a taxable income of $38,000.
d. Write a formula defining a piecewise linear function $T(I) = \cdots$ that gives tax owed in Mississippi as a function of taxable income.

25. The following table gives the 2004 Louisiana state tax rates for married persons filing jointly.

IF YOUR TAXABLE INCOME IS OVER	BUT NOT OVER	THE TAX YOU OWE IS	OF THE AMOUNT OVER
$0	$12,500	2%	$0
$12,500	$25,000	$250 + 4%	$12,500
$25,000		$550 + 6%	$25,000

a. Find the tax owed on a taxable income of $4400.
b. Find the tax owed on a taxable income of $17,000.
c. Find the tax owed on a taxable income of $56,000.
d. Write a formula defining a piecewise linear function $T(I) = \cdots$ that gives tax owed in Louisiana as a function of taxable income.

26. The following table gives the 2004 District of Columbia tax rates for married persons filing separately.

IF YOUR TAXABLE INCOME IS OVER	BUT NOT OVER	THE TAX YOU OWE IS	OF THE AMOUNT OVER
$0	$10,000	5%	$0
$10,000	$30,000	$500 + 7.5%	$10,000
$30,000		$2000 + 9.3%	$30,000

a. Find the tax owed on a taxable income of $4400.
b. Find the tax owed on a taxable income of $17,000.
c. Find the tax owed on a taxable income of $36,000.
d. Write a formula defining a piecewise linear function $T(I) = \cdots$ that gives tax owed in the District of Columbia as a function of taxable income.

27. The state of Michigan had a 2004 flat tax rate of 3.9% of taxable income.

a. Find the tax owed on a taxable income of $5700. Compare this value to the tax owed on the same amount in the state of Connecticut (found in Exercise 23).
b. Find the tax owed on a taxable income of $43,000. Compare this value to the tax owed on the same amount in the state of Connecticut (found in Exercise 23).
c. Write a formula defining a linear function $T(I) = \cdots$ that gives tax owed in Michigan as a function of taxable income.
d. For what incomes would a taxpayer in Michigan pay more tax than a comparable taxpayer in Connecticut? For what incomes would the Michigan taxpayer pay less?

2.4 FITTING LINEAR MODELS TO DATA

In Section 2.1 we saw that a population whose growth is modeled by a linear function grows with a constant rate of change—that is, with the same change in population each year. In this section we discuss the modeling of data of a sort that might be said to display an "almost constant" rate of change—with the annual changes from year to year being approximately but not exactly equal. As an example of data that might therefore be described as "almost but not quite linear," the following table shows the population of Charlotte, North Carolina, as recorded in the decade census years of 1950–1990.

YEAR	POPULATION (THOUSANDS)	CHANGE (THOUSANDS)
1950	134	
1960	202	68
1970	241	39
1980	315	74
1990	396	81

Source: U.S. Census Bureau.

The third column of this table shows (for each decade year) the change in the population during the preceding decade. We see that the population of Charlotte increased by roughly 70 to 80 thousand people during the 1950s, the 1970s, and the 1980s. It increased by somewhat less during the 1960s, but still with a change measured in roughly comparable tens of thousands. We might wonder whether this qualifies as "almost linear" population growth. The way to answer such a question is to plot the data and take a look.

The data points corresponding to this table are plotted in Fig. 2.4.1. Surely most people would agree that these points appear to lie on or near some straight line.

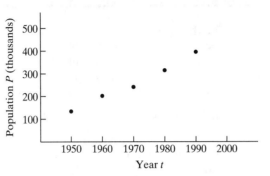

FIGURE 2.4.1 Scatter plot of the Charlotte 1950–1990 population data.

But how can we find a straight line that passes through or near each data point in the figure? One way is simply to pass a straight line through the first and last data points—those for 1950 and 1990.

EXAMPLE 1   Finding a Linear Population Model

Use the populations from 1950 and 1990 to find a linear function that describes (at least approximately) the population of Charlotte over this time period.

SOLUTION

First, we find the average rate of change of this function:

$$a = \frac{\Delta P}{\Delta t} = \frac{396 - 134}{1990 - 1950} = \frac{262}{40} = 6.55 \text{ thousand/year}.$$

Then, using the point-slope form with $t_1 = 1950$ and $P_1 = 134$ gives the linear population model

$$P(t) = 134 + 6.55(t - 1950). \tag{1}$$

Figure 2.4.2 shows the graph of $P(t)$ along with our original census population data points for Charlotte.

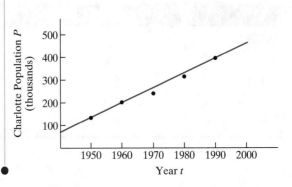

FIGURE 2.4.2 The graph of $P(t) = 134 + 6.55(t - 1950)$ approximating the 1950–1990 growth of Charlotte.

The linear function we found looks like a pretty "good fit" to the data—the 1950 and 1990 data points automatically lie on the line (why?), while the 1960 point seems to lie on the line and the 1970 and 1980 points lie just below the line. The following table shows the *discrepancies* between the actual populations and those "predicted" by the linear function in (3) for each of the 1950–1990 census years. We see that the actual 1960 population is just 2.5 thousand (only about 1%) larger than the 1960 population predicted by the linear model in (3), whereas the actual 1970 and 1980 populations are 24 thousand and 15.5 thousand, respectively, less than the corresponding linear predictions. We generally refer to these discrepancies as "errors," and will do so from here on.

t	P (ACTUAL) (THOUSANDS)	P(t) (PREDICTED) (THOUSANDS)	ERROR $P - P(t)$ (THOUSANDS)
1950	134	134	0
1960	202	199.5	2.5
1970	241	265	−24
1980	315	330.5	−15.5
1990	396	396	0

Note that the figures in each (horizontal) row of the figure satisfy the formula

$$P_{\text{actual}} = P(t) + \text{error}.$$

Smaller Errors and Fitting Data Better

Now we discuss the concept of a linear model that **best fits** given population (or other) data such as those in the following table:

t (year)	t_1	t_2	t_3	t_4
P (thousands)	P_1	P_2	P_3	P_4

In the case of the Charlotte population, there were five given data points. However, here we assume for simplicity in the general discussion that four data points are given. The final procedure will be analogous whatever the number of data points.

The table gives the actual populations P_1, P_2, P_3, and P_4 specified at $n = 4$ different times t_1, t_2, t_3, and t_4. We may ask what linear model of the form

$$P(t) = b + at \tag{2}$$

"best fits" the given data. That is, what should the numerical values of the coefficients a and b be in order that the model best fits the data? But the real question is: What does this mean? What does it mean for the model to "fit" the data well?

Figure 2.4.3 shows the errors that correspond to the discrepancies between the given data points in the tP-plane and the straight-line graph of (2). The ith error E_i is the vertical distance between the actual data point (t_i, P_i) and the corresponding point $(t_i, P(t_i))$ on the line that is "predicted" by the linear model. It

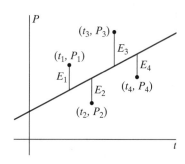

FIGURE 2.4.3 The errors in the linear model $P(t) = at + b$.

is worth emphasizing that P_i is the *actual* observed value of the population at time t_i, while $P(t_i)$ is the value *predicted* by the linear model, so

$$\text{error} = \text{actual} - \text{predicted}. \tag{3}$$

One might suspect that the linear model fits the data well if the *sum of the errors* is small. However, observe that each error E_i defined is *signed*. It is positive if the ith data point lies above the graph of $P(t) = at + b$ but is negative if the data point lies below the line. Consequently, it is possible for large positive errors to cancel out large (in absolute value) negative errors in the sum of all the errors. For instance, Fig. 2.4.3 indicates two large positive errors (data points above the line) and two large negative errors (data points below the line). Although each of these four errors is numerically large, their sum may be quite small, or even 0, just as

$$(+72) + (-54) + (+29) + (-47) = 0.$$

Thus the fact that the sum of the errors is small does *not* guarantee that all the individual errors are small.

It is therefore customary to use the sum of the squares of the errors as a measure of the overall discrepancy between the given data points and a proposed linear model.

> **DEFINITION: Sum of Squares of Errors**
>
> The phrase "**Sum of Squares of Errors**" is so common in data modeling that it is abbreviated to **SSE**. Thus the **SSE** associated with a data model based on n data points is defined by
>
> $$\text{SSE} = E_1^2 + E_2^2 + E_3^2 + \cdots + E_n^2. \tag{4}$$

Note that, however many data points are given, **the SSE is the sum of the squares of all their errors.** In plain words, if we write the actual populations in one column and the predicted populations in a second column, then the SSE is the sum of the squares of the differences between corresponding entries in the two columns:

t_i	ACTUAL P_i	LINEAR $P(t_i)$	ERROR E_i	E_i^2
1950	134	134	0	0
1960	202	199.5	2.5	6.25
1970	241	265	-24	576
1980	315	330.5	-15.5	240.25
1990	396	396	0	0

EXAMPLE 2 Finding the SSE for a Linear Model

Find the SSE for the linear population model in Example 1.

SOLUTION

Recall that that the linear model for Charlotte's population in Example 1 is

$$P(t) = 134 + 6.55(t - 1950). \tag{5}$$

The table showing the Charlotte census data now includes a final column showing the squares of the errors. The SSE is the sum of the numbers in this final column; that is,

$$\text{SSE} = 0 + 6.25 + 576 + 240.25 + 05 = 822.5.$$

● So the SSE for our Charlotte population model is 822.5.

The linear function $P(t) = 134 + 6.55(t - 1950)$ can be simplified as follows:

$$P(t) = 134 + 6.55t - 12772.5$$
$$P(t) = 6.55t - 12638.5. \quad (6)$$

So $P(t) = 6.55t - 12{,}638.5$ in (6) corresponds to choosing the numerical values $a = 6.55$ and $b = -12638.5$ in the slope-intercept linear model $P(t) = b + at$. We wonder if it is possible to find different numerical values of a and b that yield a smaller SSE than the value 822.5 found in Example 2.

EXAMPLE 3 Finding a Smaller SSE

Find a linear function with a smaller SSE for the Charlotte population data.

SOLUTION

Looking at Fig. 2.4.2 as well as at the table for Example 2, we see that the graph of $P(t) = 6.55t - 12{,}638.5$ passes right through the first and last data points, but passes 24 units above the third data point. Let's think of a line that splits the difference and passes 12 units below the first and last data point as well as 12 units above the third data point. Can you see that we can make this change by subtracting 12 from the value $b = -12638.5$ in (8), so that the new line lies 12 units lower in the tP-plane? Our new linear model is

$$P(t) = 6.55t - 12650.5 \quad (7)$$

with $m = 6.55$ as in (6) but now $b = -12650.5$ (rather than $-12{,}638.5$). The following table shows the computation of the SSE for this new altered linear model. (We started with the previous table, then recalculated the numbers in the third, fourth, and fifth columns in turn. You should verify the results for yourself.)

t_i	ACTUAL P_i	LINEAR $P(t_i)$	ERROR E_i	E_i^2
1950	134	122	12	144
1960	202	187.5	14.5	210.25
1970	241	253	−12	144
1980	315	318.5	−3.5	12.25
1990	396	384	12	144

Now the sum of the numbers in the final column is

$$\text{SSE} = 144 + 210.25 + 144 + 12.25 + 144 = 654.5$$

(as compared with 822.5 for the linear model of Example 2). Thus we have suc-
● ceeded in reducing somewhat the SSE.

So although either of the functions

$$P(t) = 6.55t - 12638.5 \quad \text{or} \quad P(t) = 6.55t - 12650.5 \qquad (8)$$

can be used as a linear model for the growth of Charlotte during the 1950–1990 period, the latter one fits the observed census data at least slightly better because it has a smaller sum of squares of errors.

But exactly *what* does the SSE really mean? Well, the SSE for either of the models in (8) is a sum of five "squared errors,"

$$\text{SSE} = E_1^2 + E_2^2 + E_3^2 + E_4^2 + E_5^2.$$

We should therefore divide by 5 to get the average of these squared errors,

$$\text{average squared error} = \frac{E_1^2 + E_2^2 + E_3^2 + E_4^2 + E_5^2}{5} = \frac{\text{SSE}}{5}.$$

But then we should take the *square root* of the average *squared* error to get the average error itself,

$$\text{average error} = \sqrt{\frac{\text{SSE}}{5}}.$$

This is for Charlotte with $n = 5$ data points. To define the average error for a model fitting n given data points we need only divide the SSE instead by n.

DEFINITION: Average Error

The *average error* in a linear model fitting n given data points is defined in terms of its SSE by

$$\text{average error} = \sqrt{\frac{\text{SSE}}{n}}. \qquad (9)$$

This formula says simply that **the average error is the square root of the average of the squares of the individual errors** (or discrepancies between predicted and actual data values).

EXAMPLE 4 Finding Average Error

Find the average error for each of the population models in (8).

SOLUTION

For the first linear model $P(t) = 6.55t - 12638.5$ in (8), we calculated its SSE = 822.5, so with $n = 5$ in (9) we find that its average error is

$$\text{average error} = \sqrt{\frac{822.5}{5}} \approx 12.826.$$

But the SSE for the second linear model $P(t) = 6.55t - 12650.5$ in (8) is only 654.5, so *its* average error is

$$\text{average error} = \sqrt{\frac{654.5}{5}} \approx 11.441.$$

These average errors—like the populations themselves—are measured in thousands. We may therefore say that the populations "predicted" by our first model for the five census years 1950, 1960, 1970, 1980, and 1990 err by an average of 12,826 people, whereas the predictions of our second model err by an average of 11,441 people. This is a tangible statement of the extent to which the second model fits the actual census data better than does the first model.

The Best Fit—The Least Possible Average Error

The question is this: What choice of the numerical coefficients a and b in the linear model $P(t) = b + at$ will minimize the average error—that is, will result in the least possible average error? This optimal linear model will be the one that we say "best fits" the given data.

Students in past generations often plotted data points on a piece of graph paper, then carefully maneuvered a ruler so that it appeared visually to come as close as possible to these points "on the average." But the modern graphing calculator comes equipped with the facility to solve "best-fitting" problems more precisely. Here we describe how this is done. You should use your graphing calculator to carry out the steps we describe.

Figure 2.4.4 shows how to enter the Charlotte population data in a calculator. The list of census years from 1950 to 1990 is stored as a list **L1**, and the corresponding list of recorded populations is stored as a list **L2**. (Note that calculator "lists" are enclosed in curly braces.) Figure 2.4.5 the resulting **STAT EDIT** menu displaying the data in table form. Here the individual items—either a year or the corresponding population—can be changed or "edited" one at a time (for instance, if an error was made in entering either list originally).

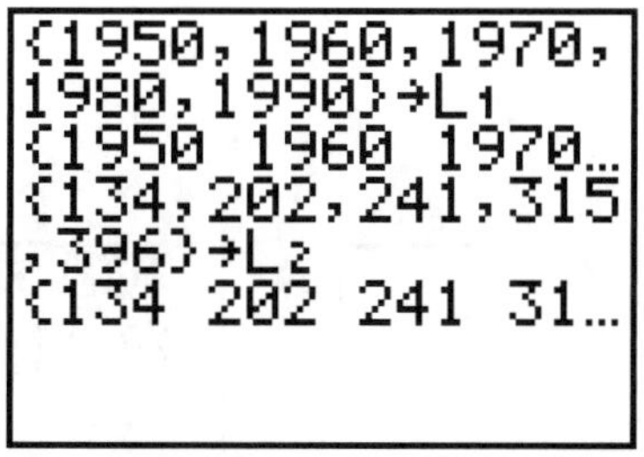

FIGURE 2.4.4 Storing the 1950–1990 population census data for Charlotte.

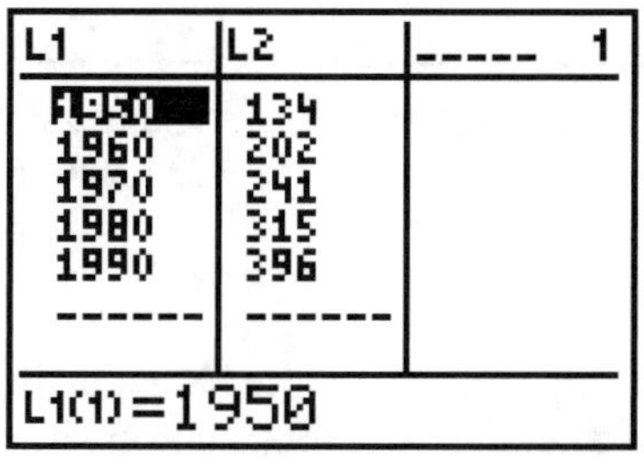

FIGURE 2.4.5 The TI-84 **STAT EDIT** menu.

Figure 2.4.6 shows the calculator's **STAT CALC** menu. Item **4: LinReg(ax+b)** on this menu is the calculator's so-called "linear regression" facility for finding the linear function **y = ax + b** that best fits the selected data. (Curve-fitting is called "regression" in statistics and calculator lingo.) When we select item **4** on this menu, the **LinReg(ax+b)** function is entered on the home calculating screen. As shown in Fig. 2.4.7, we must then enter the names of our list **L1** of input values, our list **L2** of output values, and the name **Y1** of the **Y=** menu variable where we want the resulting linear formula saved.

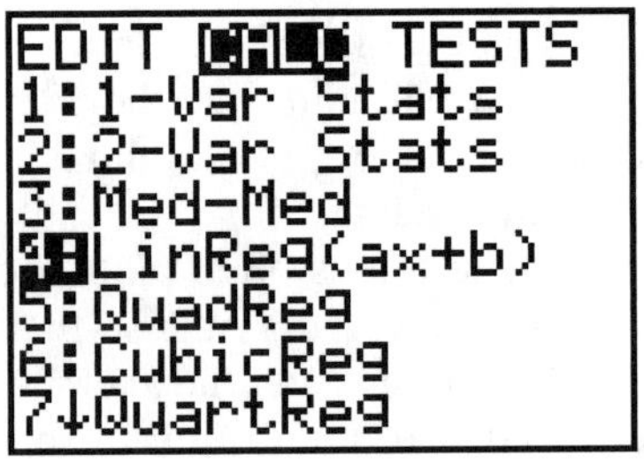

FIGURE 2.4.6 The TI-84 **STAT CALC** menu.

FIGURE 2.4.7 Fitting the **X**-data in list **L1** and the **Y**-dta in list **L2**.

Figure 2.4.8 shows the display that results when this command is entered, showing that the linear function that best fits our data is

$$y = 6.37x - 12291.3. \tag{12}$$

The **LinReg(ax+b) L1, L2,Y1** command automatically enters Equation (12) in the **Y=** menu, ready for plotting as shown in Fig. 2.4.9. With **Plot 1** turned **On** in the **STAT PLOT** menu (Fig. 2.4.10), **GRAPH** then gives the plot of the best fitting linear function shown in Fig. 2.4.11, where the original census data points are shown as small squares.

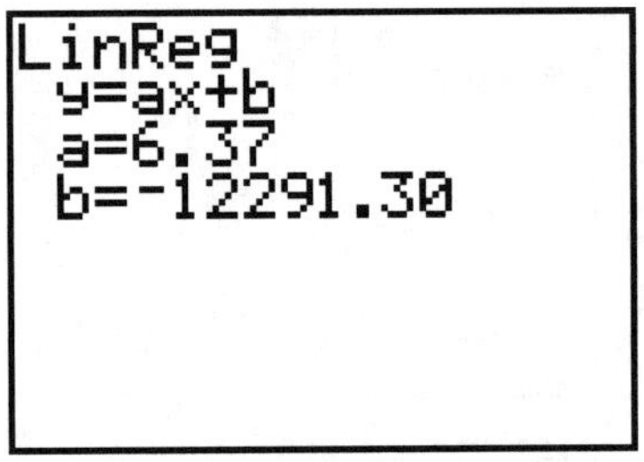

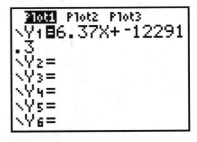

FIGURE 2.4.8 The best straight-line fit to our data points.

FIGURE 2.4.9 The best-fitting linear function ready for graphing.

The calculator automatically uses **X** as the input variable and **Y** as the output variable. For our calendar year–population situation, we use t as the input vari-

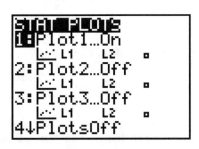
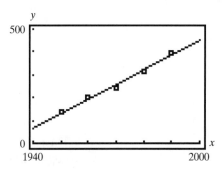

FIGURE 2.4.10 Preparing to plot the data with the best-fit straight line.

FIGURE 2.4.11 The line of best fit.

able and P as the output variable, and Equation (12) says that the linear model that best fits the 1950–1990 Charlotte census data is given by

$$P(t) = 6.37t - 12291.3 \tag{13}$$

This is the model that "best fits" the data in the sense that it has the smallest sum of the squares of the errors (SSE). While we cannot show that no model has a smaller SSE (this would involve mathematics beyond the scope of this text), we can demonstrate that the SSE for this model is smaller than either of the two we have already found. Should you want further examples, you can construct some linear models of your own and find their (larger) SSEs.

EXAMPLE 5 Finding the Average Error for the Best-Fitting Model

Find the SSE and the average error for the linear model that best fits the 1950–1990 Charlotte census data.

SOLUTION

The following table gives the actual population values and the values predicted by our best-fitting linear model, along with the errors and their squares. (You should verify these table entries.)

t (CALENDAR YEAR)	P (ACTUAL POPULATION IN THOUSANDS)	$P(t)$ (PREDICTED POPULATION IN THOUSANDS)	ERROR (ACTUAL − PREDICTED) (THOUSANDS)	ERROR SQUARED (THOUSANDS)
1950	134	130.2	3.8	14.44
1960	202	193.9	8.1	65.61
1970	241	257.6	−16.6	275.56
1980	315	321.3	−6.3	39.69
1990	396	385	11.0	121.00

Hence the SSE associated with the optimal linear model in (13) is

$$14.44 + 65.61 + 275.56 + 39.69 + 121.00 = 516.30.$$

Therefore the average error is given by

$$\text{average error} = \sqrt{\frac{516.30}{5}} \approx 10.162.$$

Thus the linear model $P(t) = 6.37t - 12291.3$ predicts 1950–1990 census year populations that differ (on the average) by 10,162 people from those actually recorded for Charlotte.

By comparison, our first linear model $P(t) = 6.55t - 12638.5$ had an SSE of 822.5 and an average error of 12.826, while our second model $P(t) = 6.55t - 12650.5$ had an SSE of 654.5 and an average error of 11.441.

Once we have a linear model that predicts the 1950–1990 census year populations for Charlotte, we can use it to predict the population of Charlotte in years for which we have no data. This is really the point of mathematical modeling since it would be foolish to use an approximate value for the population in 1970 when we have the exact value.

This issue raises several points of which you should be aware. First, your calculator will always find you a "best" linear model, whether your data "looks" linear or not. Before constructing such a model, you should consider whether it is appropriate. You can do this by plotting your data or finding the average rates of change between consecutive data points to verify that the average rate of change is more or less constant. (A set of data that increases and decreases at noticeably different rates is not a good candidate for a linear model.)

Second, you should pay attention to the type of prediction you plan to make. If you are making a prediction about a year between two data points, say 1962 (between census years 1960 and 1970), this is called **interpolation.** It is generally safe to make such a prediction because, barring some unusual circumstance, the population is not likely to fluctuate wildly over this 10-year period. (Of course, such unusual circumstances do sometimes occur. Consider the population of New Orleans on January 1, 2005, before Hurricane Katrina, compared to its population on January 1, 2006, approximately 4 months after the hurricane.)

If you make a prediction about a year earlier or later than all of your data, this is called **extrapolation.** You must be careful about predicting too far beyond (or before) the information you have because circumstances that affect the actual population may be very different from those existing when the data are collected.

Applications of Linear Modeling

Thus far, we have discussed only the modeling of linear population growth. But the world is full of other apparently linear data waiting to be modeled.

EXAMPLE 6   Finding an Appropriate Linear Model

The per capita consumption of cigarettes in 1930 and the lung cancer death rate (deaths per million males) for 1950 in the four Scandinavian countries were as follows:

COUNTRY	CIGARETTE CONSUMPTION IN 1930 c	DEATH RATE IN 1950 D
Norway	250	95
Sweden	300	120
Denmark	350	165
Finland	1100	350

It is hard to ignore the fact that higher cigarette consumption appears to be correlated with higher lung cancer death rates 20 years later.

a. Verify that this relationship is approximately linear.
b. Find the linear model that best fits this data.

SOLUTION

a. We do **MEM ClrAllLists** to purge the Charlotte population data from our calculator, then enter the c- and D-data in the lists **L1** and **L2** (Fig. 2.4.12). Then we use the command **L3=ΔList(L2)/ΔList(L1)** (as we did in Section 1.4) to determine whether the average rate of change is approximately constant. The result is shown in Fig. 2.4.13.

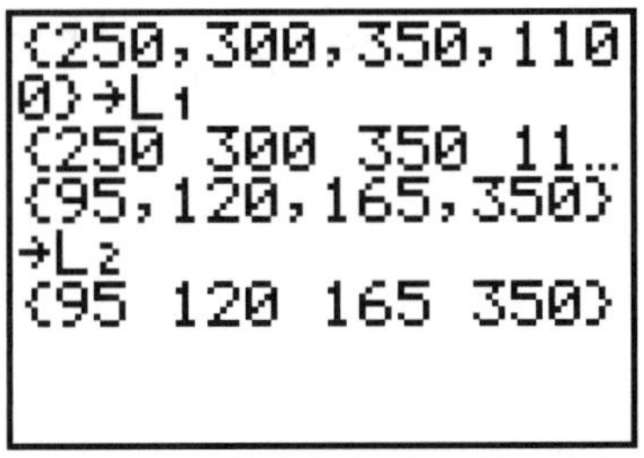

FIGURE 2.4.12 The cigarette consumption and lung cancer death date of Example 6.

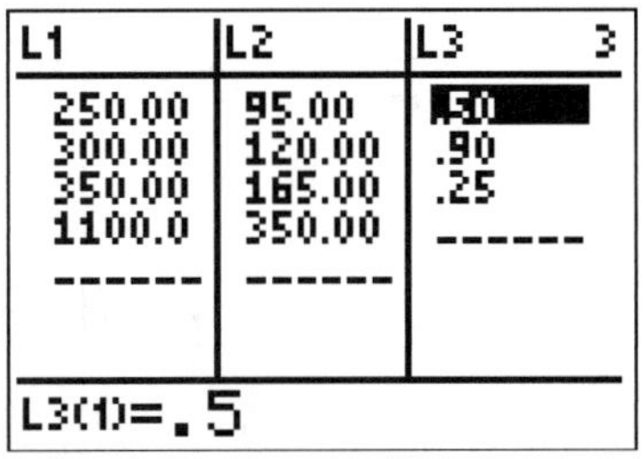

FIGURE 2.4.13 Checking the average rate of change.

At first glance, the average rates of change stored in **L3** might appear very different. But notice that each one is positive and between 0 and 1. So we conclude that the average rate of change for this table of values is approximately constant, and so we can model these data with a linear function.

b. Now we use the **STAT CALC** command **LinReg(ax+b) L1, L2, Y1**, which produces the linear function

$$y = 0.28x + 40.75, \tag{14}$$

which best fits our data (with our calculator set in two-decimal-place mode). With an appropriate window and the same **STAT PLOT** settings as before, we get the plot of this linear function and the table data points that is shown in Fig. 2.4.14.

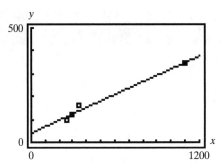

FIGURE 2.4.14 The line best fitting the cigarette consumption versus lung cancer deaths data.

Thus, using function notation and our c and D variables, we find that the linear model that best fits the given data is $D(c) = 0.28c + 40.75$.

The calculator also calculates automatically the discrepancies between the data and the predictions of the linear model in (14). These errors are stored in the **LIST** menu under the name **RESID**. If you scroll through the list indicated in Fig. 2.4.15, you see that these errors or "residuals" are -15.75, -4.74, 26.25, and 1.25.

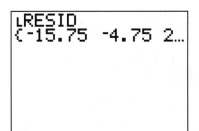

FIGURE 2.4.15 The errors or "residuals" in Example 6.

Hence the average error of our linear model is

$$\text{average error} = \sqrt{\frac{(-15.75)^2 + (-4.75)^2 + (26.25)^2 + (1.25)^2}{4}} = 15.50.$$

Thus the average error in the linear model's predictions (for the four Scandinavian countries) is 15.5 lung cancer deaths per million males.

For a comparison, the 1930 per capita cigarette consumption in Australia was 470, so the linear model predicts

$$D(470) = (0.28)(470) + 40.75 = 172.35 \approx 172$$

lung cancer deaths per million males in Australia. The actual number of such deaths in Australia in 1950 was 170 per million. (Given an average error of 15.50, this agreement is better than we have any right to expect, especially ignoring any differences there may be between Australian and Scandinavian health and lifestyles.)

Finally, let us note what the linear model $D(c) = 0.28c + 40.75$ *means*. If we substitute $c = 0$ (no cigarette consumption), we get $D(0) = 40.75$. Thus we should expect 40.75 lung cancer deaths per million males even if no cigarettes at all are smoked. Then the average rate of change (or slope) $a = 0.28$ implies an additional 0.28 death per unit increase in per capita cigarette consumption. That is, since $4 \times 0.28 = 1.12$, there should be approximately one additional death for each four-unit increase in per capita cigarette consumption. This sort of *interpretation* of a linear model frequently is more important than any specific numerical predictions of the model.

Before you begin working on the exercises, we need to discuss decimal-place settings on the calculator. When using real-world data, the values of a and b in the best-fitting linear model seldom come out "nicely." This means that if your calculator is in "float" mode, these values could be reported with as many as 10 decimal places. Because we frequently give our population values in thousands, it is convenient for interpreting the model if a and b are given as three-place decimals. Thus, the answers given in the back of the book reflect a calculator setting of three decimal places. If you use a different setting on your calculator, your answers may vary a bit from ours.

2.4 Exercises | Building Your Skills

In Exercises 1–4 the population P (in thousands) of a city in three different years is given, thereby providing three known (t, P) data points. Find the average error for the linear function $P(t) = at + b$ whose graph contains:

a. The first and third of these points.
b. The second and third of these points.

(Note that in some exercises we have "reset the clock," while in others we use actual calendar years.)

1.

t (years)	0	10	20
P (thousands)	200	270	320

2.

t (years)	0	10	20
P (thousands)	460	390	300

3.

t (years)	1980	1990	2000
P (thousands)	715	605	435

4.

t (years)	1980	1990	2000
P (thousands)	615	805	1155

90 CHAPTER 2 Linear Functions and Models

In Exercises 5–8 the population P (in thousands) of a city in four different years is given, thereby providing four known (t, P) data points. For each set of data, find:

   a. The linear model that best fits the data.
   b. The SSE and average error for this optimal linear model.

5.

t (years)	0	10	20	30
P (thousands)	360	300	300	240

6.

t (years)	0	10	20	30
P (thousands)	240	320	360	360

7.

t (years)	1960	1970	1980	1990
P (thousands)	600	800	950	1050

8.

t (years)	1960	1970	1980	1990
P (thousands)	565	485	385	265

Applying Your Skills

Exercises 9–12 give the 1950–1990 U.S. census data for a city whose average rate of growth is approximately constant.

   a. Find the linear model $P(t) = at + b$ that best fits this census data. (You may "reset the clock" or not, as you prefer.)
   b. Construct a table showing the actual populations, predicted populations, and errors for the decade years 1950–1990.
   c. Use your linear model to predict the city's population in the year 2000.

9. San Diego, California:

t (years)	1950	1960	1970	1980	1990
P (thousands)	334	573	697	876	1111

10. Buffalo, New York:

t (years)	1950	1960	1970	1980	1990
P (thousands)	580	533	463	358	328

11. Newark, New Jersey:

t (years)	1950	1960	1970	1980	1990
P (thousands)	439	405	382	329	275

12. Garland, Texas:

t (years)	1950	1960	1970	1980	1990
P (thousands)	11	39	81	139	181

13. The following table shows the percentage of a three-person American family's weekly income spent on gasoline for the indicated year.

YEAR t	1966	1976	1986	1996	2006
Percentage Spent on Gasoline P	5.5	4.4	3.0	2.7	4.2

Source: St. Louis Post-Dispatch.

a. Use a plot of the data to determine whether a linear function is a suitable model for these data.
b. If a linear model is suitable, find the linear function that best fits the data.

14. The following table shows the revenue in billions of dollars for the state of Georgia for the years 1996 through 2000.

YEAR t	1996	1997	1998	1999	2000
Revenue R (billions of dollars)	9.9	10.5	11.1	12.1	13

Source: Atlanta Journal-Constitution.

a. Use a plot of the data to determine whether a linear function is a suitable model for these data.
b. If a linear model is suitable, find the linear function that best fits the data.

15. The following table below shows the U.S. life expectancy (estimated number of years of life remaining) as a function of current age.

AGE A	0	10	20	35	50	70
Years of Life Remaining L	77.2	67.9	58.1	43.9	30.3	14.6

Source: The World Almanac and Book of Facts 2005.

a. Use a plot of the data to determine whether a linear function is a suitable model for these data.
b. If a linear model is suitable, find the linear function that best fits the data.

16. The following table shows the total number of marine species as a function of the total number of gallons of tanks for various U.S. aquariums.

Gallons g	1	1.3	2.5	2.5	4.75	5
Marine Species M	692	536	536	625	1600	500

Source: Atlanta Journal-Constitution.

a. Use a plot of the data to determine whether a linear function is a suitable model for these data.
b. If a linear model is suitable, find the linear function that best fits the data.

17. The following table gives the number of CDs (in millions) sold in the United States for the even-numbered years 1988 through 1996. A plot of these data indicates that CD sales were increasing at a reasonably constant rate over these years.

T (YEAR)	1988	1990	1992	1994	1996
S (millions of CDs)	149.7	286.5	407.5	662.1	778.9

Source: The World Almanac and Book of Facts 1998 and The World Almanac and Book of Facts 2005.

a. Find the linear model $S(t) = at + b$ that best fits these data.
b. Compare the model's prediction for the year 1995 with the actual 1995 CD sales of 722.9 million.
c. Compare the model's prediction for the year 2002 with the actual 2002 CD sales of 803.3 million.
d. What do you think may have accounted for the discrepancy between the 2002 predicted and actual values?

18. The following table gives the number of cassette tapes (in millions) sold in the United States for various years between 1994 and 2001.

T (YEAR)	1994	1995	1997	2000	2001
S (millions of cassettes)	345.4	272.6	172.6	76.0	45.0

Source: The World Almanac and Book of Facts 2005.

a. Find the linear model $S(t) = at + b$ that best fits these data.
b. Explain the meaning of the average rate of change for this linear model in terms of the situation.
c. According to this model, in what year will no more cassette tapes be sold? Is this answer reasonable?

19. The following table gives the number of athletes participating in the Winter Olympics for various years between 1924 and 2002 as a function of the number of countries represented.

Number of countries C	16	25	37	49	64	77
Number of Athletes A	258	464	1072	1272	1801	2399

Source: National Council of Teachers of Mathematics.

a. Find the linear model that best fits these data.
b. What does the average rate of change of the model mean in terms of the situation?
c. Use your model to predict the number of athletes participating if 80 countries are represented.

20. The following table gives the winning times (in seconds) for the men's 400-meter hurdles in the Summer Olympics as a function of the year.

t (years after 1920)	0	8	16	36	48	56
W (seconds)	54.0	53.4	52.4	50.1	48.12	47.64

Source: The World Almanac and Book of Facts 2005.

   a. Find the linear model $W(t)$ that best fits these data.
   b. Find the average error for the linear model, and explain what it means in terms of the situation.
   c. Given that the Summer Olympics are held only every 4 years, during which Summer Olympics does the model predict that the winning time will fall below 45 seconds?

21. The following table gives the 2004 military monthly pay scale for a Warrant Officer (W-4) based on years of service.

Y (years of service)	2	4	8	12	16	20
P (pay in dollars)	3356	3547	3872	4194	4617	4944

Source: The World Almanac and Book of Facts 2005.

   a. Find the linear model that best fits these data.
   b. According to your model, what is the pay for a Warrant Officer (W-4) with 15 years of service?
   c. Use a graph or a table of values for this model to find the years in which the model underestimates a warrant officer's pay.

22. The Consumer Price Index (CPI) is a measure of the cost to consumers of various goods and services in comparison to the cost of those items in previous years. The following table gives the CPI for all urban consumers for housing for selected years from 1987 to 1996. The years 1982–1984 serve as the base years for comparison, with the cost of housing in those years equal to 100.

T (years after 1987)	0	3	4	7	9
CPI	115.6	130.5	135.0	145.4	154.0

   a. Find the linear model that best fits these data.
   b. Use this linear model to predict the CPI for housing in 1989. Is your answer a reasonable approximation to 124.9, the actual 1989 CPI for housing? Explain your answer in terms of the plot of the data or the average rates of change between consecutive table entries.
   c. Use the linear model to predict the year in which the CPI for housing will reach 165. What assumption are you making when you make this prediction?
   d. Use the linear model to predict when the CPI for housing will reach 250. Why is it *not* a good idea to make such a prediction?

23. Thus far we have constructed linear models for data that represent a function of some independent variable. Frequently in the "real world," we are confronted with a set of data that does not actually describe a function, but that suggests an underlying relationship that might be modeled by a function. An example of such data is the relationship between height and weight for the seven infielders on the Los Angeles Dodgers roster on July 12, 1997.

Height in inches h	70	74	71	71	76	70	73
Weight in pounds W	163	170	180	145	222	185	200

a. Find the linear model that best fits these data.
b. Use the linear model to predict the weight of a major league infielder who is 6 feet tall.
c. Should you use this model to predict the weight of any American male who is 6 feet tall? Why or why not?

Chapter 2 | Review

In this chapter, you learned about linear and piecewise-linear function models. After completing the chapter, you should be able to:

- Determine whether a function described numerically, graphically or symbolically represents a linear function or a piecewise-linear function.
- Determine the average rate of change of a linear function.
- Find a linear function that models given linear data.
- Find a piecewise-linear function that models given piecewise-linear data.
- Find the best-fitting linear model for data that is approximately linear.
- Interpret the slope and y-intercept of a linear model in terms of the situation modeled.

Review | Exercises

In Exercises 1–6, decide whether, based on the table, graph, or formula, y is a linear or a piecewise-linear function of x.

1.

x	2	4	5	6	8	11
y	3	1.5	0	−1.5	−3	−4.5

2.

x	2	4	6	8	10	12
y	3	1.5	0	−1.5	−3	−4.5

3.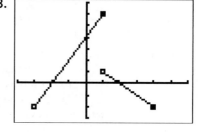

4.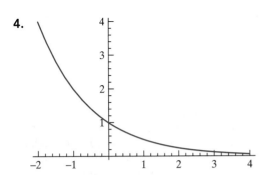

5. $3x + 4y = 2$

6. $y = x^2 + 3x$

7. Find the average rate of change for the linear function given by the following table.

x	−3	0	1.5	4	6
f(x)	−7.2	−3	−0.9	2.6	5.4

8. The population of Libertyville was 227 thousand in 2000 and was decreasing at an average rate of 2.3 thousand people per year.

  a. Find a linear population model $P(t)$ that gives the population of Libertyville as a function of the number of years after 2000.
  b. According to your model, what was the population of Libertyville in 2005?
  c. Use your model to predict the year in which the population of Libertyville falls to 200 thousand.

9. In 1989, 51% of students at Ohio State University graduated within 6 years of entering the university. By 1999, the percentage graduating within 6 years had risen to 68%.

  a. Find a linear model that gives the percentage of students graduating within 6 years as a function of the year.
  b. Explain what the average rate of change of your model means in terms of the situation.
  c. Use your model to predict the percentage of students graduating within 6 years in 2006.

10. Economists frequently use linear functions to construct models for the relationship between price and demand for a small business. Generally, we consider the selling *price p* of the item to be a linear function of the *demand x* (the number of units consumers will purchase). Pomelia has begun a home business making decorative lawn sprinklers out of copper tubing. She has found that she can sell 20 sprinklers per month if they are priced at $60 each but only 12 sprinklers per month if they are priced at $70.

  a. Find a linear function $p(x)$ giving the price of the sprinklers as a function of the number sold.
  b. How many sprinklers will she sell if the sprinklers are priced at $55?
  c. If Pomelia wants to sell at least 30 sprinklers per month, what price should she charge?
  d. Explain what the intercepts of the line mean in this situation.

11. According to Knight-Ridder Newspapers, homeowners who hire a tree service to grind and remove tree stumps from their yards can expect to pay about $95 for the first 10 inches of tree stump diameter and $1 for each additional inch thereafter.

  a. Find a piecewise-linear function $C(d)$ that gives the charge for stump grinding and removal as a function of the diameter of the stump in inches.
  b. Find $C(16)$ and explains what it means in terms of the situation.
  c. If a homeowner paid $108 for stump grinding and removal, what was the diameter of the tree stump?

12. Nambe Mills in Santa Fe, New Mexico, manufactures tableware made of a metal alloy. The pieces are sand casted and then shaped, ground, buffed, and polished. The table following gives the total grinding and polishing times (in minutes) for various pieces of Nambeware and their corresponding prices (in dollars).

Time	109.38	16.41	23.77	13.25	44.25	64.30	34.16
Price	260	39	49.50	31	89	165	75

Source: Nambe Mills.

  a. Find the best-fitting linear model for price as a function of grinding and polishing time.
  b. According to your model, what should be the price of a piece of Nambeware if its grinding and polishing time is 53.18 minutes?
  c. Interpret the slope of your model in terms of the situation.

13. (Chapter Opener Revisited) In the discussion that introduced this chapter, we looked at a plot that displayed the relationship between the duration of an Old Faithful eruption and the time to its next eruption. The plot was based on data from January, 2003, which is displayed in the following table.

Duration d (seconds)	110	147	174	231	243	273
Times to Next Eruption T (minutes)	66	73	81	92	96	101

a. It is known that the time to the next Old Faithful eruption can be predicted using the duration of an eruption and an appropriate regression model. Use the table of values to find the best-fitting linear model $T(d)$ that gives the time (in minutes) to the next eruption based on the duration (in seconds) of an eruption.

b. Use your model to predict the time to the next eruption if an eruption lasted 90 seconds.

c. Use your model to predict the time to the next eruption if an eruption lasts 260 seconds.

d. Complete the following table by finding the times to next eruption predicted by your linear model in part (a) and the corresponding errors. (Round your predicted values to the nearest whole minute.)

Duration (seconds)	110	147	174	231	243	273
Times to Next Eruption (minutes)	66	73	81	92	96	101
Predicted Value T(d) (minutes)						
Error (Actual−Predicted)						

e. According to Ralph Taylor, who has logged and analyzed much of the Old Faithful data, in recent years there have been more eruptions lasting longer than 90 seconds and fewer eruptions of shorter duration. Therefore, most predictions are made using two different regression models—one for longer durations and one for shorter durations. (Adjustments are made for the very few durations that fall in between.) The following table gives the time to next eruption predicted in the Old Faithful Visitor Center (OFVC) Logbook for each of our data points. Complete the table to find the errors for the Logbook predictions.

Duration (seconds)	110	147	174	231	243	273
Times to Next Eruption (minutes)	66	73	81	92	96	101
Time Predicted in OFVC Logbook (minutes)	65	65	84	94	94	94
Error (Actual−Predicted)						

f. Find the SSEs for your model and for the Logbook predictions. Based on the SSEs, which predictions are better? (Note that our model is based on a very small set of very linear data.)

INVESTIGATION Modeling Used Car Prices

In this chapter, you have learned that a linear function is one whose rate of change is constant. As a car ages, its market value decreases, frequently at a rate that is approximately constant. This activity will allow you to create a linear model to describe the price of a used car as a function of its age.

Select a particular make and model of a car or truck that has been in production for at least 10 years (such as a Ford Mustang, a Chevy Silverado, or a Toyota Camry). Then collect prices from a newspaper or the Internet for 10 different vehicles from at least five different model years. Try to choose vehicles that are similar in terms of features; for example, if you choose a Camry, select only SE models or only LE models. Be sure to save a copy of your data from the original source, and attach it to your report.

Once you have your prices:

1. Make a table showing the age of the vehicle in years as input and the price in dollars as output. (A vehicle of the current year is 0 years old, last year's model is 1 year old, etc.)
2. Find the best-fitting linear model $P(t)$ that gives the price of a used vehicle of your model as a function of its age.
3. Find the SSE and the average error for your model.
4. Make a graph showing the scatter plot of your data, along with the regression line. Based on your graph, is a linear function an appropriate model for your data?
5. Explain what the slope and y-intercept of the line mean in terms of the age and price of your vehicle.
6. Based on your model, how much should a 6-year-old vehicle of your make and model cost?
7. Based on your model, how old is a car of your make and model when it becomes "free"?

CHAPTER 3

NATURAL GROWTH MODELS

3.1 PERCENTAGE GROWTH AND INTEREST

3.2 PERCENTAGE DECREASE AND HALF-LIFE

3.3 NATURAL GROWTH AND DECLINE IN THE WORLD

3.4 FITTING NATURAL GROWTH MODELS TO DATA

Some time during the last day, you probably "visited" the Internet. You may have checked your e-mail, participated in a chat room, or downloaded a new ring tone for your cell phone, activities that were unavailable just a few years ago. During the 1990s the use of the Internet exploded, so that today millions of people around the world use it each day for school, work, and entertainment.

The chart in Fig. 3.0.1 shows a plot of data portraying the growth of the Internet from 1993 to 1999. The Internet Software Consortium considered a "host" to be a computer system connected directly to the Internet. We can see

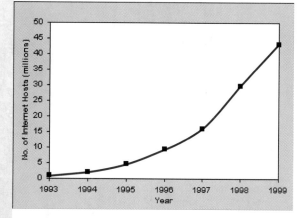

FIGURE 3.0.1 Internet growth from 1993 to 1999.
Source: Internet Software Consortium Internet Domain Survey.

from the increasing steepness of the graph that the growth of the Internet was rather slow at first, but then began to increase more rapidly. Because this function is not increasing at a constant rate, it is not a linear function. To a mathematician, a graph with this shape suggests a quite different kind of function, one that grows at the same *percentage* rate each year.

In this chapter, we study functions that model such growth and illustrate their applicability to a wide variety of real-world situations.

3.1 PERCENTAGE GROWTH AND INTEREST

Suppose that $1000 is invested on July 1, 2000, in a savings account that pays 10% annual interest. This means that, on each subsequent July 1, the amount in the account is increased by 10%—that is, "interest" equal to 1/10 of the current amount is added to the account. Thus, (1/10)($1000) = $100 in interest is added to the account on July 1, 2001, so the new amount then is $1000 + $100 = $1100. Next, (1/10)($1100) = $110 in interest is added to the account on July 1, 2002, so the new amount then is $1100 + $110 = $1210. The following table shows the resulting amount on each July 1 (after the year's interest has been added) for the first 10 years. The third column of the table shows the change in the amount from the previous year. You should use your calculator to verify all the entries in this table. That is, add 1/10 of each year's amount to that amount to calculate the subsequent year's amount.

DATE	AMOUNT	CHANGE
July 1, 2000	$1,000.00	
July 1, 2001	$1,100.00	$ 100.00
July 1, 2002	$1,210.00	$ 110.00
July 1, 2003	$1,331.00	$ 121.00
July 1, 2004	$1,464.10	$ 133.10
July 1, 2005	$1,610.51	$ 146.41
July 1, 2006	$1,771.56	$ 161.05
July 1, 2007	$1,948.72	$ 177.16
July 1, 2008	$2,143.59	$ 194.87
July 1, 2009	$2,357.95	$ 214.36
July 1, 2010	$2,593.74	$ 235.79

Observe that that the annual change increases from each year to the next. Indeed, the change in each of the last 2 years is over twice the change in the first year. Hence it is not the amount of the change that is constant from year to year, but rather the percentage of change that remains constant. Thus growth at a constant *percentage* rate—10% annually in this case—is quite different from a constant rate of change.

Many changing quantities, ranging from bank accounts to animal populations, are very different in appearance but very similar in the way they change. In equal units of time, such as during any 1-year period, from one year to the next, they grow by the same *percentage*. In this section we review the language that is needed to measure and analyze the growth of such quantities.

Percentage Increase

For a brief review of percentages, recall first that 1 **cent** is 1/100 of a dollar. This helps us to remember that the word **percent** means "one hundredth." Thus **1 percent of something is 1/100 of it**. Then p percent of it is p times 1 percent of it, which is the same as $p/100$ of it. So, if we're talking about a quantity denoted by A, then

$$p \text{ percent of } A = \frac{p}{100} \times A, \tag{1}$$

We often use the abbreviation % for the word "percent." Thus **the symbol % simply stands for the number 1/100**. That is,

$$\% = \frac{1}{100} = 0.01. \tag{2}$$

So, whenever we see the symbol % we can replace it with the number 0.01 if we wish. For instance,

$$7\% = 7 \times \% = 7 \times 0.01 = 0.07.$$

This fact is all we need to clear up the air of mystery that surrounds percentages among some people.

The phrase "percent of" simply means "% times." Hence

$$p\% \text{ of } A = p \times \frac{1}{100} \times A = p \times 0.01 \times A. \tag{3}$$

For those who remember (3), there really is no mystery to percentages. It may help to regard the symbol % as a combination of the division symbol / and the two zeroes in the quantity 1/100 in (2).

EXAMPLE 1 | Using Percentages

Find each of the following:

   a. 6% of 100
   b. 15% of 270
   c. 7.6% of 385

SOLUTION

   a. 6% of $100 = 6 \times \frac{1}{100} \times 100 = 0.06 \times 100 = 6$

b. 15% of 270 = 15 × $\frac{1}{100}$ × 270 = 0.15 × 270 = 40.5

c. 7.6% of 385 = 7.6 × $\frac{1}{100}$ × 385 = 0.076 × 385 = 29.26

(Whenever you see such calculations here, you should use your calculator to check them to make sure you know what's going on.)

We can simply replace the word **percent** with the symbol **% = 0.01** when "translating" an appropriate sentence into a mathematical equation. The following table includes some additional pairs of corresponding words and symbols that should be familiar:

WORD	SYMBOL
percent	%
of ("times")	×
and ("plus")	+
is ("equals")	=

Often we need to find the result of increasing some quantity by a given percentage. **To increase A by $p\%$ means to increase A by adding $p\%$ of A to A itself.**

EXAMPLE 2 — Percentage Increase

Suppose a shirt is priced at $26.50. If this price is increased by 6%, find the new price of the shirt.

SOLUTION

$$26.50 + (0.06)(26.50) = 26.50 + 1.59 = 28.09,$$

so the new price of the shirt is $28.09.

Interest and Constant Percentage Growth

We now begin our study of quantities that grow by equal percentages during equal time intervals. A typical such quantity is the amount A invested in a savings account that draws interest that is compounded annually. If the **annual interest rate** is $p\%$, this means that at the end of each year, $p\%$ of the amount at the beginning of the year is added to the account as "interest." **That is, the amount in the account at the beginning of the year is increased by $p\%$ at the end of the year.** Let's consider a specific example as a way of developing an appropriate function rule for the amount in the account after t years.

Suppose that $500 is invested in an account paying 6% interest, compounded annually. We would like to determine the amount A in the account after any given number of years t, that is, $A(t)$. For the sake of having a specific number to talk about, let's suppose we are interested in $A(20)$, the amount in the account after 20 years.

We can proceed as we did in Example 2, and we find the following:

- After 1 year, we have

$$A(1) = \$500 + 0.06(\$500) = \$500 + \$30 = \$530. \qquad (4)$$

- After 2 years, we have

$$A(2) = \$530 + 0.06(\$530) = \$530 + \$31.80 = \$561.80. \qquad (5)$$

- After 3 years we have (rounded to the nearest cent)

$$A(3) = \$561.80 + 0.06(\$561.80) = \$561.80 + \$33.71 = \$595.51. \qquad (6)$$

We could continue on in this manner all the way to 20 years, but this is just not very efficient. Let's look carefully at what we have done.

In (4), we calculated that the amount A in the account in dollars after 1 year is $A(1) = \$500 + 0.06(\$500)$. If we factor that expression, we see that

$$A(1) = \$500(1 + 0.06) = \$500(1.06).$$

Similarly, factoring (5), we obtain $A(2) = \$530(1 + 0.06) = \$530(\$1.06)$. But $\$530$ is $A(1)$, so we can replace $\$530$ with $\$500(1.06)$ to get

$$A(2) = [\$500(1.06)](1.06) = \$500(1.06)^2.$$

Again, factoring (6), we find $A(3) = \$561.80(1 + 0.06) = \$561.80(1.06)$. Since $\$561.80$ is $A(2)$,

$$A(3) = [\$500(1.06)^2](1.06) = \$500(1.06)^3.$$

Do you see the pattern that is emerging?

- After **1** year, the amount in the account is $A(\mathbf{1}) = 500(1.06) = 500(1.06)^1$ dollars.
- After **2** years, the amount in the account is $A(\mathbf{2}) = 500(1.06)^2$ dollars.
- After **3** years, the amount in the account is $A(\mathbf{3}) = [500(1.06)](1.06) = 500(1.06)^3$ dollars.

You can probably now guess that the answer to our original question, "How much is in this account after 20 years?" is found by calculating $A(20) = 500(1.06)^{20}$. That is, the exponent on the multiplier 1.06 matches the number of years the money has been in the account. We see in Fig. 3.1.1 the calculator computation of this value, $\$1603.57$.

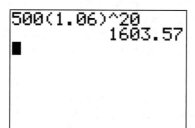

FIGURE 3.1.1 Calculating $500(1.06)^{20}$.

Therefore, after t years, the amount in this account will be $A(t) = 500(1.06)^t$ dollars. In particular, we see that the amount A is a function of the number of years t that the money remains in the account.

Because there is nothing special about the amount we start with or the rate of interest we are earning, we can find a similar function to represent any such annual compound interest situation.

If A_0 denotes the amount originally deposited in the account and the interest rate, written in decimal form, is r, then the amount in the account after the first year is given by

$$A(1) = A_0 + A_0 r = A_0(1 + r).$$

During the second year the amount $A(1)$ is itself multiplied by $(1 + r)$, so at the end of 2 years the amount in the account is

$$A(2) = A(1)(1 + r) = [A_0(1 + r)](1 + r) = A_0(1 + r)^2.$$

During the third year the amount $A(2)$ is multiplied by $(1 + r)$, so at the end of 3 years the amount in the account is

$$A(3) = A(2)(1 + r) = [A_0(1 + r)^2](1 + r) = A_0(1 + r)^3.$$

Again we see the same pattern as before. In each case, the **final exponent** on the right equals the **input value** (time in years) on the left.

Therefore, when A_0 dollars are invested in an account with an annual compound interest rate whose decimal equivalent is r, then the amount in the account at the end of t years is

$$A(t) = A_0(1 + r)^t \text{ dollars}. \tag{7}$$

This function rule is frequently written as $A(t) = P(1 + r)^t$ because the initial amount A_0 is referred to as the *principal* in investment situations.

EXAMPLE 3 Annual Compound Interest

Suppose you deposit $1000 in a savings account that draws 5% interest compounded annually. How long will you have to wait until you have $1300 in the account?

SOLUTION

If we substitute $P = 1000$ and $r = 0.05$ in (7), we get the function $A(t) = 1000(1.05)^t$, which gives the amount in the account after t years. In order to answer this "Here's the output, what's the input?" question, we can calculate this amount with the successive values $n = 1, 2, 3, \ldots$ until our money reaches $1300. An easy way to do this is to store our function in **Y1** and then look at the table of function values for **TblStart = 0** and **ΔTbl = 1**. The calculator screen in Fig. 3.1.2 shows the resulting table.

We see that we had "almost" reached $1300 after 5 years, but it required 6 years to actually exceed our goal of $1300. So the answer to the question asked is 6 years. (But if we had asked after what year is the amount in the account ● *closest* to $1300, the answer would have been 5 years.)

X	Y1
0.00	1000.0
1.00	1050.0
2.00	1102.5
3.00	1157.6
4.00	1215.5
5.00	1276.3
6.00	1340.1

X=0

FIGURE 3.1.2 Table of function values for $A(t) = 1000(1.05)^t$.

Natural Growth Models

The U.S. Constitution as adopted in 1783 decrees that a census (or count) of the people shall be conducted every 10 years. The third column of the following table records the results from the first six census counts in the decade years 1790, 1800, 1810, 1820, 1830, and 1840. The fourth column shows the results of an annual growth rate of 3%, starting with the initial 1790 population $P_0 = 3.9$ million. Note the close correspondence, in these early decades of our nation's history, between the actual population figures and those predicted by the "3% mathematical model" $P(t) = 3.9 \times (1.03)^t$.

t (YEARS)	YEAR	CENSUS POPULATION (MILLIONS)	$P(t) = 3.9 \times (1.03)^t$ (ROUNDED, MILLIONS)
0	1790	3.9	3.9
10	1800	5.3	5.2
20	1810	7.2	7.0
30	1820	9.6	9.5
40	1830	12.9	12.7
50	1840	17.1	17.1

The world is full of quantities that, like savings accounts and the early U.S. population, appear to grow at a constant percentage rate per unit of time. For instance, this is so common for many other populations—of people, animals, insects, bacteria—that such growth is called **natural growth**. If a population starts at time $t = 0$ with initial population numbering P_0 and thereafter grows "naturally" at an annual rate of r, then the number of individuals in the population after t years is given by

$$P(t) = P_0(1 + r)^t. \tag{8}$$

Observe that this formula is the same as Equation (7), except that here we write P for population instead of A for amount. Regardless of the letters we use as variables, a function in this form denotes a quantity growing at a constant percentage rate. A graphing calculator typically requires that **X** (rather than **T**) be used for the independent variable in its **Y=** functions menu. But you should realize that the formula $P(x) = P_0(1 + r)^x$ defines exactly the same function as (8) because it says to do precisely the same thing with x or t (whichever letter we use to denote the independent variable). The independent variable by any other name is (like a rose?) still the independent variable.

We will call the percentage form of r in either (7) or (8) the **annual growth rate** because this value indicates the percentage by which the quantity is growing each year. We will call the value $1 + r$ the **annual growth factor** because this is the value by which our amount is multiplied each year.

EXAMPLE 4 Using a Natural Growth Model

Suppose the U.S. population, starting at 3.9 million in 1790, had continued indefinitely to grow at a constant 3% annual rate.

a. Use your model to predict the U.S. population in 1890.

b. According to your model, when would the country's population have reached 100 million?

SOLUTION

a. Using the model $P(t) = 3.9 \times (1.03)^t$ with $t = 0$ in 1790, the year 1890 is $t = 100$, so $P(100) = 3.9 \times (1.03)^{100} = 75$ million people (rounded to the nearest million).

b. Here we want to find the value of t such that $P(t) = 3.9 \times (1.03)^t = 100$ million. In Fig. 3.1.3 we plotted the graphs **Y1 = 3.9*1.03^X** and **Y2 = 100** in the viewing window $0 < x, y < 150$ and then used the calculator's intersection-finding facility to determine the indicated coordinates (109.8, 100) of the point of intersection. Thus it would take about 109.8 years (starting in 1790) for the U.S. population to reach 100 million.

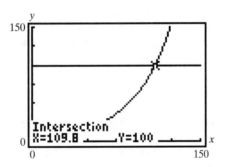

FIGURE 3.1.3 $P = 100$ when $t = 109.8$

Let's assume for the sake of discussion that each calculated population occurs at the midpoint (July 1) of the corresponding year. The 109 years after July 1, 1790, would be July 1, 1899 (because 1790 + 109 = 1899). The population of 100 million would then occur about 0.8 year later, early in the year 1900.

A prediction based on an assumed formula is one thing, and actual reality may be another. The results in Example 4 could have been calculated (perhaps using a slide rule instead of a modern calculator) early in the nineteenth century, as soon as enough census data were available to suggest a 3% annual growth rate. But how could one then have been sure that the U.S. population would continue to grow at this same 3% constant annual rate? Did it? One should never merely accept mathematical predictions without wondering about the possibility of a reality check. We cannot be sure of the validity of a mathematical model until we have checked its predictions against real-world facts.

In this case, sitting here more than a century later, we *know* how it turned out. Figure 3.1.4 exhibits a spreadsheet (available at this book's web site as **USpop.xls**) showing both the population predicted by the "3% mathematical model" and the actual population recorded in U.S. census data at 10-year intervals through the year 1900. In the graph, predicted populations are plotted as a smooth curve and the actual census populations are plotted as square dots.

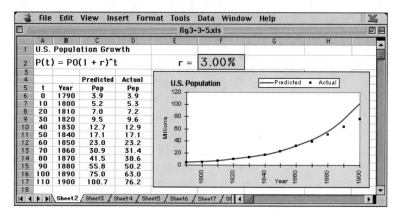

FIGURE 3.1.4 Predicted and actual U.S. populations for 1790-1900.

We observe that the assumed 3% rate of growth appears to have been maintained initially—perhaps through the first half of the nineteenth century—but the actual U.S. population growth seems to have slowed appreciably during the second half of the century. Indeed, in the year 1900 the actual population of the United States was about 76 million rather than the 100 million that is predicted by the model $P(t) = 3.9 \times (1.03)^t$.

EXAMPLE 5 Interpreting a Natural Growth Model

The revenue for the state of Georgia from 1990 to 2001 can be modeled approximately by the natural growth function $R(t) = 6.8(1.068)^t$, where t is years after 1990 and $R(t)$ is measured in billions of dollars.

(Source for data: *Atlanta Journal-Constitution*.)

a. According to this model, what was the State of Georgia's revenue in 1990?

b. At what annual rate was Georgia's revenue growing over this time period?

c. Use the model to predict the year in which Georgia's revenue reaches $10 billion.

d. Use the model to predict Georgia's revenue in the year 2010.

SOLUTION

a. In a natural growth model $A(t) = A_0(1 + r)^t$, A_0 represents the initial amount. So our initial amount, 6.8 (billion dollars) represents the revenue for Georgia in 1990.

b. Similarly, $(1 + r) = 1.068$ is our growth *factor*. Therefore, the growth rate $r = 1.068 - 1 = 0.068$. Changing this decimal form to percent form gives us an annual growth rate of 6.8%.

c. With $A(t)$ stored in **Y1**, 10 stored in **Y2**, and a viewing window of $-1 < x, y < 20$, the **intersect** command on the calculator gives the intersection point shown in Fig. 3.1.5.

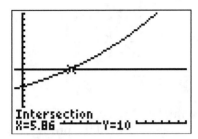

FIGURE 3.1.5 $R = 10$ when $t = 5.86$.

Assuming, as we usually do, that the years referred to begin on January 1, 5.86 years takes us to the latter part of 1995. So, the revenue for the state of Georgia is predicted to have reached $10 billion in 1995.

d. Since $A(t)$ is already stored in **Y1** and our calculator window is (just) large enough for the input we are interested in (year 2010, $t = 20$), we can use the **value** command to evaluate $A(20)$.

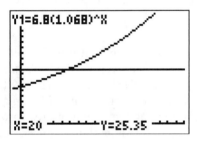

FIGURE 3.1.6 When $t = 20$, $R = 25.35$.

As Fig. 3.1.6 indicates, our model predicts a 2010 Georgia revenue of $25.35 billion.

Iteration

Modern calculators and computers provide especially simple ways to carry out repetitive calculations like the one in Example 3 (in which $1000 is deposited in a savings account that draws 5% interest compounded annually). Indeed, this is exactly what modern computers were invented (in the 1940s and 1950s) to do. Suppose we enter

$$1000 \rightarrow A$$

in our calculator to record the initial deposit A, and then enter the command

$$A*(1 + 0.05) \rightarrow A$$

to calculate and store (as the value of **A**) the amount in the account after 1 year if 5% interest is compounded annually.

Now comes the punch line—if we simply press the **ENTER** key again, the last command line is executed again! That is, the value 1050.00 currently stored as **A**

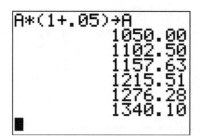

FIGURE 3.1.7 Example 3 by iteration.

is itself multiplied by the growth factor 1.05, and the result stored as the new value of **A**. Having started the process, each press of the **ENTER** key carries out the computation for another year and stores the result. So, in order to find out how much is in the account after 6 years, we need only type the line **A*(1+0.05)** → **A** *once*, and then press the **ENTER** key five times in succession (keeping careful count of how many times we've entered the command). Figure 3.1.7 shows the result, and provides the same conclusion as in Example 3.

This is our first example of *iteration*—doing the same thing again and again. Specifically, the successive results $A_0, A_1, A_2, A_3, \ldots$ are calculated *iteratively* when the same formula is used repeatedly to calculate each new result from the preceding one. In Example 3, this *iterative formula* is the formula

$$A_n = A_{n-1}(1 + 0.05), \tag{9}$$

which says that each amount is multiplied by $(1 + 0.05)$ to get the next amount. Doing this over and over is called *iterating* the formula. So Fig. 3.1.7 shows the result of iterating (9) six times, starting with $A_0 = 1000$.

In this situation iteration provides an alternate strategy for calculating yearly amounts in annual compound interest problems. Iteration has a long history in mathematics; iterative methods were sometimes the only ones available to solve a problem and were used long before calculators and computers were available.

When we repeat the iteration of Fig. 3.1.7, we "plow" the output back into the right-hand side of the equation $A_{new} = A_{old}(1 + 0.05)$. Indeed, the word *iterate* apparently stems from the Latin verb *iterare* with the meaning "to plow again." The concept of iteration is as old as the word. Two thousand years ago the Babylonians introduced the iteration

$$x_{n+1} = \frac{1}{2}\left(x_n + \frac{A}{x_n}\right), \tag{10}$$

which can be used to calculate more and more accurate approximations to the **square root** $\sqrt{A}$ of a given positive number A. One starts with an initial guess x_0 and then uses (10) to calculate the successive iterates $x_1, x_2, x_3, \ldots$. It happens that the initial guess x_0 need not be especially accurate—any nonzero guess will do for a start. For instance, $x_0 = A/2$ is a convenient starting point, and then the **Babylonian square root algorithm** can be implemented with the calculator commands

$$2 \to \mathbf{A}$$

$$\mathbf{A}/2 \to \mathbf{X}$$

$$(1/2)(\mathbf{X} + \mathbf{A}/\mathbf{X}) \to \mathbf{X}$$

to approximate the square root of the number $A = 2$. The first two commands here initialize the variables **A** and **X**, and the last one is the iterative command.

SECTION 3.1 Percentage Growth and Interest **109**

```
            1.000000
(1/2)(X+A/X)→X
            1.500000
            1.416667
            1.414216
            1.414214
            1.414214
```

FIGURE 3.1.8 Approximating the square root of 2.

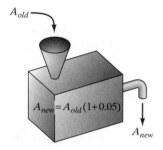

FIGURE 3.1.9 Iteration (9) as an input-output process.

The results shown in Fig. 3.1.8 indicate that the six-place value $\sqrt{2} \approx 1.414214$ is reached quickly. With any positive number A and any initial guess x_0 the successive approximations generated by the iteration in (10) eventually "stabilize" in this way, agreeing to the number of decimal places displayed. In Exercises 28 and 29, you can explore this historic method for calculating square roots.

Input-Output Processes

The iterative formula in (9) or (10) can be visualized as an **input-output process** as illustrated in Fig. 3.1.9. We think of a "black box" with a hidden mechanism inside that performs the computation described by the iterative formula. Specifically, when an **input** $A = A_{old}$ is fed into the box, the iteration is performed and the **output** A_{new} is produced. We often find it useful to interpret a complicated iteration as a "simple" input-output process—simple in that for many purposes we need not think explicitly of what is happening inside the box. It may only be important that, whatever the input, the same process always produces the corresponding output.

3.1 Exercises | Building Your Skills

In Exercises 1–4, find each number.

1. 5% of 200
2. 3.5% of 50
3. 2.25% of 18
4. 4.19% of 58.35

In Exercises 5–8, calculate the new price (rounded off to the nearest cent) if

5. The old price of $110 is increased by 13%.
6. The old price of $1720 is increased by 4.2%.
7. The old price of $69.50 is increased by 3.9%.
8. The old price of $250 is increased by 8.25%.
9. Explain why the following method works to calculate the amount to be paid in a restaurant if you wish to add a 15% tip to the original bill amount A. First you write down the amount A. Then you move the decimal point in A one unit to the left and write this amount under A. Finally divide this second amount by 2 and write this amount down also. The total amount (including tip) to be paid is then the sum of the three amounts you have written down.

10. The town of Bridgeport had 20 thousand people in the year 2000 and was growing at a rate of 4% per year. Find a natural growth function $P(t)$ that gives the population of Bridgeport as a function of t, years after 2000.

11. The town of Jackson had 130 thousand people in the year 2000 and was growing at a rate of 6.3% per year. Find a natural growth function $P(t)$ that gives the population of Jackson as a function of t, years after 2000.

12. A small company currently has total sales of $80,000 and projects that its sales will grow at a rate of 2% per year. What are the company's projected sales in 10 years?

13. A self-employed tax preparer currently has an advertising budget of $500 and plans to increase this budget by 1.5% per year. In how many years will her advertising budget rise to $600?

Applying Your Skills

14. A 1996 presidential primary candidate proposed a "flat tax" under which a family of four would pay as its federal income tax each year an amount equal to 17% of the portion of its taxable income in excess of $36,000. Suppose that a family's 1996 taxable income was $50,000.

    a. Under this proposal, how much would be owed as income tax?
    b. Suppose that this family's actual federal income tax bill for 1996 was $6015 plus 28% of the amount of their taxable income over $40,100. How much income tax did the family owe?
    c. Under which method of assessing taxes would the family pay less tax?

15. Repeat Exercise 14 for a family whose taxable income was $200,000.

16. Suppose that $400 is deposited in an account paying 6% interest compounded annually.

    a. Find a natural growth function $A(t)$ that gives the amount in the account after t years.
    b. How much is in the account after 5 years?
    c. How long will it take until the account has $1000? Give the year after which the amount is **at least** $1000.

17. Suppose that $2000 is deposited in an account paying 3.45% interest compounded annually.

    a. Find a natural growth function $A(t)$ that gives the amount in the account after t years.
    b. How much is in the account after 8 years?
    c. How long will it take until the account has $3200? Give the year after which the amount is **at least** $3200.

18. How long (rounded off accurate to the nearest year) does it take to triple an initial deposit of $1000 if the annual interest rate is 8%?

19. Suppose that $1000 is deposited in an account.

    a. Complete column B to show how long it will take until the initial deposit is doubled for various interest rates. (Give the year after which the amount in the account is **closest to** $2000.)
    b. Complete column C by finding the product of the values in column A and column B for each row.

COLUMN A % INTEREST RATE	COLUMN B NUMBER OF YEARS TO DOUBLE	COLUMN C COLUMN A × COLUMN B
4		
6		
8		
9		

c. Based on your results, what might the **rule of 72** say?

20. If you invest $1000 at 12% annual interest, show that you will have $1973.82 after 6 years and $2210.68 after 7 years. What initial deposit (accurate to the nearest cent) would lead to precisely $2000 (accurate to the nearest dollar) after 6 years? To answer this question, you will need to try several different initial values, finding successively better estimates of the needed initial deposit; obviously you need to start with a bit more than $1000.

21. The population of Jacksonville, Florida, was 635 thousand in 1990 and grew at an average annual rate of 1.49% throughout the 1990s. Assuming that this rate of growth continues,

    a. Find a natural growth function $P(t)$ that gives the population of Jacksonville as a function of t, years after 1990.
    b. Use your function model to predict the year in which the population of Jacksonville will grow to 800 thousand.
    c. Use your function model to predict the population of Jacksonville in 2003. How does your prediction compare to the actual 2003 population of 733,781?

22. The population of Lincoln, Nebraska, was 192 thousand in 1990 and grew at an average annual rate of 1.64% throughout the 1990s. Assuming that this rate of growth continues,

    a. Find a natural growth function $P(t)$ that gives the population of Jacksonville as a function of t, years after 1990.
    b. Use your function model to predict the population of Lincoln in 2003. How does your prediction compare to the actual 2003 population of 235,594?
    c. Use your function model to predict the year in which the population of Lincoln will grow to 350,000.

23. The population of Corpus Christi, Texas, was 257 thousand in 1990 and grew at an average annual rate of 0.75 % throughout the 1990s. Assuming that this rate of growth continues,

    a. Find a natural growth function $P(t)$ that gives the population of Corpus Christi as a function of t, years after 1990.
    b. Use your function model to predict the population of Corpus Christi in 2003. How does your prediction compare to the actual 2003 population of 279,208?
    c. Use your function model to predict the year in which the population of Corpus Christi will grow to 400 thousand.

24. Data from the *St. Louis Post-Dispatch* indicate that the cost in dollars of a gallon of gasoline from 1966 to 2006 can be modeled approximately by the natural growth function $G(t) = 0.32(1.055)^t$, where t is years after 1966.

  a. According to this model, what was the price of a gallon of gas in 1966?
  b. At what annual percentage rate was the price of a gallon of gas growing over this time period?
  c. Use the model to predict the price of a gallon of gas in 2016.
  d. Use the model to predict the year in which the price of a gallon rises to $5.

25. The number of people (in millions) enrolled in health maintenance organizations (HMOs) for the 20 years following 1976 can be given approximately by the natural growth model $P(t) = 6.0(1.12)^t$, where t is years after 1976. (Model based on data from *World Almanac and Book of Facts 2005*.)

  a. According to this model, how many people were enrolled in HMOs in 1976?
  b. At what annual percentage rate was the number of people enrolled in HMOs growing over this time period?
  c. Use the model to predict the number of people enrolled in HMOs in 1999. How does your prediction compare to 81.3 million, the actual number of people enrolled in HMOs in 1999?
  d. Use the model to predict the number of people enrolled in HMOs in 2000. How does your prediction compare to 80.9 million, the actual number of people enrolled in HMOs in 2000?
  e. How do you explain the discrepancy in part (d)?

26. According to the *Atlanta Journal-Constitution*, the market research firm NPD Group "reported a dramatic 53% increase in sales of suits, dress pants, sport coats, and jackets to young men" in 2005 over 2004 sales of $645 million dollars.

  a. Assuming that this trend continues, find a natural growth function $S(t)$ that gives the sales as a function of t, years after 2004.
  b. Based on your model, what were the 2005 sales of suits, dress pants, sport coats, and jackets to young men?
  c. Use your model to predict the year in which sales will reach $2 billion? ($2 billion is $2000 million.)

27. Data from *USA Today* indicate that in 1994, Brazil's exports were valued at $46.4 billion and increased at an average annual rate of 8.51% over the next 10 years. ("Exports" refers to goods and services, and are measured in constant 2000 U.S. dollars.)

  a. Assuming that this trend continues, find a natural growth function $E(t)$ that gives the value of Brazil's exports as a function of t, years after 1994.
  b. Based on your model, what was the value of Brazil's exports in 2004?
  c. Use your model to predict the year in which Brazil's exports will double.

28. Apply the Babylonian square root algorithm to approximate the (principal) square root of 19. Try different initial guesses and discuss how the accuracy of the initial guess affects the number of iterations required to get six-place accuracy.

29. What happens with the Babylonian algorithm if a negative rather than a positive initial guess is used?

3.2 PERCENTAGE DECREASE AND HALF-LIFE

In Section 3.1, we considered natural growth functions that describe quantities that increase at a constant annual percentage rate. Natural growth functions can also be used to describe quantities that *decrease* at a constant percentage rate. (Oddly enough, we refer to this entire class of functions as natural "growth" functions, regardless of whether the quantity involved is growing or declining.)

Percentage Decrease

Some department store "clearance centers" operate on the principle that the price of an article is reduced by an additional 10% each week it remains in stock. Suppose that such a clearance center has a pair of shoes initially priced at $150. What function describes the price of the shoes in terms of the number of weeks it remains in stock?

Recall that to increase A by $p\%$ means to increase A by adding $p\%$ of A to A itself. Similarly, **to decrease A by $p\%$ means to decrease A by subtracting $p\%$ of A from A itself.**

So for our clearance shoes:

- After 1 week, the price P would be

$$P(1) = \$150 - 0.10(\$150) = \$135. \tag{1}$$

- After 2 weeks, the price P would be

$$P(2) = \$135 - 0.10(\$135) = \$121.50. \tag{2}$$

- After 3 weeks, the price P would be

$$P(3) = \$121.50 - 0.10(\$121.50) = \$109.35. \tag{3}$$

We could continue on in this manner, finding the price of the shoes as long as they remain in stock, or we could use an iterative formula as we did in Section 3.1. But our goal here is to determine a function that describes the price of the shoes.

In (1), we calculated that the price P of the shoes after 1 week is $P(1) = \$150 - 0.10(\$150)$. If we factor that expression, we see that

$$P(1) = \$150(1 - 0.10) = \$150(0.90).$$

Similarly, factoring (2), we get $P(2) = \$135(1 - 0.10) = \$135(0.90)$. But $135 is $P(1)$, so we can replace $135 with $150(0.90) to get

$$P(2) = [\$150(0.90)](0.90) = \$500(0.90)^2.$$

And again, factoring (3), we find $P(3) = \$121.50(1 - 0.10) = \$121.50(0.90)$. Since $121.50 is $P(2)$,

$$P(3) = [\$150(0.90)^2](0.90) = \$150(0.90)^3.$$

We see the same kind of pattern here that we observed in increasing natural growth functions. That is,

- after **1** week, the price of the shoes is $P(\mathbf{1}) = 150(0.90) = 500(0.90)^\mathbf{1}$ dollars;

- after **2** weeks, the price of the shoes is $P(2) = 150(0.90)^2$ dollars;
- after **3** weeks, the price of the shoes is $P(3) = 150(0.90)^3$ dollars.

Therefore we conclude that, after t weeks, the price of the shoes will be $P(t) = 150(0.90)^t$ dollars.

EXAMPLE 1   Finding an Input Value

If you are willing to pay only $80 for the shoes from the clearance center, how long must you wait to buy them?

SOLUTION

Since the price of the shoes is reduced only once a week, a table displaying function values at the end of each week will indicate when the price of the shoes is $80 or less. Entering $P(t)$ into **Y1** and 80 into **Y2**, we set **TblStart=0** and **ΔTbl =1**. Figure 3.2.1 shows the resulting table.

FIGURE 3.2.1 Finding when $P(t) \leq 80$.

X	Y1
0.000	150.00
1.000	135.00
2.000	121.50
3.000	109.35
4.000	98.415
5.000	88.574
6.000	79.716

X=0

We see that if the shoes are still in stock after 6 weeks, the price will be $79.72 (to the nearest cent). Since the price is below $80, you can now happily purchase the shoes. (Of course, someone else may have been willing to pay more, and may have bought them earlier!)

Declining Natural Growth Models

In general, suppose that A is a quantity that is decreasing at a constant percentage rate r (expressed as a decimal) during each unit of time t. If A_0 denotes the initial value of the quantity, then the amount after the first unit of time (hour, day, week, year, whatever) is given by

$$A(1) = A_0(1 - r).$$

During the second unit of time, the amount $A(1)$ is itself multiplied by $(1 - r)$, so that

$$A(2) = A(1)(1 - r) = [A_0(1 - r)](1 - r) = A_0(1 - r)^2.$$

During the third unit of time, the amount $A(2)$ is multiplied by $(1 - r)$, so that

$$A(3) = A(2)(1 - r) = [A_0(1 - r)^2](1 - r) = A_0(1 - r)^3.$$

Again we see the same pattern as before. In each case, the **final exponent** on the right equals the **input value** (time) on the left.

Therefore, when a quantity whose initial value is A_0 is decreasing during each unit of time at a rate whose decimal equivalent is r, then the amount of the quantity after t units of time is

$$A(t) = A_0(1 - r)^t. \tag{4}$$

EXAMPLE 2 Finding a Declining Natural Growth Function

Suppose you invested $1200 in a dot-com stock in 1990, and rather than growing (as you expected), the value of that investment decreased at an annual percentage rate of 4%.

a. Find a natural growth function that describes the value of the investment as a function of t, the number of years after 1990.
b. What was the value of your investment in 1995?
c. If you have decided to sell the stock when its value declines to $500, in what month and year will you sell the stock?

SOLUTION

a. If we substitute $A_0 = \$1200$ and $r = 0.04$ in (4), we get the natural growth function $A(t) = \$1200(0.96)^t$ giving the value of the investment—that is, the amount in the account after t years—as a function of t.

b. In order to answer this "Here's the input, what's the output?" question, we need to evaluate our function when $t = 5$. Thus $A(5) = \$1200(0.96)^5 = \$978.45$ (correct to the nearest cent).

c. In this case, we are given the output ($500) and asked to find the appropriate input (time). Letting **Y1=1200(0.96)^X** and **Y2=500**, and using the window indicated in Fig. 3.2.2, we find that the **intersect** command gives intersection point shown in Fig. 3.2.3.

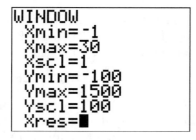

FIGURE 3.2.2 The window for solving **1200(0.96)^X = 500**.

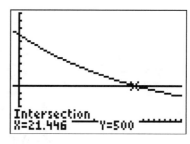

FIGURE 3.2.3 $A = 500$ when $t = 21.446$.

Thus, you need to sell the stock after 21.446 years. Since (0.446 year)(12 months/year) = 5.352 months, we have completed 5 months and are in the sixth; you need to sell the stock in June 2011.

If you prefer to learn only one formula for natural growth functions, you should remember the one from Section 3.1, $A(t) = A_0(1 + r)^t$. When you need to create a decreasing natural growth function, you can think of its growth rate as negative (because adding the opposite of a number is equivalent to subtracting the number itself).

We saw in Chapter 2 that once we have "built" one or more functions, we can answer many different types of input-output questions, such as,

- What is the value of a function for a given input?
- For what input does a function have a given output?
- For what input is the function's value double its initial value?
- For what input do two functions have the same output?

We can also answer such questions for the natural growth functions we created in this chapter (and have done so in Examples 1 and 2). Indeed, we will be doing this for all the different types of functions we create throughout this book. This is the key to understanding the "big idea" of mathematical modeling: the questions do not change—it is only the functions that model the situation that are different.

EXAMPLE 3 | When Are Two Populations Equal?

In 2000, the population of Baltimore, Maryland, was 651 thousand and was declining at an average annual rate of 1.14%. At the same time, the population of Fort Worth, Texas, was 541 thousand and was increasing at an average annual rate of 2.64%. In what month and year will the populations of Baltimore and Fort Worth be equal?

SOLUTION

Our first task is to find a natural growth model for the population of each city. Since Baltimore's initial population is 651 thousand and its annual rate of decline is 0.0114, the population is given by

$$B(t) = 651(1 - .0114)^t = 651(0.9886)^t,$$

where $t = 0$ in 2000 and $B(t)$ is given in thousands.

Fort Worth's initial population is 541 thousand and its annual rate of increase is 0.0264; its population is given by

$$F(t) = 541(1 + .0264)^t = 541(1.0264)^t,$$

where $t = 0$ in 2000 and $F(t)$ is given in thousands.

The question we need to answer then is "What value of t makes $B(t) = F(t)$?" With $B(t)$ stored in **Y1** and $F(t)$ stored in **Y2**, and using the window $-1 < x < 15$, $-10 < y < 800$, we use the intersect feature of the calculator. Figure 3.2.4 shows the result.

We see (as we expect) one increasing function (the population of Fort Worth) and one decreasing function (the population of Baltimore). You should always verify that the graph your calculator displays reflects what you know about the function or functions you are graphing. It is easy to make an error in

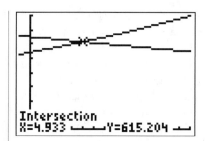

FIGURE 3.2.4 $B(t) = F(t)$ when $t = 4.933$.

entering a function into the calculator—if something looks wrong to you, check it before reporting an answer.

In this case, however, all is fine, and the intersection point (4.933, 615.204) tells us that 4.933 years after January 1, 2000, the populations of Baltimore and Fort Worth are equal. Since 0.933×12 is 11.196, the populations are equal in December 2004.

The intersection point also tells us the answer to a question we didn't ask. That is, what are the populations of the cities when they are equal? The second coordinate gives us these equal populations. So our models predict that on some day in December 2004 Baltimore and Fort Worth will each have a population of
● 615.204 thousand, or 615,204 people.

Half-Life

When discussing an increasing natural growth function, it makes sense to ask when the initial amount doubles. With a decreasing natural growth function, the quantity is getting smaller, so the amount will never be double the initial amount. However, we are frequently interested in answering a similar question—when does the quantity decline to *half* the initial amount?

For a quantity described by a decreasing natural growth function, the length of time it takes for the quantity to decrease to half its initial value is called its **half-life**. While we do not typically talk about the half-life of an investment gone bad or a city's population, this idea is very common in discussing the decay of radioactive elements or the metabolism of drugs in the human body.

EXAMPLE 4 | Finding Half-Life

A laboratory has a 50-gram sample of bismuth-210, a radioactive element that decays at a daily rate of approximately 12.94%.

    a. What is the half-life of bismuth-210?
    b. How long will it take until only 2 grams of the sample remain?

SOLUTION

    a. The natural growth function describing the number of grams in the sample is $B(t) = 50(0.8706)^t$. To determine the half-life, we need to figure out when $B(t) = 25$. We will again use the intersect command from the calculator's **CALC** menu, with $B(t)$ stored in **Y1** and 25 stored in **Y2**. In an appropriate window, the calculator displays the intersection point shown in Fig. 3.2.5. (For many students, the most difficult part of these problems is finding an appropriate window. We leave you to discover a "good" window here on your own, as practice for future situations. Just keep trying windows until you find one where the intersection of the two graphs shows clearly.)

The intersection point indicates that the half-life of bismuth-10 is (correct to the nearest day) 5 days.

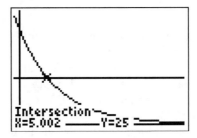 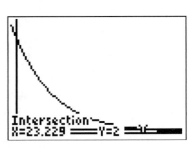

FIGURE 3.2.5 $A = 25$ when $t = 5.002$.

FIGURE 3.2.6 $B = 2$ when $t = 23.229$.

b. Although the wording is much different, this question is the same kind of "Here's the output, what's the input?" question that we answered in part (a). (We really are asking you the same questions over and over again!) So if we replace 25 with 2 in **Y2** and use **intersect** once again, the result is shown in Fig. 3.2.6. Thus, it takes a bit over 23 days for only 2 grams of the sample to remain.

There are two important points to note about this example. First, although we gave you a specific number of grams in the sample, this value is not required for the calculation of the half-life. The half-life of a substance does not depend on the initial amount given, but rather on properties inherent in the substance itself. If a particular amount had not been given in Example 4, we could have chosen any amount we like (our favorite number perhaps?) or the arbitrary amount A_0. In this case, we would be need to solve the equation

$$0.5\, A_0 = A_0(0.8706)^t.$$

If we divide both sides of this equation by A_0, we obtain the equivalent equation

$$0.5 = (0.8706)^t.$$

We can solve this equation graphically, as we did before, with 0.8706^t in **Y1**, and 0.5 in **Y2**. (You should try this to see that we get exactly the same half-life.)

Second, we notice that while it took only 5 days for the first 25 grams of the sample to decay, the next 23 grams (so that only 2 grams remain) took approximately 18 days to decay. Why is this? This sample is not decaying by the same *amount* every 5 days—that would make it a linear function. It is decaying by the same *percentage* (50% of what is there) every 5 days—that's what makes it a natural growth function. Because half of our sample "disappears" every 5 days, we keep halving and halving and halving indefinitely. Thus, at least in theory, all of the sample never goes away. Of course, at some point, whatever scale we are using to measure the sample is not sensitive enough to recognize that there is any sample left. At this point, we can say that the weight of the sample is "essentially" zero. What we mean by this is that neither we nor the scale we are using can distinguish between the weight of the sample and the real number 0.

So for a declining natural growth function, as our input values increase, our output values get closer and closer to 0 but never actually "get there." Thus the graph of such a function gets closer and closer to the x-axis but never actually touches it. (Our calculator has the same difficulty as the scale weighing the sample—if you go out "far enough," it appears that the output values are all 0. But in reality they are not; it is just the physical limitations of the calculator that prevent us from seeing the difference.)

You probably have had no occasion to think about the half-life of bismuth-210 in your daily life, but the same mathematical principles apply to how your body eliminates certain chemicals from your blood. Metabolism of prescription medications, over-the-counter remedies, and even caffeine behaves like a decreasing natural growth function. (Surprisingly, alcohol—when measured by the drink—is eliminated from the body in a linear fashion, typically at the rate of about 1 drink per hour.)

EXAMPLE 5 | The Half-Life of Caffeine

In a healthy adult, the amount of caffeine in the bloodstream decreases at an hourly rate of approximately 11.5%.

a. What is the half-life of caffeine in a healthy adult?

b. A can of Red Bull energy drink contains about the same amount of caffeine as a cup of regular coffee, 100 mg. If you drink a Red Bull or a cup of coffee, how much caffeine will remain in your bloodstream after 2 hours?

Sources: American Journal of Clinical Pathology and www.redbullusa.com.

SOLUTION

a. Since a specific initial value is not given here, the natural growth function describing the amount of caffeine in the bloodstream is $A(t) = A_0(0.885)^t$. We need to determine when the amount of caffeine falls to half the initial amount, that is, to $0.5A_0$. Thus we need to solve the equation $0.5A_0 = A_0(0.885)^t$. As before, we can find an equivalent equation, $0.5 = 0.885^t$, and solve this equation for t. Figure 3.2.7 shows the graphical solution of this equation. Once again, you should verify this result in your calculator.

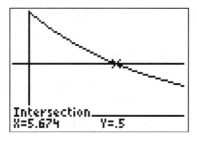

FIGURE 3.2.7 $A = 0.5$ when $t = 5.674$.

So the half-life of caffeine in a healthy adult is about 5.7 hours.

b. Here we do have a specific initial amount, so our natural growth function is $A(t) = 100(0.885)^t$. We are asked to determine how much caffeine is in the bloodstream after 2 hours—that is, to find the output for an input value of 2. We evaluate $A(2) = 100(0.885)^2 = 78.4335$; to the nearest milligram, 78 mg of caffeine remain after 2 hours.

What we have presented here is a simplified view of how a body metabolizes caffeine. There are many variables involved—a person's gender, weight, and medical condition and how quickly he or she consumes the caffeine. But it is worth noting that natural growth models, whether increasing or decreasing, occur in many applications that at first glance seem quite different. What unites all of these situations is that they involve quantities that either increase or decrease at a constant percentage rate.

3.2 Exercises

Building Your Skills

In Exercises 1–6, calculate the new price (rounded off to the nearest cent) if

1. the old price of $82 is decreased by 7%.
2. the old price of $640 is decreased by 2%.
3. the old price of $69.50 is decreased by 1.5%.
4. the old price of $1080 is decreased by 6.3%.
5. the old price of $535 is decreased by 0.4%.
6. the old price of $192 is decreased by 0.25%.
7. Suppose that a quantity whose initial value is 120 is reduced by 25%, and then the result is increased by 25%. Is the final value 120, more than 120, or less than 120? Why?
8. Suppose that a quantity whose initial value is 200 is increased by 15%, and then the result is decreased by 15%. Is the final value 200, more than 200, or less than 200? Why?
9. The town of Bilston had 35 thousand people in the year 2000 and was declining at a rate of 3% per year. Find a natural growth function $P(t)$ that gives the population of Bilston as a function of t, years after 2000.
10. The town of Independence had 210 thousand people in the year 2000 and was declining at a rate of 4.1% per year. Find a natural growth function $P(t)$ that gives the population of Independence as a function of t, years after 2000.
11. A manufacturing company currently loses $20,000 a year on defective items. It is implementing a plan by which it hopes to reduce this amount by 1% each year for the next 10 years. If the company is successful, how much will it lose on defective items 5 years from now?
12. A small accounting firm is spending $60,000 a year on clerical staff. The president of the firm believes that they can reduce this amount by 5% per year over the next 5 years by hiring temporary workers. Assuming the president is correct, how much would the firm spend on clerical staff in 3 years?

Applying Your Skills

13. Suppose that you invest $1000 in shares of a new stock you heard about from a friend. Instead of gaining value (as you hoped), the stock loses 2% of its value each month.

a. Find a natural growth function $V(t)$ that gives the value of the stock after t months.
b. What is the stock worth after 6 months?
c. If you are unwilling to lose more than $200 on this venture, in how many months must you sell the stock (assuming that it continues to lose value at this rate)?

14. The population of Birmingham, Alabama, was 266 thousand in 1990 and was decreasing at an average annual rate of approximately 1% throughout the 1990s. Assuming that this rate of decline continues,

   a. find a natural growth function $P(t)$ that gives the population of Birmingham as a function of t, years after 1990.
   b. use your function model to predict the year in which the population of Birmingham will fall to 220 thousand.
   c. Use your function model to predict the population of Birmingham in 2003. How does your prediction compare to the actual 2003 population of 236,620?

15. The population of Toledo, Ohio, was 333 thousand in 1990 and was decreasing at an average annual rate of 0.59% throughout the 1990s. Assuming that this rate of decline continues,

   a. find a natural growth function $P(t)$ that gives the population of Toledo as a function of t, years after 1990.
   b. use your function model to predict the population of Toledo in 2003. How does your prediction compare to the actual 2003 population of 308,973?
   c. use your function model to predict the year in which the population of Toledo will fall to 280 thousand.

16. The population of Cincinnati, Ohio, was 331 thousand in 2000 and decreased at an average annual rate of 1.53% over the next 3 years. At the same time, the population of Plano, Texas, was 222 thousand and grew at an average annual rate of 2.92%. Assuming that these trends continue, in what month and year will the populations of Cincinnati and Plano be the same?

17. The population of Boston, Massachusetts, was 589 thousand in 2000 and decreased at an average annual rate of 0.4% over the next 3 years. At the same time, the population of Sacramento, California, was 407 thousand and grew at an average annual rate of 3.02%. Assuming that these trends continue, in what month and year will the populations of Boston and Sacramento be the same?

18. Canada's Office of Consumer Affairs reported that in 1982 the average household spending for food constituted 15.1% of household expenditures. From 1982 to 2003 this amount decreased at an average annual rate of 1.53%.

   a. Assuming that this trend continues, find a natural growth function $F(t)$ that gives percentage expenditure on food as a function of t, years after 1982.
   b. Use your model to predict the year in which percentage expenditure on food decreases to 10%.
   c. Based on your model, what percentage of household expenditures did the average household spend on food in 1996?

19. According to the American Cancer Society, death rates from diseases other than cancer were falling from 1950 to 2003, while cancer death rates remained virtually the same. Measured per hundred thousand, the death rate from heart

disease in 1950 was 586.8. This amount decreased at an average annual rate of 1.74% over the next 53 years.

  a. Assuming that this trend continues, find a natural growth function $D(t)$ that gives heart disease deaths per hundred thousand as a function of t, years after 1950.
  b. Based on your model, how many heart disease deaths per hundred thousand were there in 1967?
  c. Use your model to predict the year in which heart disease deaths per hundred thousand fall to 300.

20. As indicated in Example 5, the amount of caffeine in the bloodstream decreases at an hourly rate of approximately 11.5%. A double shot of espresso contains 80 mg of caffeine.

  a. Find a natural growth model $C(t)$ that gives the amount of caffeine in the bloodstream as a function of time t in hours.
  b. How much caffeine is in the bloodstream after 3 hours?
  c. How long does it take for the caffeine level in the blood to fall to 10 mg?

21. A laboratory has a 30-gram sample of radon-222, which decays at a rate of 16.6% per day.

  a. Find a natural growth model $R(t)$ that gives the amount of radon-222 as a function of t, time in days.
  b. How long does it take for the sample to decay to 5 grams?
  c. How much radon-222 remains after 2 weeks?

22. A laboratory has a 800-mg sample of cesium-137, which decays at an annual rate of 2.28%.

  a. Find a natural growth model $C(t)$ that gives the amount of cesium-137 as a function of t, time in years.
  b. How much of the sample remains after 5 years?
  c. What is the half-life of cesium-137, correct to the nearest year?

23. Many people use the over-the-counter medication ibuprofen for relief from muscle strains and joint pain. The amount of ibuprofen in an adult's bloodstream decreases at an hourly rate of approximately 29%, and a normal adult dose is 400 mg.

  a. Find a natural growth model $I(t)$ that gives the amount of ibuprofen in the bloodstream as a function of time t in hours.
  b. What is the half-life of ibuprofen?
  c. How much ibuprofen is in a person's bloodstream after 3 hours?

3.3 NATURAL GROWTH AND DECLINE IN THE WORLD

In Sections 3.1 and 3.2, we discussed natural growth models for which we knew the annual growth rate. However, in many situations we suspect (or assume) constant-percentage growth but know the actual population at only a couple of different times. The following example illustrates several different methods that can then be used to determine the population's growth rate r.

EXAMPLE 1 Finding the Growth Rate *r*

The city of Bethel had a population of 25 thousand in 1990 and 40 thousand in 2000. What was the city's percentage rate of growth (rounded off accurate to one decimal place in percentage points) during this decade?

SOLUTION

We take $t = 0$ in 1990 to get started. Using the natural growth function $P(t) = P_0(1 + r)^t$ with $P_0 = 25$ (thousand), we have

$$P(10) = 25(1 + r)^{10} = 40 \tag{1}$$

since the population 10 years later in 2000 was 40 thousand. We need to solve this equation for r. Various methods are available.

Graphical Method We graph the functions **Y1 = 25 (1+X)^10** and **Y2 = 40**, with x instead of r denoting the unknown in Equation (1), in the viewing window $0 < x < 0.10, 0 < y < 60$. (For sake of investigation, we estimate initially that the unknown rate $r = x$ is less than $10\% = 0.10$.) Figure 3.3.1 shows this plot; the rising curve is the curve $y = 25(1 + x)^{10}$. We used the calculator's intersection-finding facility to solve automatically for the intersection point and obtained the point $(0.048, 40)$.

FIGURE 3.3.1 The graphs $y = 25(1 + x)^{10}$ and $y = 40$.

Thus $x = 0.048$, and the growth rate $r = 4.8\%$ describes the growth of Bethel during this decade. Therefore, the mathematical model

$$P(t) = 25 \times 1.048^t \tag{2}$$

gives the population (in thousands) of Bethel as a function of years after 1990.

Numerical Method Another approach is to solve Equation (1) by the method of tabulation. We use our calculator to tabulate values of the function **Y1 = 25*(1 + X)^10** (which we previously stored). Figure 3.3.2 shows the result when we construct a table with **TblStart = 0** and **ΔTbl = 0.001** and then scroll down in this table until the population figures (in the **Y1** column) approach the target of 40 (thousand).

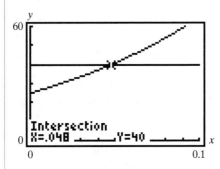

FIGURE 3.3.2 P is closest to 40 when $r = 0.048$.

The second column entry that is closest to 40 corresponds to the first column entry of $x = 0.048 = 4.8\% = r$ for the annual percentage growth rate. This is the same growth rate we found graphically and leads to the same mathematical model shown in Equation (2).

Symbolic Method Finally, we illustrate an entirely algebraic approach. In order to solve the equation

$$25(1 + r)^{10} = 40,$$

we divide each side by 25 and obtain the equation

$$(1 + r)^{10} = \frac{8}{5}$$

Taking the tenth root, or the 1/10 power, of each side yields

$$1 + r = \left(\frac{8}{5}\right)^{1/10} = 1.048.$$

What we have found here is the growth *factor*—the value by which we multiply each "old" population of Bethel to get each "new" population of Bethel. The question asks us to find the growth *rate* r. So,

$$r = \left(\frac{8}{5}\right)^{1/10} - 1.$$

The calculator then gives $(8/5)\^(1/10) - 1 = 0.048$ (approximately), so once again we see that $r \approx 4.8\%$. Pay careful attention to the parentheses in this expression. If you forget either set of parentheses, you will not get the correct growth rate!

As we have seen, it is "the rule rather than the exception" when a choice of different methods is available to solve a given problem. The preceding example illustrates the **rule of three**, which advocates the consideration of **graphical** and **numerical** methods as well as the **symbolic** methods of ordinary algebra. Of course, plotting the graph of a function is **graphical**, calculating a table of values is **numerical**, and solving an equation algebraically (as we did here) is **symbolic**. It is important that you develop, through experience and practice, some judgment as to which of the available methods of approach is likely to work best (or seem easiest to you) in a given situation.

In this situation many students prefer to find the correct natural growth function symbolically, particularly when the growth rate is not specifically requested. Suppose that we are given both the initial value A_0 (when $t = 0$) and a "new value" V after a specified "elapsed time" of N years (or other appropriate time unit). Then an easy way to "build" the natural growth function $A(t)$ is to write

$$A(t) = (\text{initial value})\left(\frac{\text{new value}}{\text{initial value}}\right)^{t/\text{elapsed time}}, \tag{3}$$

that is,

$$A(t) = A_0\left(\frac{V}{A_0}\right)^{t/N}. \tag{4}$$

Here's the reasoning behind this formula. We are given that the amount A increased from A_0 to V in time N. Thus we multiply A_0 by the factor V/A_0 to get the new value $A_0(V/A_0) = V$ after passage of the time period N. We think of t/N such time periods having passed in time t. If we started with A_0 and multiplied by the factor V/A_0 once for each of these elapsed time periods—that is, a total of t/N times—then we would get the result shown in (4). This is so because multiplication t/N times by V/A_0 amounts to multiplication by the "aggregate factor" $(V/A_0)^{t/N}$ that we see in (4).

EXAMPLE 2 Finding a Natural Growth Model

The city of Greendale had a population of 46 thousand in 1996 and 41 thousand in 2000.

a. Assuming natural growth, find a model that gives the population of Greendale as a function of years after 1996.

b. What was the city's population in 2005?

c. Use your model to predict the year in which the population of Greendale's falls to 30 thousand.

SOLUTION

a. With $t = 0$ in 1996, we have $A_0 = 46$, $V = 41$, and $N = 2000 - 1996 = 4$. Substitution of these values in (4) yields the natural growth model

$$P(t) = 46\left(\frac{41}{46}\right)^{t/4} \quad \text{(thousands)}$$

which gives the population of Greendale t years after the year 1996.

b. In parts (b) and (c), we are once again back to our usual input-output questions. Here we are given input ($t = 9$ years) and asked for output (population). Since part (c) is the "backward" question, for which we will want to use the **intersect** command, we will go ahead and store $P(t)$ in **Y1**. This allows us to use the **value** command to evaluate $P(9)$. Figure 3.3.3 shows our function, with the required parentheses; Fig. 3.3.4 shows $P(9)$.

FIGURE 3.3.3 The function modeling Greendale's population.

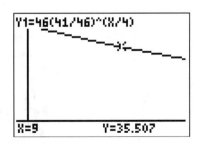

FIGURE 3.3.4 Finding the population when $t = 9$.

According to our model, the 2005 population of Greenwood was 35.507 thousand, or 35,507 people.

c. The window we used in part (b), $-1 < x < 15$, $-5 < y < 50$, is not quite large enough in the x-direction to show clearly the intersection of the graph of **Y1** with the line **Y2 = 30**. So in a new window, with **Xmax=20**, the **intersect** command gives the result shown in Fig. 3.3.5.

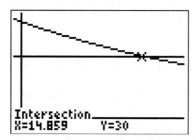

FIGURE 3.3.5 Finding when the population is 30 (thousand).

Our model predicts that by 2010 the population of Greenwood will have decreased to 30 thousand.

■

Notice that in Example 2, the calculator performs perfectly well without "knowing" explicitly either the annual growth rate r or the annual growth factor $1 + r$. If, for whatever reason, we wanted to determine $1 + r$, we could apply the law of exponents

$$a^{m/n} = (a^m)^{1/n} = (a^{1/n})^m.$$

In this case, then,

$$P(t) = 46\left(\frac{41}{46}\right)^{t/4} = 46\left[\left(\frac{41}{46}\right)^{1/4}\right]^t.$$

Since $\left(\frac{41}{46}\right)^{1/4} = 0.9716$ (correct to four decimal places), the growth factor for the population of Greendale is approximately 0.9716.

Because the growth factor is less than 1, we know that the population is declining rather than growing. (Of course, we knew this from the data we were given.) If we subtract 1 from the growth factor to get the growth rate, we find that $r = -0.0284 = -2.84\%$, with the negative sign indicating that the population was decreasing at the rate of 2.84% per year.

Exponential Models

We have used the function

$$A(t) = A_0(1 + r)^t \tag{5}$$

as a mathematical model for the *natural growth* of a quantity that starts (at time $t = 0$) with initial amount A_0 and thereafter grows at an annual percentage rate whose decimal equivalent is r.

There are situations (like Example 2) in which the growth rate r itself is not of specific interest, and it may then be easier to simply think of the "base constant"

$$b = 1 + r \tag{6}$$

that appears (raised to the tth power) in (5). Then (5) takes the simpler-looking form

$$A(t) = A_0 \cdot b^t. \tag{7}$$

> **DEFINITION: Exponential Function**
>
> An exponential function is a function of the form
>
> $$f(x) = a \cdot b^x \tag{8}$$
>
> with **base** b and **exponent** x (its independent variable).

Note that b is a *constant* raised to a *variable power*. By contrast, in an ordinary power function like $7x^3$, x is a *variable* raised to a *constant power*.

Equation (8) then says that **a natural growth function is an exponential function.** Frequently, the verbal description of a natural growth function gives the base constant b rather than the growth rate r; in this situation, we can write the function directly if we use the exponential form $A(t) = A_o \cdot b^t$.

EXAMPLE 3 | Using a Constant Base

Suppose that a population of rabbits initially has 10 rabbits and is doubling every year.

    a. Find an exponential function that models the growth of this population.
    b. How many rabbits will there be in 5 years?
    c. What is the annual rate of growth of this population?

SOLUTION

    a. When we say that a quantity is doubled, we mean that it is multiplied by 2. Since the base constant b is the factor by which the population is multiplied each year, b is 2 for this population. With an initial population of 10, our function model is then

$$P(t) = 10 \cdot 2^t.$$

    b. Because the population is doubling, we can easily find its value when $t = 5$. We don't even need the calculator for this computation. We can use inputs from 0 to 5, and then double the population as we go from one year to the next, as shown in the following table. (Of course, we wouldn't want to do this if we were asked about, say, 20 years, and we would be unable to do it "in our head" if we were asked about a fractional number of years.)

t	P(t)
0	10
1	20
2	40
3	80
4	160
5	320

c. The base constant b is just another "name" for the growth factor $1 + r$. For this population $b = 2$, so we have $1 + r = 2$, and thus $r = 1$. To give r as a percent, we write $1 = 1.00 = 100\%$. Therefore, a population that is doubling every year is growing at an annual rate of 100%.

A population that *triples* every year is described by the function $P(t) = P_0 \cdot 3^t$; one that *quadruples* every year is described by $P(t) = P_0 \cdot 4^t$; and one that *quintuples* annually is described by $P(t) = P_0 \cdot 5^t$. Obviously the larger the base constant b, the faster the exponential function b^t grows in value as t increases.

There's no reason that it—whatever ***it*** is—has to happen in a single year. Our rabbit population of Example 3 could double every 3 years, rather than every year. In that case, we can modify the table we constructed to reflect this new situation:

t	P(t)
0	10
3	20
6	40
9	80
12	160
15	320

You can see that we multiply the initial population of 10 by

- 2^1 when $t = 3$
- 2^2 when $t = 6$
- 2^3 when $t = 9$

and so on. Notice that each exponent is determined by dividing t by 3. That is, instead of multiplying by 2^t, we are multiplying by $2^{t/3}$. So the function describing this population is $P(t) = 10 \cdot 2^{t/3}$.

Natural Growth Model: Multiplication by b Every N years

If a quantity with initial value A_0 grows naturally and is multiplied by the number b every N years (or other appropriate unit of time), then it is described by the function

$$A(t) = A_0 \cdot b^{t/N} \tag{9}$$

This formula is equivalent to the earlier formula in (4). To see this, suppose the value of the quantity after the initial period of N years is denoted by V. Then the initial amount A_0 has been multiplied by the factor $b = V/A_0$ to get V. But substitution of $b = V/A_0$ in (9) gives the same function $A(t) = A_0(V/A_0)^{t/N}$ shown in (4).

EXAMPLE 4 A Population That Triples

Suppose that Quasimodo has four bats in his belfry, and the bats are tripling every 5 years.

  a. Find a function $B(t)$ that gives the number of bats as a function of the time in years.
  b. How long will it take for the population of bats to grow to 50 bats?
  c. How many bats will there be in Quasimodo's belfry in 7 years?

SOLUTION

  a. With an initial population of four bats and the population tripling every 5 years, the function model is $B(t) = 4 \cdot 3^{t/5}$.
  b. Storing $B(t)$ in **Y1** and 50 in **Y2** (Fig. 3.3.6), we use the **intersect** command to obtain the screen shown in Fig. 3.3.7.

FIGURE 3.3.6 Storing the functions to solve $B(t) = 50$.

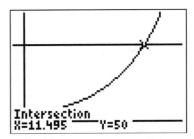

FIGURE 3.3.7 The intersection point (11.495, 50).

So after approximately $11\frac{1}{2}$ years the population of bats will grow to 50.

  c. Here we can use the **value** command, or we can type the function evaluation directly on the home screen of the calculator, as shown in Fig. 3.3.8.

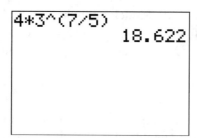

FIGURE 3.3.8 Evaluating $B(7)$.

After 7 years, there will be 18 or 19 bats in the belfry (depending on whether we ignore the "fractional" bat or round to the nearest whole bat).

Warning We emphasize once again that the parentheses enclosing the exponent 7/5 in Example 4(c) are vital (as are the ones in our function definition as well)! If we omitted them we would get

4*3^7/5

1749.600

instead. You should recognize immediately that this cannot be the correct value. Since $t = 7$, the population has tripled once (to 12 bats at $t = 50$, but has not tripled again (to 36 bats at $t = 10$).

The calculator, of course, finds the correct value for the calculation you requested. The calculator "understands" the ordinary order of operations you learned in elementary algebra. That is, *unless parentheses indicate otherwise*, raising to a power comes before multiplying or dividing (which then occur in order from left to right). Thus the command **4*3^7/5** means to first raise 3 to the seventh power, then multiply this value by 4, and finally divide the result by 5. Without parentheses this calculation is **4*(3^7)/5** instead of **4*3^(7/5)**. Most people dislike superfluous parentheses because they complicate typing (and also make the eyes glaze over) and therefore try not to use them unless actually necessary to tell the calculator precisely what to do. But a good practice is "When in doubt, use parentheses" to make sure the calculator does what you intend.

EXAMPLE 5 | Radioactive Decay

A nuclear reactor accident at the state's engineering school has left its campus contaminated with three times the maximal amount S of radiation that is safe for human habitation. Two and one half months after the accident the campus radiation level has declined to 75% of its original level. Assuming natural decline of this radiation level,

a. how long must students and faculty members wait before it is safe for them to return to campus;

b. what is the half-life of this substance?

SOLUTION

a. Let us write $A(t)$ for the amount of radiation still present after t months. If we measure A as a multiple of the maximal safe amount S, then we won't need to know exactly what S is (who knows this sort of thing offhand?). For in terms of these "safe units" (SU) we're given that $A(0) = 3$ SU and we want to find when $A(t) = 1$ SU, so it's safe for folks to come back to campus.

In addition to the initial amount $A_0 = 3$, we're given that the amount of radiation is multiplied by the factor $b = 75\% = 0.75$ every $N = 2.5$ months. Therefore Equation (8) gives

$$A(t) = 3 \times 0.75^{t/2.5}.$$

Notice that here time is measured in months rather than in years. To find when $A = 1$, we define **Y1 = 3*0.75^(X/2.5)** and **Y2 = 1** with the idea of seeing where the amount graph crosses the horizontal line $y = 1$. Figure 3.3.9 shows a plot in the viewing window defined by $0 < x < 12, 0 < y < 4$. Automatic intersection-finding yields the intersection point (9.5471, 1). Thus we see that A falls to 1 SU in just over $9\frac{1}{2}$ months. With human lives at stake, we probably ought to add a margin for safety and wait at least 10 months, maybe a full year, before reoccupying the campus.

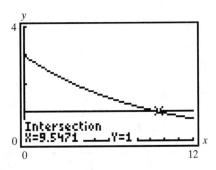

FIGURE 3.3.9 Solving the equation $3 \times 0.75^{t/2.5} = 1$.

b. Recalling that the half-life of a substance is the amount of time for only half it to remain, we replace **Y2 = 1** with **Y2 = 1.5** (half of our initial 3 SU). For the same window as before, **intersect** gives the intersection point (6.024, 1.5). Therefore, the half-life of the radioactive substance is approximately 6 months.

SUMMARY

We say that a quantity **grows** or **declines** *naturally* provided that

$$A(t) = A_0 \cdot b^t$$

for constants A_0 and b both greater than zero. Recalling the relation $b = 1 + r$ between the growth rate r and the (positive) base constant b, we see that as t increases, $A(t)$

- **increases** if $r > 0$ (and hence $b > 1$),
- **decreases** if $r < 0$ (and hence $b < 1$).

This means that, as we scan the graph of $A(t)$ from left to right, the curve

- **rises** if $b > 1$ (natural growth),
- **falls** if $0 < b < 1$ (natural decline).

Letting $A_0 = 1$, we can see typical "growth curves" for values of $b > 1$ in Fig. 3.3.10. Figure. 3.3.11 shows some typical "decay curves" for several values of $b < 1$.

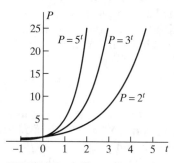

FIGURE 3.3.10 Natural growth curves.

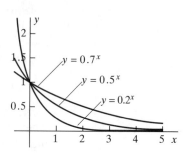

FIGURE 3.3.11 Natural decay curves.

Many quantities grow or decline naturally—that is, are multiplied by equal factors in equal times—and are therefore described by exponential functions.

3.3 Exercises | Building Your Skills

Each of the tables in Exercises 1–4 gives data from a function that is exponential. Determine the factor by which y is multiplied each year, and find a function of the form $y = a \cdot b^x$ that fits these points exactly.

1.

x	0	1	2	3	4
y	10	20	40	80	160

2.

x	0	1	2	3	4
y	5	15	45	135	405

3.

x	0	1	2	3	4
y	100	20	4	0.8	0.16

4.

x	0	1	2	3	4
y	80	40	20	10	5

In Exercises 5–12, write a natural growth or decay function model $p(t)$ with t in years.

5. $P(0) = 100$ and $P(1) = 175$.

6. $P(0) = 400$ and $P(1) = 155$.

7. $P(0) = 120$ and $P(4) = 40$.

8. $P(0) = 120$ and $P(3) = 180$.

9. $P(0) = 75$ and P doubles every year.
10. $P(0) = 125$ and P triples every year.
11. $P(0) = 100$ and P doubles every $2\frac{1}{2}$ years.
12. $P(0) = 50$ and P triples every $3\frac{1}{4}$ years.
13. Find the half-life for the function model in Exercise 6.
14. Find the half-life for the function model in Exercise 7.
15. Find the annual growth or decay rate for the function model in Exercise 7.
16. Find the annual growth or decay rate for the function model in Exercise 8.

Applying Your Skills

17. The population of Amsterdam, New York, was 20.7 thousand in 2000 and 18 thousand in 2003. Assuming natural growth,
    a. find its annual percentage decrease;
    b. find its predicted population in the year 2010;
    c. find the year in which the population falls to half the initial amount.

18. The population of Crestwood, Missouri, was 11.2 thousand in 1990 and 11.9 thousand in 2000. Assuming natural growth,
    a. find its annual percentage increase;
    b. find the predicted population in the year 2010;
    c. find the year in which the population doubles.

19. The population of Nevada was 1.2 million in 1990 and 2 million in 2000. Assuming natural growth,
    a. find its annual percentage increase;
    b. find the predicted population in the year 2010;
    c. find the year in which the population reaches 3 million.

20. The population of North Dakota was 642.2 thousand in 2000 and 636.7 thousand in 2003. Assume natural growth;
    a. find its annual percentage decrease;
    b. find the predicted population in the year 2010;
    c. find the year in which the population falls to 500 thousand.

21. The population of the world's more developed regions was 1.002 billion in 1965 and 1.176 billion in 1985, while the population of the world's less developed regions was 2.356 billion in 1965 and 3.706 billion in 1985.
    a. Using a natural growth model, in what year do you expect the population of the world's more developed regions to double?
    b. Using a natural growth model, in what year do you expect the population of the world's less developed regions to double?

22. The Giving USA Foundation reports that individual Americans gave 11.19 billion dollars to charity in 1964 and 92.52 billion dollars in 1994.
    a. Assuming natural growth, find the annual percentage rate of growth in individual American's charitable giving over this time period.

b. Assuming this trend continued, predict the amount of money given to charity by individual Americans in 2004.
   c. Compare your prediction in part (b) with the actual value, $187.92 billion.

23. According to *USA Today*, spending on medical equipment is rising rapidly because of the aging of the baby boomer population and the improving quality of health care in countries like India and China. In 2001, U.S. medical device revenue was $47.5 billion and was projected to be $80.2 billion in 2006.
   a. What annual percentage increase in revenue does this represent?
   b. Use a natural growth model to predict the U.S. medical device revenue in 2010.
   c. In what year does your model predict U.S. medical device revenue of $100 billion?

24. In 1979, the typical microcomputer contained 29 thousand transistors, while in 1993, a typical microcomputer contained 3.1 million transistors.
   a. Assuming natural growth, find a function giving the number $N(t)$ of transistors in a typical microcomputer t years after 1979.
   b. Find the annual growth rate r in part (a), expressed as a percentage.
   c. At this rate, how many months does it take to double the number of transistors in a typical microcomputer?
   d. Assuming this annual rate of increase continues, how many transistors (rounded off accurate to the nearest million) did the typical microcomputer contain in the year 2001?

25. A naturally growing bacteria population $P(t)$ numbers 49 at 12 noon.
   a. Write a formula giving $P(t)$ after t hours if there are 294 bacteria at 1 PM.
   b. How many bacteria are there at 1:45 PM?
   c. At what clock time (to the nearest minute) that afternoon are there 20 thousand bacteria?

26. How long does it take a naturally growing bacteria population to triple if it doubles in 1.5 hours?

27. The number of bacteria in a culture increased sixfold in 10 hours. Assuming natural growth, how long did it take their number to double?

28. The English language evolves naturally in such a way that 77% of all words disappear (or are replaced) every 1000 years.
   a. Of a basic list of words used by Chaucer in the year 1400, what percentage would have been in use in the year 2000?
   b. How long will it take until only 1% of Chaucer's words are still in use?

29. A survey by the market research firm NPD Group, as reported in the *San Antonio Express News*, found that over the 20-year period from 1985 to 2005, the percentage of Americans who said they find overweight people less attractive dropped from 55% to 24%.
   a. Find a natural growth model that gives the percentage of Americans who find overweight people less attractive as a function of years after 1985.
   b. Use your model to predict the percentage of Americans who find overweight people less attractive in 2008.
   c. In what year will only 5% of Americans find overweight people less attractive?

29. An accident at a nuclear power plant has left the surrounding area polluted with radioactive material that decays naturally. The initial amount of radioactive material present is 15 SU (safe units), and 5 months later it is still 10 SU.

   a. Write a formula giving the amount $A(t)$ of radioactive material (in SU) remaining after t months.
   b. What amount of radioactive material will remain after 8 months?
   c. How long (in total number of months or fraction thereof) will it be until $A = 1$ SU so it is safe for people to return to the area?

30. On April 26, 1986, the worst accident in the history of nuclear power occurred at the Chernobyl plant in Ukraine. An estimate of the amount of radioactive cesium-137 released from the plant is 2.7 million curies. If cesium-137 has a half-life of 30 years, how much of the amount released remained on April 26, 2006, the twentieth anniversary of the accident?

31. The National Center for Health Statistics reported that 557,271 Americans died of cancer in 2002, and that number declined to 556,902 deaths in 2003.

   a. Find a natural growth model that gives the number of Americans who died of cancer as a function of years after 2002.
   b. What was the annual percentage decrease in the number of Americans dying of cancer?
   c. If this trend continues, how many years will it take for American cancer deaths to fall below 500,000?

32. Thousands of years ago ancestors of the American Indians crossed the Bering Strait from Asia and entered the Western Hemisphere. Since then, they have fanned out across North and South America. The single language that the original Indian settlers spoke has since split into many Indian "language families." Assume that a language family develops into 1.5 language families every 6000 years. There are now 150 Indian language families in the Americas. About when did the first American Indians arrive?

33. On the April 24, 1999, edition of the Car Talk radio show, Tom and Ray presented a puzzler concerning a fellow who visited a nursery for advice concerning a new lawn. The following scenario is an adaptation of that puzzler. Mike wants to have a lush, new lawn for the Fourth of July. Since it is already May 30, it is too late for grass seed, and sod is too expensive for his budget. Matt, the nursery worker, suggests a new product—a small plug of grass that doubles in size every day. After consulting a diagram of the yard, Matt does some calculations, and reports his conclusion. If Mike plants only one plug on June 1, he will have the lawn he desires on June 30. Being a bit nervous, Mike decides to buy two, just to be on the safe side. If Mike plants two grass plugs on June 1, on what day will he have his new lawn?

3.4 FITTING NATURAL GROWTH MODELS TO DATA

In Section 3.1 we discussed briefly the data in the following table, which shows the U.S. population as recorded during the first six census counts, in the decade years 1790, 1800, 1810, 1820, 1830, and 1840.

CHAPTER 3 Natural Growth Models

t	YEAR	U.S. CENSUS POPULATION (MILLIONS)	CHANGE (MILLIONS)
0	1790	3.9	
10	1800	5.3	1.4
20	1810	7.2	1.9
30	1820	9.6	2.4
40	1830	12.9	3.3
50	1840	17.1	4.2

We saw that the U.S. population was growing at approximately 3% per year over these 50 years, and we modeled the population with the natural growth function

$$P(t) = 3.9 \times (1.03)^t. \tag{1}$$

But now we wonder whether there is a natural growth function that will fit these data even better.

Recall that in Section 2.4, we used the calculator's linear regression feature to find the best-fitting linear model—the one that had the smallest SSE and average error. This is the criterion we would like to continue to use for judging which model best fits a set of data. So let us first consider the SSE and average error for our "3% mathematical model" for the early U.S. population.

The following table compares the actual population figures with those predicted by this 3% natural growth model, and gives both the errors and their squares.

t	ACTUAL POPULATION	$P(t) = 3.9 \times (1.03)^t$	ERROR E	E^2
0	3.9	3.90	0.00	0.0000
10	5.3	5.24	0.06	0.0036
20	7.2	7.04	0.16	0.0256
30	9.6	9.47	0.13	0.0169
40	12.9	12.72	0.18	0.0324
50	17.1	17.10	0.00	0.0000

Population values and errors in millions.

The sum of the squares of the errors shown in the table is

SSE = 0.0000 + 0.0036 + 0.0256 + 0.0169 + 0.0324 + 0.0000 = 0.0755.

Since there are $n = 6$ data points, the average error in this natural growth model is given by

$$\text{average error} = \sqrt{\frac{\text{SSE}}{n}} = \sqrt{\frac{0.0755}{6}} = \sqrt{0.0126} \approx 0.1122. \tag{2}$$

Thus the average discrepancy between the actual census population and the population predicted by the natural model in (1) is about 0.112 million, or 112,000 persons.

We could now search around, trying different natural growth models, to see if we can find one with a smaller SSE and average error. We could adjust either the initial amount or the annual growth rate or both, each time checking to see if we get a better model. However, happily for us, the calculator finds not only linear regression, but also many other kinds, including exponential regression. For a given set of data, exponential regression will supply a "best-fitting" function of the form

$$y = a \cdot b^x. \tag{3}$$

In a typical data-modeling situation, the numbers a and b in (3) are not known in advance. Indeed the question ordinarily is "What numerical values for a and b yield the best natural growth model for the given data?" As before, the "optimal" model is the one that best fits the actual data, and the better of two different models is the one giving the lesser average error.

The Best Fitting Natural Growth Model

Figure 3.4.1 shows 1790–1840 U.S. population data entered in a calculator. The t-values 0, 10, 20, 30, 40, 50 corresponding to the decade years from 1790 to 1840 are stored as list **L1**, and the corresponding list of census population figures is stored as list **L2**. Figure 3.4.2 shows the resulting **STAT EDIT** menu displaying the data in table form.

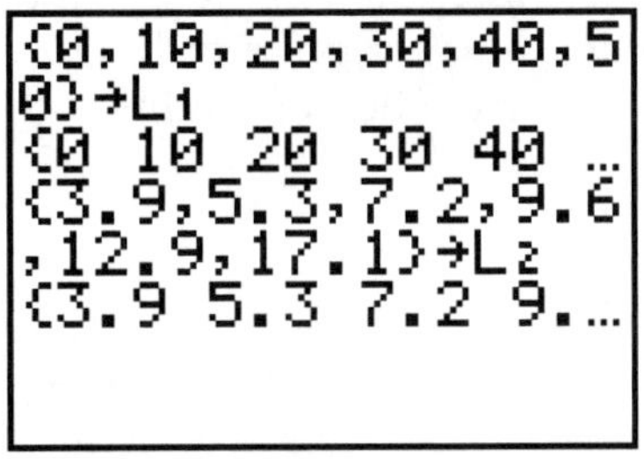

FIGURE 3.4.1 Storing the 1790–1840 U.S. census data.

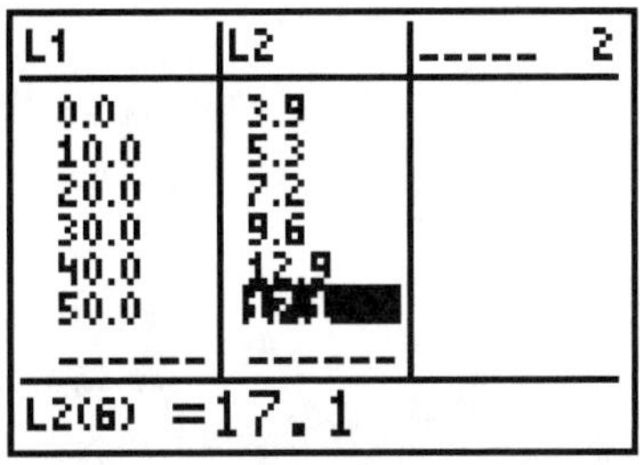

FIGURE 3.4.2 TI-83 **STAT EDIT** screen showing the U.S. population data in table form.

In the calculator's **STAT CALC** menu, item **0: ExpReg** (below item 9) is the calculator's so-called "exponential regression" facility for finding the natural growth (or "exponential") curve $y = a \cdot b^x$ that best fits the selected data. As we did with linear regression, we will tell the calculator where our inputs are stored **(L1)**, where our outputs are stored **(L2)**, and where to save our natural growth function rule **(Y1)**. So the command we enter is **ExpReg L1, L2, Y1**. Figure 3.4.3 shows the display that results when this command is entered.

The displayed results say that the natural growth curve best fitting the given data is

$$y = 3.9396 \times (1.0300)^x. \tag{4}$$

138 CHAPTER 3 Natural Growth Models

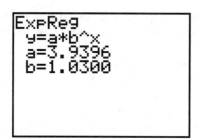

FIGURE 3.4.3 The optimal natural growth function.

In terms of time t and population P rather than the calculator's variables x and y, the natural growth function best fitting the actual 1790–1840 U.S. population growth is given by

$$P(t) = 3.9396 \times (1.0300)^t. \tag{5}$$

Observe that this optimal population function corresponds to a population that grows at a 3.00% annual rate, starting with an initial population of 3.9396 million when $t = 0$. Note also that this best-fitting initial population, about 3.94 million, is different from the actual initial population of 3.9 million. So when we attempt to best-fit a given list of population data, we are free to modify both the initial population and the rate of growth in order to fit the data as closely as possible.

The **ExpReg L1, L2,Y1** command automatically enters Equation (4) in the **Y=** menu. With **Plot 1** turned **On** in the **STAT PLOT** menu, **GRAPH** then gives the plot of the best-fitting natural growth curve shown in Fig. 3.4.4, where the original census data points are shown as small squares.

Here we are less interested in the graph than in a table of values that we can use to compute the average error in our optimal fit. Our calculator now provides the table of (rounded off) values of **Y1 = 3.9396*1.03^X** shown in Fig. 3.4.5.

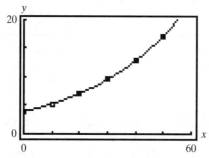

FIGURE 3.4.4 The optimal natural growth fit.

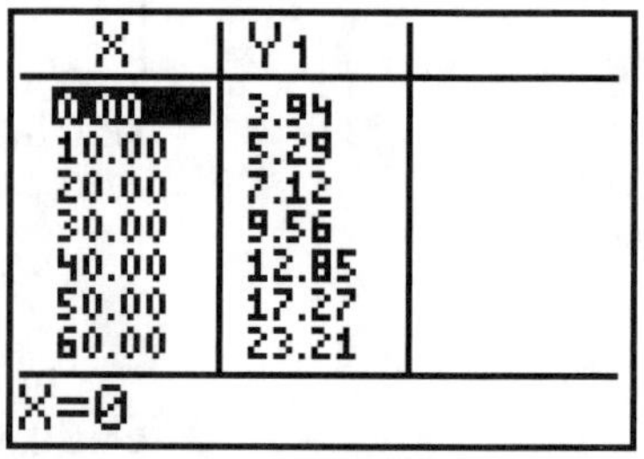

FIGURE 3.4.5 Values of the optimal natural growth function $P(t) = 3.9396 \times 1.0300^t$.

When we combine these values with our original U.S. population data, we get the following table. It compares the original census data with the population figures predicted by the optimal natural growth model $P(t) = 3.9396 \times (1.0300)^t$.

SECTION 3.4 Fitting Natural Growth Models to Data

t	ACTUAL POPULATION	$P(t) = 3.94 \times (1.03)^t$	ERROR E	E^2
0	3.9	3.94	−0.04	0.0016
10	5.3	5.29	0.01	0.0001
20	7.2	7.12	0.08	0.0064
30	9.6	9.56	0.04	0.0016
40	12.9	12.85	0.05	0.0025
50	17.1	17.27	−0.17	0.0289

Population values and errors in millions.

As usual, the final column shows the *squares* of the errors. Hence the SSE associated with the optimal natural growth model in (18) is

$$\text{SSE} = 0.0016 + 0.0001 + 0.0064 + 0.0016 + 0.0025 + 0.0289 = 0.0411.$$

Since there are $n = 6$ data points, the average error in the optimal model is given by

$$\text{average error} = \sqrt{\frac{\text{SSE}}{n}} = \sqrt{\frac{0.0411}{6}} = \sqrt{0.0069} \approx 0.0828.$$

Thus the average discrepancy between the actual census population and the population predicted by the natural model in (5) is about 0.083 million, or 83 thousand persons.

In this case, the average error of 83 thousand for our "best-fitting" model is less than the average error of 112 thousand in the 3% natural growth model $P(t) = 3.9 \times (1.03)^t$. However, contrary to one's natural expectation, the so-called "best-fitting model" given by a calculator's exponential-regression facility is not always an exponential model with the least possible average error. The example that follows demonstrates this surprising fact.

EXAMPLE 1 Comparing Natural Growth Models

One of the indicators of China's growing economy is its trade volume with other nations. The table gives this trade volume in billions of dollars for selected years from 1980 to 2004.

YEAR	TRADE VOLUME (BILLIONS OF DOLLARS)
1980	38
1985	70
1990	115
1995	281
2000	474
2004	1155

Source: Atlanta Journal-Constitution.

140 CHAPTER 3 Natural Growth Models

a. Use the first and last data points to find a natural growth model giving trade volume as a function of years after 1980.
b. Find the SSE and average error for the model in (a).
c. Use exponential regression to find the best-fitting natural growth model giving trade volume as a function of years after 1980.
d. Find the SSE and average error for the model in (c).

SOLUTION

a. Using the formula

$$A(t) = A_0 \left(\frac{V}{A_0}\right)^{t/N}$$

from Section 3.3, we obtain the natural growth function $A_1(t) = 38\left(\frac{1155}{38}\right)^{t/24}$, where A_1 is measured in billions and t is years after 1980.

b. As before, we enter our inputs in **L1**, our outputs in **L2**, and our natural growth function $A(t)$ in **Y1**. We will use our calculator's table to find the squares of the errors as follows:

1. In the **STAT EDIT** menu, highlight **L3**. At the bottom of the screen **L3 =** appears. Hit **VARS**, arrow over to **Y-VARS**, hit **ENTER**, select Function, then **ENTER** again to select **Y1**. The bottom line of the **STAT EDIT** menu now reads **L3 = Y1**. Type **(**, then **2ND** and **1** for **L1**, then **)**. The bottom line of the **STAT EDIT** menu now reads **L3 = Y1(L1)**. Hit **ENTER**, and the calculator displays in **L3** (Fig. 3.4.6) the output values predicted by the natural growth model.

L1	L2	L3
0.000	38.000	38.000
5.000	70.000	77.393
10.000	115.00	157.62
15.000	281.00	321.02
20.000	474.00	653.81
24.000	1155.0	1155.0

L2(1)=38

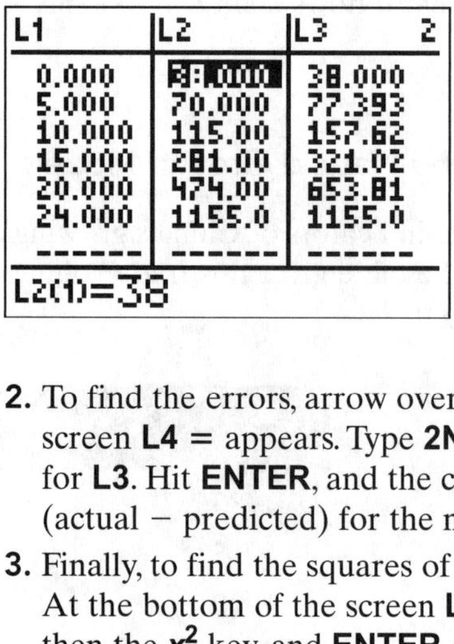

FIGURE 3.4.6 Values predicted by the natural growth model $A_1(t) = 38\left(\frac{1155}{38}\right)^{t/24}$.

2. To find the errors, arrow over and highlight **L4**. At the bottom of the screen **L4 =** appears. Type **2ND** and **2** for **L2**, then **−**, and **2ND** and **3** for **L3**. Hit **ENTER**, and the calculator displays in **L4** the errors (actual − predicted) for the natural growth model.

3. Finally, to find the squares of the errors, arrow over and highlight **L5**. At the bottom of the screen **L5 =** appears. Type **2ND** and **4** for **L4**, then the x^2 key, and **ENTER**. The calculator displays in **L5** the squares of the errors for this model. Figure 3.4.7 shows the errors in **L4** and their squares in **L5**. (Notice that the errors for the first and last data points are 0 because we built the function to contain those exact points.)

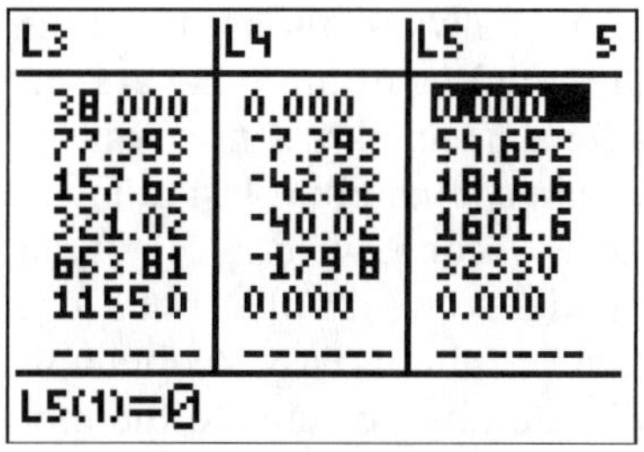

FIGURE 3.4.7 Predicted values, error, and the squares of the errors for $A_1(t) = 38\left(\dfrac{1155}{38}\right)^{t/24}$.

So the SSE for this model is given by

$$\text{SSE} = 0 + 54.652 + 1816.617 + 1601.630 + 32330.207 + 0 \\ = 35803.106,$$

where we have scrolled down through **L5** to obtain three decimal places for each value. Therefore, the average error is $\sqrt{\dfrac{35803.106}{6}} = 77.248$, correct to three decimal places.

c. We already have our data stored in **L1** and **L2**, so we execute the command **ExpReg L1, L2,Y1**, and obtain the natural growth function **Y1 = 34.218*1.149^X**. Thus, the best-fitting exponential model for the China trade volume data is $A_2(t) = 34.218(1.149)^t$, again with A_2 given in billions and t as years after 1980.

d. Following exactly the same steps as in part (b), we obtain Fig. 3.4.8, showing the predicted values, the errors, and their squares.

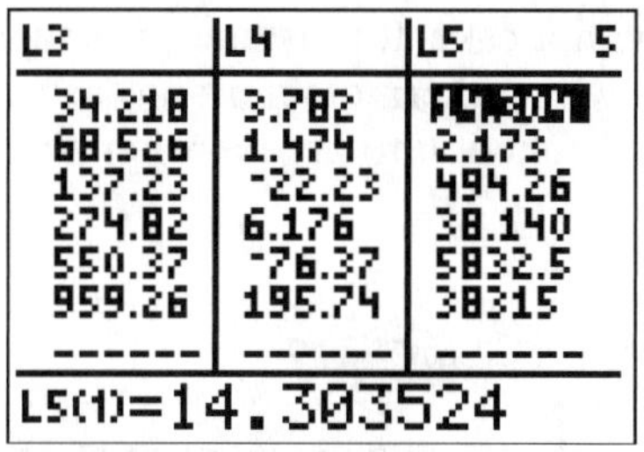

FIGURE 3.4.8 Predicted values, error, and the squares of the errors for $A_2(t) = 34.218(1.149)^t$.

So,

$$\text{SSE} = 14.304 + 2.173 + 494.255 + 38.140 + 5832.463 + 38315.034 \\ = 44696.369, \text{ and the average error is } \sqrt{\dfrac{44696.369}{6}} = 86.310.$$

Note that this average error, for the best-fitting function $A(t) = 34.218(1.149)^t$, exceeds the average error of 77.248 for our original exponential model $A(t) = (38)(1155/38)^{t/24}$, as found in part (b).

We see that, unlike the linear regression from Section 2.4, exponential regression does not always provide us with the function with the smallest possible SSE. (This is because of the way the calculator does exponential regression, which we can't explain without ideas from the next chapter.)

So if it's not always "the best there is," why do we bother? First, it is a model that is quite good at fitting our data—it is indeed the "best-fitting" exponential model in a certain technical sense (which we will describe in Chapter 4). Frequently it *is* better than a "first point-last point" model like the one we created in part (a) of Example 1. Second, we don't have the mathematics that's needed to find the "perfect" model, so we'd rather not spend lots of time and energy trying different models when one that is easy to obtain will do very nicely.

We will continue to refer to the models that we obtain by regression as the models that "best fit" our data (even though that term is sometimes misleading). This gives us the vocabulary by which we will recognize when we are asked to use regression to find a function model.

Newton's Law of Cooling

Suppose that a hot object with initial temperature T_0 is placed in a relatively cool medium with constant temperature A. For example, we might put a hot cake (just out of the oven) in a cool room with air temperature A, or we might put a hot rock in a large water tank with water temperature A. Then **Newton's law of cooling** says that the difference

$$u(t) = T(t) - A \qquad (6)$$

between the temperature of the object at time t and the temperature A is a naturally declining quantity. That is,

$$u(t) = a \cdot b^t \qquad (7)$$

with appropriate values of the positive parameters a and b, where $b < 1$. In the case of the hot cake, for instance, we might measure its falling temperature several times with a thermometer and then attempt to fit the resulting data with an exponential function to predict when the cake will be cool enough to serve.

EXAMPLE 2   When Is the Cake Cool?

Suppose a cake is baked in an oven at 350°F. At 1 PM it is taken out of the oven and placed to cool on a table in a room with air temperature 70°F. We plan to slice and serve it as soon as it has cooled to 100°F. The temperature of the cake is measured every 15 minutes for the first hour, with the following results.

Time t (PM)	1:00	1:15	1:30	1:45	2:00
Temperature of cake T (°F)	350	265	214	166	143

When will the cake be cool enough to eat?

SOLUTION

To use Newton's law we must replace the cake temperature T with the difference $u = T - 70$ that we see in Equation (6), with $A = 70°F$ (room temperature). Measuring time t in hours and subtracting 70 from each temperature entry, we find that the given data take the form

SECTION 3.4 Fitting Natural Growth Models to Data **143**

t	0	0.25	0.5	0.75	1
u	280	195	144	96	73

If we can fit these data with an exponential model $u(t) = a \cdot b^t$, then we can find a model for the temperature of the cake and then solve graphically for the time when $T = 100°F$ as desired.

We begin by entering the list of t-values as calculator list **L1** and the list of u values as list **L2** (Fig. 3.4.9). Then the **STAT CALC** command **ExpReg L1, L2,Y1** produces the result shown in Fig. 3.4.10.

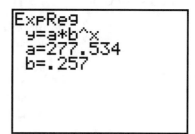

FIGURE 3.4.9 Entering the data of Example 5.

FIGURE 3.4.10 The best exponential fit $u(t) = 277.534(0.257)^t$.

So the best-fitting exponential function is given by

$$u(t) = 277.534(0.257)^t \qquad (8)$$

(on replacing **X** with t and **Y** with u). Finally, remembering Equation (6), we add $A = 70°F$ (the room temperature) to both sides to obtain the cake's temperature function

$$T(t) = 70 + 277.534(0.257)^t \quad (°F) \qquad (9)$$

It's always important to keep your eye on the cake. Here, literally *no one* cares what is the average error in the natural decay function (8). Instead, we all want to know when we can eat our cake. That is, accepting (9) as our mathematical model for the cooling of our cake, for what value of time t is it true that

$$70 + 277.534(0.257)^t = 100°F? \qquad (10)$$

In Fig. 3.4.11 we have graphed the functions **Y1 = 70+277.534*(0.257)^X** and **Y2 = 100** and asked for the coordinates of the point of intersection.

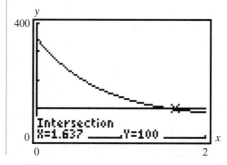

FIGURE 3.4.11 Solving Eq. (10).

144 CHAPTER 3 Natural Growth Models

The *x*-value of 1.637 tells us that our cake reaches an edible temperature of 100°F after 1.637 hours (starting at 1 PM). Since $0.637 \times 60 = 38.220$, this means we can slice and eat the cake at about 2:38 PM.

In previous sections of this chapter we saw many different quantities whose growth can be modeled with natural growth functions. We can use exponential regression to find the best-fitting model whenever we are given a set of data whose behavior exhibits natural growth.

3.4 Exercises | Building Your Skills

Each of the tables in Exercises 1–8 gives data from a function that is approximately exponential.

  a. First use the exponential regression facility of your calculator to find the exponential model of the form $y = a \cdot b^x$ that best fits these points. Find a and b accurate rounded off to three decimal places.
  b. Calculate the average error in this optimal exponential model.

1.

x	0	1	2	3	4
y	5	7	10	15	23

2.

x	1	2	3	4	5
y	35	75	170	385	865

3.

x	0	2	4	6	8
y	15	30	60	110	220

4.

x	1	1.5	2	2.5	3
y	22	23	24	25	27

5.

x	0	1	2	3	4
y	40	27	17	11	7

6.

x	1	3	5	7	9
y	215	40	7	1	0

7.

x	0.5	0.75	1	1.25	1.5
y	80	65	55	45	40

8.

x	0	5	10	15	20
y	1000	775	600	465	360

Applying Your Skills

In each of Exercises 9–12 the 1960–1990 population census data for a U.S. city is given.

    a. Find the exponential model $P = a \cdot b^t$ (with $t = 0$ in 1960) that best fits these census data.
    b. What was the city's average annual percentage rate of growth during this 30-year period?
    c. Use your model to predict the city's population in 2000.

9. San Diego

Year	1960	1970	1980	1990
Population (thousands)	573	697	876	1111

10. Phoenix

Year	1960	1970	1980	1990
Population (thousands)	439	584	790	1008

11. Cleveland

Year	1960	1970	1980	1990
Population (thousands)	876	751	574	506

12. Buffalo

Year	1960	1970	1980	1990
Population (thousands)	533	463	358	328

13. San Antonio

Year	1960	1970	1980	1990
Population (thousands)	588	654	786	935

14. Raleigh

Year	1960	1970	1980	1990
Population (thousands)	94	123	150	212

In Exercises 15–21, use the best-fitting exponential model to determine the answer to each question. (For problems using calendar years, "reset the clock" so that $t = 0$ in the year for which you first have data.)

15. In a certain lake the intensity I of light beneath the surface is a naturally declining function of the depth x in feet. The following table gives I as a percentage of the intensity of light at the surface:

Depth x (feet)	0	5	10	15
Intensity I (%)	100	60	35	20

How deep would you have to dive in this lake so that the light intensity there will be 1% of the surface intensity?

16. The atmospheric pressure p (in pounds per square inch, lb/in^2) is a naturally declining function of the altitude x above sea level. The following table gives p as a function of x in thousands of feet:

Altitude x (thousands of feet)	0	4	8	12
Pressure p (lb/in^2)	14.7	12.6	10.9	9.3

    a. Suppose that you (like most people) cannot survive without special conditioning at an air pressure of less than 7.5 lb/in^2. Then how high could you safely fly in an airplane without pressurization?

    b. What is the air pressure atop Mt. Everest (at 29,029 feet)?

17. The amount money earned by winning the Kentucky Derby has grown quite a bit since the Derby's first running in 1875. The following table gives the net amount earned by the winner as a function of the year of the race.

Year t	1875	1900	1925	1950	1975
Winnings ($)	2850	4850	52,950	92,650	209,600

Source: Churchill Downs Incorporated (www.kentuckyderby.com).

    a. In what year would you expect the prize money to grow to $1,000,000?

    b. Use your model to predict the prize money in 2007.

18. As cassette tapes and compact discs became more popular, the sales of vinyl singles declined according to the accompanying table.

Year t	1988	1989	1992	1994	1996
Millions of units s	65.6	36.6	19.8	11.7	10.1

Source: The World Almanac and Book of Facts 1998.

Suppose that vinyl singles will be discontinued when their sales fall below 2 million. In what year will this occur?

19. The success of a major league baseball team is frequently tied very closely to the quality of its pitching staff. The relief pitchers (the "bullpen") come in the game to pitch when a pitcher tires or gets in trouble. The table gives the ERA (earned run average) and the number of games saved by the bullpen of the Atlanta Braves over various seasons between 1969 and 2005.

ERA	2.60	3.73	5.11	3.06	4.22
Games saved	57	41	24	49	32

Source: Atlanta Journal-Constitution.

If the Braves bullpen had an ERA of 6.50 (a very bad ERA), how many games would you expect them to save in a season?

20. The gross domestic product (GDP) of a country is a very common measure of the vitality of its economy. China's economy has been growing rapidly, as shown by the increase in its GDP in the accompanying table.

Year t	1980	1985	1990	1995	2000	2004
GDP of China (billions of dollars)	266	306	388	706	1072	1647

Source: Atlanta Journal-Constitution.

    a. According to your model, what will the GDP of China be in 2010?
    b. When will the GDP of China rise to 2000 million (2 trillion) dollars?

21. A BlackBerry is an all-in-one mobile device that includes phone, e-mail, Web browser, and organizer features. The number of BlackBerry subscribers has grown according to the following table.

Year t	2001	2002	2003	2004	2005
Thousands of subscribers S	165	321	534	1070	2510

Source: USA Today.

    a. At what average annual rate did the number of BlackBerry subscribers grow during this period?
    b. If this trend continues, how many BlackBerry subscribers would you expect there to be in 2010?

22. Suppose a hot cake like the one of Example 2 is taken out of an oven at 175°C and immediately set out on a cool porch where the temperature is 10°C. During the next hour the following temperature readings of the cake are taken:

Time (PM)	5	5:20	5:40	6
Temperature (°C)	175	125	85	60

The cake will be brought in and served when it has cooled to 35°C. When do you expect this to be?

23. A pitcher of buttermilk initially at 25°C is set out on the 10°C porch of Exercise 22. During the next half hour the following temperature readings are taken:

Time (PM)	1	1:10	1:20	1:30
Temperature (°C)	25	20	17	15

The buttermilk will be brought in and served when it has cooled to 12°C. When do you expect this to be?

Chapter 3 | Review

In this chapter, you learned about exponential functions and models. After completing the chapter, you should be able to

- Determine whether a relation described numerically, graphically, or symbolically represents an exponential function.
- Find the output value of an exponential function for a given input value.
- Find the input value of an exponential function for a given output value.
- Solve an equation or inequality involving an exponential function

- Find an exponential function model that fits given exponential data.
- Find the best-fitting exponential model for data that are approximately exponential.
- Find the annual percentage rate of growth or decay for an exponential model.

Review Exercises

In Exercises 1–7, the given information (table, graph, or formula) gives y as a function of x. Determine whether each function is linear, exponential, or neither.

1.

x	−3	−2	0	4	7	10
y	4.5	2	0	8	24.5	50

2.

x	−2	−1	1	3	4	5
y	32	16	4	1	.5	.25

3.

x	−4	−3	−1	2	4	5
y	29	26	20	11	5	2

4.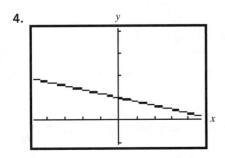

5.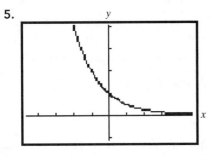

6. $y = 3x^2$

7. $y = 3 \cdot 2^x$

8. Suppose that a child who lost her first tooth in 1998 was left $1.00 by the Tooth Fairy. Let us assume that the Tooth Fairy leaves the same amount for each tooth, if payments are adjusted for inflation. Then, based on the Consumer Price Index, the Tooth Fairy should have left her mother $0.19 for a tooth in 1963 and given her grandmother $0.09 for a tooth in 1938. Find the best-fitting exponential function for these data, and use it to predict how much the Tooth Fairy should leave for the little girl's daughter if she loses a tooth in 2023.

Source for conversion factors: Robert Sahr, Political Science Department, Oregon State University.

9. An accident at a nuclear power plant has left the surrounding area polluted with a radioactive element that decays naturally. The initial radiation level is 10 times the maximum amount S that is safe, and 100 days later it is still 7 times that amount. How long (to the nearest day after the original accident) will it be before it is safe for people to return to the area?

10. Suppose that, as of January 1, 1999, a state's annual Medicaid expenditures are $1 billion and increasing at 6% per year, while its annual Medicaid tax income is $2 billion and is increasing at 3% per year. If these rates continue, during what calendar year will the state's Medicaid budget go into the red (with expenditures exceeding income)?

11. According the BP Statistical Review of World Energy 2006, the United States produced 7733 thousand barrels of oil a day in 2000 and 6830 thousand barrels daily in 2005. Use an exponential model to determine the average annual rate at which U.S. daily oil production was declining over this time period.

12. The following table gives the gross revenue R in millions of dollars for the Broadway season that begins in the indicated year t. Find the best-fitting exponential model for these data and use it to predict the year in which gross revenue will grow to 1000 million (1 trillion) dollars.

t (year)	1984	1989	1992	1996	2001	2005
R (millions)	209	282	328	499	643	862

Source: League of American Theatres and Producers.

13. (Chapter Opener Revisited) In the discussion that introduced this chapter, we looked at a plot that displayed the growth of the Internet over the years from 1993 to 1999. The plot was based on data from the Internet Software Consortium giving the number of Internet hosts, which is displayed in the following table.

Calendar year	1993	1994	1995	1996	1997	1998	1999
t (years after 1993)	0	1	2	3	5	5	6
H (millions)	1.3	2.2	4.9	9.5	16.1	29.7	43.2

a. Find a natural growth function $H_1(t)$ that models the number of Internet hosts as a function of years after 1993 and agrees with the data given for 1993 and 1998.

b. At what average annual rate was the number of Internet hosts growing over this time period? What is the sum of squares of errors SSE_1 in the model $H_1(t)$ considering the given data for the years through 1993 through 1998 but ignoring the data for 1999?

c. Use your model to predict the number of Internet hosts in 2004. What was the error in your approximation if the actual value was 317.6 million?

d. Your prediction is so much larger than the actual value because the number of Internet hosts did not continue to grow at the rate you found in part (b). The data in the following table give the number of Internet hosts for the years 1998 to 2002.

Calendar year	1998	1999	2000	2001	2002
t (year after 1993)	5	6	7	8	9
H (millions)	29.7	43.2	72.4	109.6	147.3

Source: ISC Internet Domain Survey.

Find a natural growth function $H_2(t)$ that models the number of Internet hosts as a function of years after 1993 and agrees with the data given for 1998 and 2002.

e. At what average annual rate was the number of Internet hosts growing over this time period? What is the sum of squares of errors SSE_2 in the model $H_2(t)$ considering the given data for the years through 1998 through 2002?

f. Use your model to predict the number of Internet hosts in 2004. Find the error in your approximation, given that the actual value for 2004 was 317.6 million.

g. The two exponential functions $H_1(t)$ and $H_2(t)$ in parts (a) and (d) agree at the point $t = 5$ corresponding to 1998. Consider the function $H(t)$ that equals $H_1(t)$ for $t \leq 5$ but equals $H_2(t)$ for $t \geq 5$. Can you see why $H(t)$ might be called a "piecewise-exponential model" for the number of Internet hosts from 1993 to 2002? What is the SSE of this model considering the data from 1993 to 2002. Can you find SSE quickly—without doing any more real computation—using your answers to parts (b) and (e)?

h. What is the average error in the piecewise-exponential model $H(t)$ of (g) for the data from 1993 to 2002?

INVESTIGATION Population Projections for U.S. Cities

In this activity, you will create function models based on U.S. Census data for one of the 100 largest cities in the United States. U.S. Census data are available at www.census.gov.

From the census data for 1990, select the *xy*th largest city, where *x* and *y* are the last two digits of your student identification number. This is your city. (For example, if your student ID number ends in 37, you should find the data for the 37th largest city in the United States. If your ID number ends in 00, choose the 100th largest city.) Record the populations for your city in 1950, 1960, 1970, 1980, and 1990.

1. Find the best-fitting linear model for your population data, using years after 1950 as the independent variable.
2. Calculate the SSE for your linear model and the average error in the approximation.
3. Use your linear model to predict your city's population in 2000.
4. Find the best-fitting exponential model for your population data.
5. Calculate the SEE for your exponential model and the average error in the approximation.
6. Use your exponential model to predict your city's population in 2000.
7. On the basis of having the smallest average error, which model is the best predictor of your city's population?
8. Use census data to determine your city's actual population in 2000. Which model came closer to the actual population?

CHAPTER 4
CONTINUOUS GROWTH AND LOGARITHMIC MODELS

4.1 COMPOUND INTEREST AND CONTINUOUS GROWTH

4.2 EXPONENTIAL AND LOGARITHMIC FUNCTIONS

4.3 EXPONENTIAL AND LOGARITHMIC DATA MODELING

In 1850 when Horace Greeley said, "Go West, young man," was he starting a trend or merely reporting one? From the beginning of this nation's history, there has been westward expansion. One way to measure this westward expansion is to consider the center of population of the United States. The center of population is determined by considering the land mass of the United States as a weightless rigid plane that would balance at that central point if each person were given equal weight and each person's influence on the central point was proportional to his or her distance from it.

The chart in Fig. 4.0.1 shows the westward location of the center of population (in degrees of West longitude). We can see that from 1830 to 1860, the center of population was moving west more rapidly than it was from 1860 to 1900. From 1900 to 1940, the center of population moved

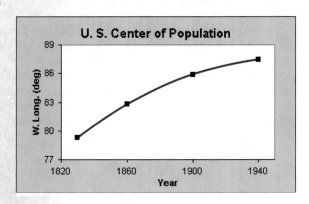

FIGURE 4.0.1 West longitude location (in degrees) of the U.S. center of population for the years 1830, 1860, 1900, and 1940.

Source: The World Almanac and Book of Facts 1998.

west even less rapidly. (From 1950 onward, Alaska and Hawaii are included in the calculation, so later years should be graphed separately.)

This pattern of growth is the opposite of what we observed with the natural growth functions of Chapter 3. Using the data of Fig. 3.0.1, we saw that the growth of the Internet was slow at first, but then increased more rapidly. We saw that this natural growth graph curved upward, while our center of population graph curves downward.

A quantity like the center of population that increases at a decreasing rate suggests a different kind of function to a mathematician, one that is, in some sense, the "opposite" of a natural growth function.

In this chapter we will study such functions and their relationship to exponential functions.

4.1 COMPOUND INTEREST AND CONTINUOUS GROWTH

In Section 3.1 we discussed the growth of the amount in an investment account that earns interest annually. If the amount initially invested is A_0 dollars and the annual interest rate (in decimal form) is r, then the amount in the account is multiplied by the factor $(1 + r)$ each year. It follows that the amount A_n the account contains after t years is given by

$$A_t = A_0(1 + r)^t. \tag{1}$$

In particular, the amount in the account after 1 year is

$$A_1 = A_0(1 + r). \tag{2}$$

These formulas describe computations in which interest is earned annually; that is, the current amount A in an account is updated **once** each year by the addition of rA in "earned interest" for the year.

Compounding Interest Semiannually

However, banks often update the amount in an account by the addition of earned interest more frequently than once per year, such as semiannually, quarterly, monthly, weekly, or even daily. This is called *compounding* of interest.

If the "interest period" is less than a full year, then less interest is earned during that period. For instance, the interest earned in a half-year would be half the interest rA earned in a full year. If the initial amount is A_0, then the addition of the interest $\frac{1}{2} rA_0$ earned during the first half-year yields the amount

$$A = A_0 + \tfrac{1}{2} rA_0 = A_0\left(1 + \frac{r}{2}\right)$$

in the account after 6 months. Thus the addition of interest to an account after a half-year corresponds to the multiplication of the amount by the factor $(1 + \frac{r}{2})$

rather than by the annual interest factor $(1 + r)$. At the end of the second half-year, the current amount $A_0(1 + \frac{r}{2})$ in the account is again multiplied by the factor $(1 + \frac{r}{2})$. Hence the amount in the account after a full year is

$$A_1 = A_0\left(1 + \frac{r}{2}\right) \cdot \left(1 + \frac{r}{2}\right) = A_0\left(1 + \frac{r}{2}\right)^2. \qquad (3)$$

So the addition of interest *twice* per year corresponds to multiplication by the factor $(1 + \frac{r}{2})^2$, the *square* of the factor $(1 + \frac{r}{2})$ corresponding to a single half-year interest period. In t years, the amount in the account is multiplied by this factor t times, that is, by

$$\left(1 + \frac{r}{2}\right)^2 \cdot \left(1 + \frac{r}{2}\right)^2 \cdots \left(1 + \frac{r}{2}\right)^2 = \left[\left(1 + \frac{r}{2}\right)^2\right]^t = \left(1 + \frac{r}{2}\right)^{2t}.$$

Consequently, the amount in the account after t years of interest paid semiannually is

$$A_t = A_0\left(1 + \frac{r}{2}\right)^{2t}. \qquad (4)$$

Compare this formula with Formula (1). Whereas Formula (1) describes the result of interest at the annual rate r compounded annually, Formula (4) describes the result of annual interest r **compounded semiannually**.

EXAMPLE 1 | Interest Compounded Semiannually

Compare the results after 1 year and after 10 years when $1000 is invested

  a. at 10% annual interest compounded annually;
  b. at 10% annual interest compounded semiannually.

SOLUTION

  a. With annual compounding, the amount after 1 year is

  $$\$1000 \cdot (1 + 0.10) = \$1100,$$

  and

  $$\$1000 \cdot (1 + 0.10)^{10} = \$2593.74$$

  is the amount after 10 years.

  b. For semiannual compounding with $r = 0.10$ and $t = 1$, Formula (3) gives

  $$\$1000 \cdot \left(1 + \frac{0.10}{2}\right)^2 = \$1000 \cdot (1 + 0.05)^2 = \$1102.50,$$

  after 1 year if the interest is compounded semiannually. Then Formula (4) with $r = 0.10$ and $t = 10$ gives

  $$\$1000 \cdot (1 + 0.05)^{20} = \$2653.30$$

  for the amount after 10 years.

Observe that compounding the interest semiannually makes our investment grow more quickly. Semiannual compounding yields $2.50 more than annual compounding in 1 year but almost $60 more after 10 years.

Compounding Even More Frequently

Suppose now that interest is compounded quarterly, that is, four times rather than twice per year. Then the interest earned in a quarter would be one-fourth the interest rA earned in a full year. If the amount at the beginning of a quarter is A, then the addition of the interest $\frac{1}{4}rA$ at the end of the quarter yields the amount

$$A + \frac{1}{4}rA = A\left(1 + \frac{r}{4}\right).$$

Thus the amount in the account is multiplied by factor $(1 + \frac{r}{4})$ at the end of each quarter. This happens four times in a full year, so the amount in the account after 1 year is

$$A_1 = A_0\left(1 + \frac{r}{4}\right)^4. \tag{5}$$

So quarterly compounding of interest corresponds to multiplication of the amount in the account by $(1 + \frac{r}{4})^4$ each year. In t years, the amount in the account is multiplied by this factor t times, that is, by

$$\left(1 + \frac{r}{4}\right)^4 \cdot \left(1 + \frac{r}{4}\right)^4 \cdots \left(1 + \frac{r}{4}\right)^4 = \left[\left(1 + \frac{r}{4}\right)^4\right]^t = \left(1 + \frac{r}{4}\right)^{4t}.$$

Consequently, the amount in an account after t years at annual interest rate r **compounded quarterly** is

$$A_n = A_0\left(1 + \frac{r}{4}\right)^{4t}. \tag{6}$$

EXAMPLE 2 | Interest Compounded Quarterly

Now find the amount after 1 year and after 10 years in the account of Example 1 if the 10% annual interest is compounded quarterly.

SOLUTION

With $A_0 = \$1000$ and $r = 0.10$, Formula (5) gives

$$\$1000 \cdot (1 + 0.025)^4 = \$1103.81$$

for the amount after 1 year—$1.31 more than the amount obtained by compounding semiannually. Formula (6) with $r = 10\% = 0.10$ and $t = 10$ gives

$$\$1000 \cdot (1 + 0.025)^{40} = \$2685.06$$

for the amount after 10 years—almost $32 more than the amount obtained by compounding semiannually.

The general case of compound interest corresponds to compounding at the decimal rate r at equal intervals n times during each year. Then the interest earned in each such "interest period" is $\frac{rA}{n}$. If the amount at the beginning the period is A, then the addition of the interest $\frac{rA}{n}$ at the end of the period yields the amount

$$A + \frac{rA}{n} = A\left(1 + \frac{r}{n}\right).$$

Thus the amount in the account is multiplied by factor $\left(1 + \frac{r}{n}\right)$ at the end of each interest period. This happens n times in a full year, so the amount in the account after 1 year is

$$A_1 = A_0\left(1 + \frac{r}{n}\right)^n. \tag{7}$$

It follows that the amount in the account is multiplied by $\left(1 + \frac{r}{n}\right)^n$ each year. In t years, the amount in the account is multiplied by this factor t times, that is, by

$$\left(1 + \frac{r}{n}\right)^n \cdot \left(1 + \frac{r}{n}\right)^n \cdots \left(1 + \frac{r}{n}\right)^n = \left[\left(1 + \frac{r}{n}\right)^n\right]^t = \left(1 + \frac{r}{n}\right)^{nt}.$$

Consequently, the amount in an account after t years at annual interest rate r **compounded n times annually** is

$$A_t = A_0\left(1 + \frac{r}{n}\right)^{nt}. \tag{8}$$

Observe that each of the previous (numbered) formulas in this section is a special case of formula (8). In particular,

- $n = 1$ gives the formula $A_t = A_0(1 + r)^t$ for interest compounded annually;
- $n = 2$ gives the formula $A_t = A_0\left(1 + \frac{r}{2}\right)^{2t}$ for interest compounded semi-annually; and
- $n = 4$ gives the formula $A_t = A_0\left(1 + \frac{r}{4}\right)^{4t}$ for interest compounded quarterly.

EXAMPLE 3 Interest Compounded Monthly

Finally, find the amount after 1 year and after 10 years in the account of Example 1 if the 10% annual interest is compounded monthly.

SOLUTION

With $A_0 = \$1000$, $r = 0.10$, and $n = 12$, Formula (7) gives

$$\$1000 \cdot \left(1 + \frac{0.10}{12}\right)^{12} = \$1104.71$$

for the amount after 1 year—90 cents more than the amount obtained by compounding quarterly. Formula (8) with $n = 12$ and $t = 10$ gives

$$\$1000 \cdot \left(1 + \frac{0.10}{12}\right)^{120} = \$2707.04$$

for the amount after 10 years—almost $19 more than the amount obtained by compounding quarterly. ●

As Good As It Gets?

Examples 1–3 suggest that perhaps we can earn more and more on an investment by insisting that our interest be compounded more and more frequently. The natural question is: How good does it get? For simplicity of investigation, we assume 100% annual interest, so $r = 1$ in the following example.

EXAMPLE 4 | Compounding More and More Frequently

Suppose that the amount $A_0 = 1$ (1 dollar) is invested in an account that draws 100% annual interest. Calculate the amount in this account after 1 year if interest is compounded

- **a.** monthly
- **b.** weekly
- **c.** daily
- **d.** hourly
- **e.** every minute
- **f.** every second

SOLUTION

If we substitute $A_0 = 1$ and $t = 1$ (year) in (8), we find that the amount in the account after 1 year is given by

$$A_1 = \left(1 + \frac{1}{n}\right)^n \qquad (9)$$

if n is the number of times interest is compounded during the year. We need only substitute

- $n = 12$ to compound interest monthly,
- $n = 52$ to compound weekly,
- $n = 365$ to compound daily,
- $n = 365 \times 24 = 8760$ to compound hourly,
- $n = 8760 \times 60 = 525{,}600$ to compound every minute, and
- $n = 525{,}600 \times 60 = 31{,}536{,}000$ to compound every second.

Figure 4.1.1 illustrates an efficient way to substitute these successive values of n into (9).

```
                8760.00000
(1+1/K)^K
                   2.71813
60*K→K
              525600.0000
(1+1/K)^K
                   2.71828
60*K→K
```

FIGURE 4.1.1 Compounding 100% interest more and more often.

The following table shows the results we obtain in this way.

100% INTEREST COMPOUNDED	AMOUNT AFTER 1 YEAR ($)
(a) Monthly	2.61304
(b) Weekly	2.69260
(c) Daily	2.71457
(d) Hourly	2.71813
(e) Each minute	2.71828
(f) Each second	2.71828

In actual practice, of course, the amount in the account will be rounded to two decimal places, indicating dollars and cents.

Apparently we have crept up on a greatest possible "limiting value" of 2.71828 dollars. Indeed, this same five-decimal-place value results if still larger values of n are used, corresponding to compounding even more than once per second. [Although a typical nine-place calculator encounters round-off error that causes inaccuracies in the five-place result if values much larger than the nine-digit value $k = 31,536,000$ are substituted in (8).]

In this example, we started with an investment of $1. Hence, if a bank compounds our 100% annual interest every hour, minute, or second—the table says that whichever of these it is makes no difference—then after 1 year our $1.00 initial investment has grown to $2.72. But if we had instead invested $1000, then the figures in the second column of the table would be multiplied by 1000. Thus, after 1 year we would have $2718.13 if the bank had compounded our interest hourly, $2718.28 (15 cents more) if it had compounded each minute or second. But this is as good as it gets—starting with an investment of $1000, we can earn no more than $2718.28 in a year at 100% interest, even if the bank compounds interest 10 (or 100, or 1000) times per second.

Even though ordinary financial computations are unlikely to require such high accuracy, a "high-precision" computer program can calculate with 15-place accuracy (say) the limiting value obtained by computing the value $(1 + 1/n)^n$ with larger and larger values of n without limit. The result is the number

$$e = 2.7\,1828\,1828\,45\,90\,45\ldots, \qquad (10)$$

which is generally regarded as the most important special number in mathematics (even more important than the number $\pi = 3.14159\,26\,5358\,9793\ldots$). In summary, our numerical investigation indicates that

$$\left(1 + \frac{1}{n}\right)^n \approx e \approx 2.71828 \qquad (11)$$

if n is sufficiently large.

We henceforth reserve the letter e to denote the number in (10), just as π is universally reserved to denote the famous number *pi* that we use in calculating areas and perimeters of circles. On a graphing calculator, the number e has its own key—typically the key beneath the π-key (as in Fig. 4.1.2).

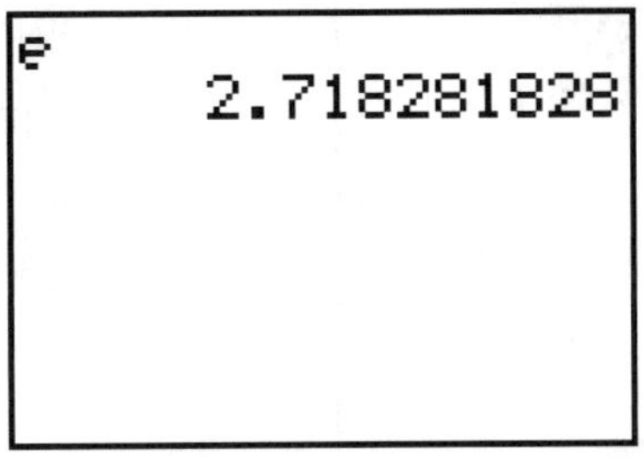

FIGURE 4.1.2 Pressing the [2nd] [÷] key on a TI-84 calculator yields the number e.

```
e
   2.718281828
```

The number e, like π or $\sqrt{2}$, is known to be *irrational*—it cannot be expressed as a terminating or repeating decimal. But can you see how the spaces printed in (10) make it possible for almost anyone, even without a great memory, to remember the first 9 (or even 15) decimal places of e?

Continuously Compounded Interest

Banks sometimes advertise that they compound your interest not quarterly, not monthly or even daily, but *continuously*. The implication is that they thereby pay you "all the interest the law allows." What does this really mean?

As a practical matter it means compounding interest so frequently that compounding it still more frequently would yield no further return on our investment. As we saw in Example 4 and the subsequent discussion, this means that an annual interest rate of $r = 100\%$ would result in our original investment A_0 being multiplied by the special number e each year. According to Section 3.3, multiplication by the constant e each year corresponds to the exponential function with base e. Consequently, the function

$$A(t) = A_0 e^t \tag{12}$$

gives the amount in the account after t years if 100% annual interest is compounded continuously.

The exponential function e^t in (12), with the special base e, is fundamental and is called **the natural exponential function**. It has its own key (usually denoted by e^x) on any scientific or graphing calculator.

EXAMPLE 5   Interest Compounded Continuously

Suppose you invest $1000 at 100% interest compounded continuously. What will your account be worth after 5 years?

SOLUTION

Substitution of $A_0 = 1000$ and $t = 5$ in (12) gives

$$A(5) = 1000 e^5$$

after 5 years. The e^x-key on our calculator gives the resulting amount $148,413.16 (Fig. 4.1.3).

158 CHAPTER 4 Continuous Growth and Logarithmic Models

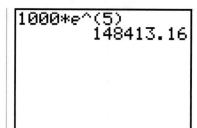

FIGURE 4.1.3 Pressing the [e^x] key puts **e^(** on the screen, so we need only type 5) and press the ENTER key.

"Amazing, but true!" Which is why nobody's ever going to offer you 100% annual interest compounded continuously.

So how do we calculate the result of continuous compounding with a more realistic annual interest rate like 6%? Well, suppose that in Example 4 we would have invested our dollar at annual interest rate r (rather than $100\% = 1$). Then our first step in the solution would have been to write

$$A_1 = \left(1 + \frac{r}{n}\right)^n \tag{13}$$

for the resulting amount after 1 year if this interest rate r is compounded n times annually. Since we saw in (11) that $(1 + \frac{1}{n})^n \approx e$ if n is sufficiently large, it may be reasonable to guess that

$$\left(1 + \frac{r}{n}\right)^n \approx e^r \tag{14}$$

if n is sufficiently large.

EXAMPLE 6 Approximating e^r

Verify (14) numerically with $n = 1$ million and $r = 0.1, 0.2, 0.3, \ldots, 1.9, 2.0$.

SOLUTION

With x in place of r we want to verify that

$$\left(1 + \frac{x}{1000000}\right)^{1000000} \approx e^x \tag{15}$$

for $x = 0.1, 0.2, 0.3, \ldots$. Figure 4.1.4 shows the left- and right-hand-side functions in (15) defined on a graphing calculator.

FIGURE 4.1.4 The functions in (15).

Figures 4.1.5 and 4.1.6 show the result of tabulating both functions.

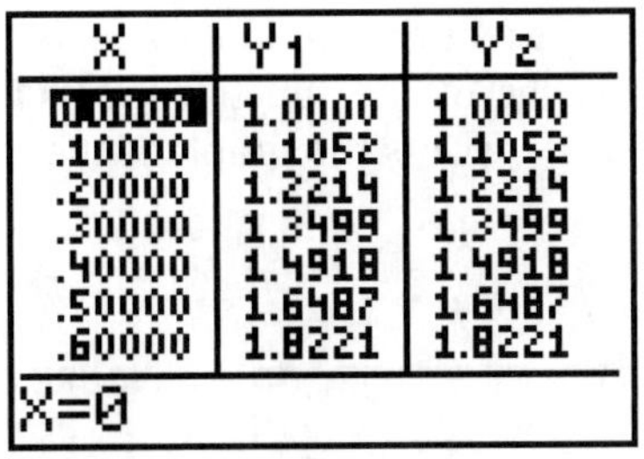

X	Y1	Y2
0.0000	1.0000	1.0000
.10000	1.1052	1.1052
.20000	1.2214	1.2214
.30000	1.3499	1.3499
.40000	1.4918	1.4918
.50000	1.6487	1.6487
.60000	1.8221	1.8221

X=0

FIGURE 4.1.5 Resulting table with **TblStart=0** and **ΔTbl=0.1.**

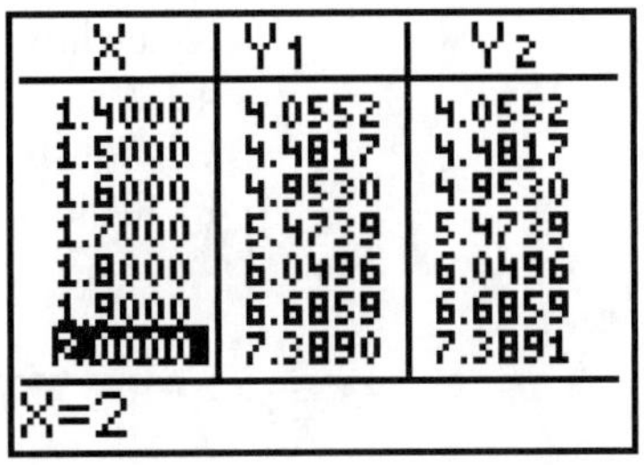

X	Y1	Y2
1.4000	4.0552	4.0552
1.5000	4.4817	4.4817
1.6000	4.9530	4.9530
1.7000	5.4739	5.4739
1.8000	6.0496	6.0496
1.9000	6.6859	6.6859
2.0000	7.3890	7.3891

X=2

FIGURE 4.1.6 Further corroboration of (15).

- The agreement we see in these tables can hardly be coincidental.

Equation (14) implies that the 1 dollar grows to e^r dollars in 1 year at interest rate r compounded continuously. Thus a dollar is multiplied by the factor e^r each year, and hence by the factor

$$(e^r)^t = e^{rt}$$

in t years. The following principle tells what happens to an arbitrary initial investment of A_0 dollars.

PRINCIPLE: Continuously Compounded Interest

If the initial amount A_0 is invested at annual interest rate r compounded continuously, then the resulting amount after t years is given by

$$A(t) = A_0 e^{rt}. \tag{16}$$

Thus the natural exponential function (with base e) is the key to all problems involving interest compounded continuously.

EXAMPLE 7 Finding an Input

Suppose you deposit $A_0 = 1000$ dollars in a savings account that draws 12% annual interest compounded continuously. How long will you have to wait until the amount in the account doubles?

SOLUTION

Equation (16) with $r = 0.12$ says that the amount in the account after t years is given by

$$A(t) = 1000\, e^{0.12t} \tag{17}$$

Therefore, we need to solve the equation

$$1000\, e^{0.12t} = 2000 \tag{18}$$

to see how long it takes for the amount to double. Let's do this graphically. Figure 4.1.7 shows the graphs of the functions **Y1=1000*e^(0.12*X)** and **Y2=2000**. Obviously, the rising curve is the graph of the exponential function on the left-hand side.

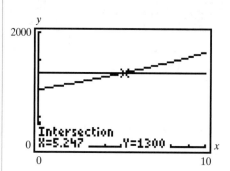

FIGURE 4.1.7 Solving Equation (18).

We have already employed our calculator's automatic intersection-finding facility; it yields the solution $x = 5.776$. Thus it takes 5.776 years $\approx$ 5 years, 40.35 weeks for our original investment to double in value.

Effective Annual Yield

Suppose that you are offered two investment opportunities, one earning 9% interest compounded annually, the other earning 8.75% compounded monthly. How do you decide which of the two investments will earn more money? The easiest way is to compute each investment's effective annual yield.

> **DEFINITION:** The **effective annual yield** for an investment is the percentage rate that would yield the same amount of interest if interest were compounded annually.

EXAMPLE 8 | Finding Effective Annual Yield

Find the effective annual yield for an investment paying 8.75% compounded monthly.

SOLUTION

According to our compound interest formula (8), after n years the investment will be worth

$$A_t = A_0\left(1 + \frac{0.0875}{12}\right)^{12t}. \tag{19}$$

Recall that the formula for annual compound interest is $A_t = A_0(1+r)^t$, so in order to determine what annual interest rate r would yield the same interest, we need to rewrite our expression in (19),

$$A_t = A_0\left(1 + \frac{0.0875}{12}\right)^{12t}$$

$$= A_0\left(\left(1 + \frac{0.0875}{12}\right)^{12}\right)^t$$

$$\approx A_0(1 + 0.0911)^t$$

- Thus, the effective annual yield for this investment is (approximately) 9.11%.

Our earlier question was "Which is better, 8.75% compounded monthly or 9% compounded annually?" We can see, then, that 8.75% compounded monthly yields a bit more than 9% compounded annually, so the interest rate offered is not the whole story.

It is common to refer to the stated rate as the *nominal* rate. (*Nominal* comes from the Latin *nomen*, which means *name*.) Note that since determining effective annual yield converts a nominal interest rate to an annual compound rate, the effective annual yield for 9% interest compounded annually is just 9%. In the case of annual compound interest, the nominal rate and the effective annual yield are the same.

EXAMPLE 9 | Which Is Better?

Determine which is the better investment, 4.75% compounded quarterly or 4.5% compounded continuously.

SOLUTION

In order to determine which investment is better, we will determine the effective annual yield for each.
   First, 4.75% compounded quarterly:

$$A_n = A_0\left(1 + \frac{0.0475}{4}\right)^{4n}$$

$$= A_0\left(\left(1 + \frac{0.0475}{4}\right)^4\right)^n$$

$$\approx A_0(1 + 0.0484)^n.$$

The effective annual yield is (approximately) 4.84%.
   Second, 4.5% compounded continuously:

$$A_n = A_0 e^{0.045n}$$

$$= A_0(e^{0.045})^n$$

$$\approx A_0(1 + 0.0460)^n.$$

The effective annual yield is (approximately) 4.60%.
- So the better investment is the one with 4.75% interest compounded quarterly.

In this case, the higher interest rate beats the more frequent compounding. When choosing among several different investments, it is important to be "comparing apples to apples." Once you obtain the effective annual rate for each, it is easy to see which offers the best return on your investment.

If you are borrowing money rather than investing it, the actual cost of financing, expressed as a yearly rate, is called the *annual percentage rate* or *APR*. In order for consumers to be able to compare different loans, lenders are required by the Federal Truth in Lending Act to report both the rate that is used to calculate monthly payments and the APR. In the case of a mortgage payment, the APR is generally higher than the advertised interest rate on the loan because APR takes into account not only interest but also origination fees, points, mortgage insurance, and any other fees. When borrowing money, the lower the APR, the better is the deal.

Continuous Growth and Decay

It is not just money that grows at a continuous compound rate. While the mathematical model $A(t) = 100e^{0.03t}$ could represent an investment of $100 that earns interest at a 3% compounded continuously, it could just as easily represent the population of a city of 100,000 people that is growing at a continuous annual rate of 3%, or the growth of a colony of 100 bacteria with a continuous hourly rate of 3%. The mathematics is the same. It is your job to pay attention to how the population is being measured (in thousands, hundreds, or single units), to whether the continuous growth is being considered over years, days, or seconds, and to what you are measuring (dollars, people, or bacteria).

> **DEFINITION:** **Continuous Growth with Rate r**
> If the growth of a quantity is described by the function $A(t) = A_0 e^{rt}$, then we say that it **grows continuously** and has **continuous growth rate r**.

In Chapter 3, we saw that the function model $A(t) = A_0(1 + r)^t$ represents **natural growth** if r is positive and **natural decay** if r is negative. Similarly, the function $A(t) = A_0 e^{rt}$ represents **continuous growth** if r is positive and **continuous decay** if r is negative.

EXAMPLE 10 | Radioactive Decay

Carbon-14 is a radioactive substance found in plants and animals, including humans. It begins to decay at death at a continuous annual rate of (approximately) 0.012%. Determine the percentage of carbon-14 remaining in a mummy that is 5000 years old.

SOLUTION

If we let $A_0 = 100$ represent the 100% of the carbon-14 that is present at death, then the percentage remaining after t years is given by

$$A(t) = 100e^{-0.00012t}.$$

After 5000 years, we have $A(5000) = 100e^{-0.00012 \times 5000} \approx 34.99$. Thus, about 35% of the original amount of carbon-14 remains.

4.1 Exercises | Building Your Skills

In Exercises 1–4, calculate the amount in an account after n years if A_0 dollars is initially invested and interest at the annual rate r is compounded (a) quarterly, (b) monthly, (c) weekly, (d) daily, (e) continuously.

1. $A_0 = 1000$, $r = 4\%$, $n = 1$
2. $A_0 = 2000$, $r = 8\%$, $n = 1$
3. $A_0 = 5000$, $r = 2\%$, $n = 50$
4. $A_0 = 5000$, $r = 1\%$, $n = 100$

In Exercises 5–8, find how long (in years, rounded to three decimal places) it takes an amount A_0 initially invested in an account to double in value if interest at the indicated annual rate r is compounded continuously.

5. $r = 3\%$
6. $r = 5\%$
7. $r = 6\%$
8. $r = 2\%$

In Exercises 9–12, find the effective annual yield for each nominal interest rate.

9. 3.5% compounded monthly
10. 4% compounded semiannually
11. 3% compounded continuously
12. 3.75% compounded continuously
13. Solve $e^x = 2$ graphically. Give your answer accurate to three decimal places.
14. Solve $e^x = 8$ graphically. Give your answer accurate to three decimal places.
15. What is the relation between the answers to Exercises 13 and 14?
16. Which is the better investment—one earning 5% compounded quarterly, or one earning 4.65% compounded continuously?
17. Which is the better investment—one earning 4.5% compounded semiannually, or one earning 4.25% compounded continuously?

Applying Your Skills

18. On January 1, 2000, an investor put $1000 in an account paying 3% interest compounded monthly.
    a. Find a function that gives the amount in the account after t years.
    b. In what month and year will the amount grow to $1500?
    c. How much is in the account on January 1, 2006?

164 CHAPTER 4 Continuous Growth and Logarithmic Models

19. You deposit $3000 in an account earning 4.25% compounded weekly on January 1, 2004.

   a. Find a function that gives the amount in the account after t years.
   b. In what month and year will the amount grow to $10,000?
   c. How much is in the account on January 1, 2014?

20. You deposit $2500 in an account earning 3.75% compounded continuously on January 1, 2005.

   a. Find a function that gives the amount in the account after t years.
   b. How much is in the account on January 1, 2020?
   c. In what month and year will the amount triple?

21. On January 1, 2000, an investor put $4800 in an account paying 5.2% interest compounded continuously.

   a. Find a function that gives the amount in the account after t years.
   b. How much is in the account on January 1, 2025?
   c. In what month and year will the amount double?

22. Rank the following investments from best to worst in terms of the effective annual yield:

   Account A: 7% compounded daily;
   Account B: 7.1% compounded monthly;
   Account C: 7.05% compounded continuously.

23. Rank the following investments from best to worst in terms of the effective annual yield:

   Account A: 6.2% compounded daily;
   Account B: 6.3% compounded monthly;
   Account C: 6.25% compounded continuously.

24. In the year 2000, the total output G of goods and services in the world was 45 trillion dollars and was growing at a continuous annual rate of 4.6%.

   a. Find a function $G(t)$ that gives the total output of goods and services in the world as a function of t, the number of years after 2000.
   b. According to your model, what will the total output of goods and services be in 2010?
   c. According to your model, when will the total output of goods and services double?

25. At 8 AM, there is a colony of 300 bacteria, and it is growing at a continuous hourly rate of 10%.

   a. Find a function $P(t)$ that gives the number of bacteria as a function of the time, with $t = 0$ being 8 AM.
   b. How many bacteria are there at noon? Round your answer to the nearest whole number of bacteria.
   c. At what time (to the nearest minute) on that same day does the colony of bacteria grow to 800?

26. Under certain conditions, *Escherichia coli* bacteria grow at a continuous hourly rate of about 20%.

   a. If a colony of these bacteria contains 1000 bacteria at 8 AM, find a function $E(t)$ that gives the number of bacteria as a function of the time.

b. How long will it take the colony to grow to 5000 bacteria? Give your answer in hours and minutes, correct to the nearest minute.

c. How many bacteria will be present in at 2 PM? Round your answer to the nearest whole number of bacteria.

27. An archaeologist has discovered a mummy and determined that only 32% of the original amount of carbon-14 remains in the mummy. How old, to the nearest year, is the mummy? (Recall from Example 10 that the continuous rate of decay for carbon-14 is 0.012%.)

28. A laboratory has a 100-mg sample of radium-226, which decays at the continuous yearly rate of .044%.

a. How much of the sample remains after 5 years?

b. How many years will it take until there are only 10 grams of the sample remaining? Round your answer to the nearest whole year.

29. A credit card for college students advertises an annual percentage rate (APR) of 9.9%, with finance charges calculated using simple interest on the monthly balance. The minimum payment on the credit card is 3% of the outstanding balance or $10, whichever is greater. Now suppose that your credit card balance is $500 on January 1, 2001, and that you make only the minimum payment on the first of each month thereafter. If you make no further purchases on the card, when will you pay off the credit card? How much interest will you have paid over that time period? What continuous annual compound rate of interest will you have paid on this debt?

30. This exercise is a numerical investigation of the fact that $(1 + \frac{r}{n})^n \approx e^r$ if n is sufficiently large. Calculate the value of $(1 + \frac{2}{n})^n$ (rounded off accurate to three decimal places) with $n = 10, 100, 1000, \ldots, 1000000$. Does it appear that $(1 + \frac{2}{n})^n$ approaches the value of e^2 as n gets larger and larger?

4.2 EXPONENTIAL AND LOGARITHMIC FUNCTIONS

In Section 4.1 we introduced the **natural exponential function** defined by

$$f(x) = e^x \tag{1}$$

using the very special base number $e \approx 2.71828$. In order to investigate this important function, let us recall the following *laws of exponents*.

THEOREM 1 Laws of Exponents with General Base a

If the base numbers a and b are positive, and x and y are any real numbers, then

$$a^x a^y = a^{x+y}, \tag{2}$$

$$a^{-x} = \frac{1}{a^x}, \tag{3}$$

166 CHAPTER 4 Continuous Growth and Logarithmic Models

$$\frac{a^x}{a^y} = a^{x-y}, \tag{4}$$

$$(a^x)^y = a^{xy}, \tag{5}$$

$$a^0 = 1, \tag{6}$$

$$(ab)^x = a^x b^x. \tag{7}$$

In case you do not remember *why* (6) is true, observe that

$$a^0 = a^{1-1}$$
$$= a^1 a^{-1} \quad [\text{using}(2)]$$
$$= a \cdot \frac{1}{a} \quad [\text{using}(3)]$$
$$= \frac{a}{a} \quad [\text{algebra}]$$
$$a^0 = 1.$$

EXAMPLE 1 | Using Laws of Exponents

Use the laws of exponents to simplify the expression $\frac{(2^2)^3 3^{-3}}{2^{-2}}$.

SOLUTION

$$\frac{(2^2)^3 3^{-3}}{2^{-2}} = \frac{2^6 3^{-3}}{2^{-2}} = \frac{2^6 2^2}{3^3} = \frac{2^8}{3^3} = \frac{256}{27} \approx 9.4815.$$

Since we will be looking at base e exponential functions, it's worth restating the laws of exponents using the special number $e \approx 2.71828$ as the base a.

THEOREM 2 Laws of Exponents with Base e

If x and y are any real numbers, then

$$e^x e^y = e^{x+y}, \tag{8}$$

$$e^{-x} = \frac{1}{e^x}, \tag{9}$$

$$\frac{e^x}{e^y} = e^{x-y}, \tag{10}$$

$$(e^x)^y = e^{xy}, \tag{11}$$

$$e^0 = 1. \tag{12}$$

SECTION 4.2 Exponential and Logarithmic Functions

EXAMPLE 2 Simplifying an Exponential Expression

Simplify $\sqrt{e^{6x-1}}$.

SOLUTION

$$\sqrt{e^{6x-1}} = (e^{6x-1})^{1/2} = e^{3x-1/2} = \frac{e^{3x}}{\sqrt{e}} \approx \frac{e^{3x}}{1.6487} \approx 0.6065\, e^{3x}.$$

Common Logarithms

You may have seen "common" base 10 logarithms (if only briefly) in a previous mathematics course.

> **DEFINITION: Common Logarithm**
> The common logarithm of the positive number x is the power to which 10 must be raised in order to obtain the number x. It is denoted by $\mathbf{log_{10}\, x}$. Thus
> $$y = \log_{10} x \quad \text{means that} \quad 10^y = x. \tag{13}$$
> Frequently we omit the subscript 10 and simply write $\log x$ for the common logarithm of the positive number x.

The definition of logarithms seems inherently confusing to almost everyone. Sometimes it helps to phrase it as a question. If we ask what is the common logarithm of x, we're asking what number to put in the "question mark" blank:

$$\log x = ? \quad \text{provided that} \quad 10^? = x. \tag{14}$$

This is not so hard to figure out if the given number x is an (integer) power of 10.

EXAMPLE 3 Finding Common Logarithms Algebraically

Find each of the following.

    a. $\log 100$
    b. $\log 10000$
    c. $\log 10$
    d. $\log 0.1$
    e. $\log 0.001$

SOLUTION

a. $\log 100 = 2$ because $10^2 = 100$.
b. $\log 10000 = 4$ because $10^4 = 10000$.
c. $\log 10 = 1$ because $10^1 = 10$.
d. $\log 0.1 = -1$ because $10^{-1} = \dfrac{1}{10} = 0.1$.
e. $\log 0.001 = -3$ because $10^{-3} = \dfrac{1}{10^3} = 0.001$.

We can see that in each case the logarithmic and exponential equations are just different versions of the same relationship. ●

However, we need a calculator to find the value of $\log x$ if x is *not* an integer power of 10. The common logarithm key on a graphing calculator is the one marked **log** (what else?). However, it is important to realize that calculators use function notation, so the logarithm of x is denoted by **log(x)** with parentheses included. Figure 4.2.1 illustrates the use of a calculator to find some typical logarithms. When we press the log key, the calculator displays **log(** on the screen and waits for us to enter the **x)** with the value of x whose logarithm we want to find.

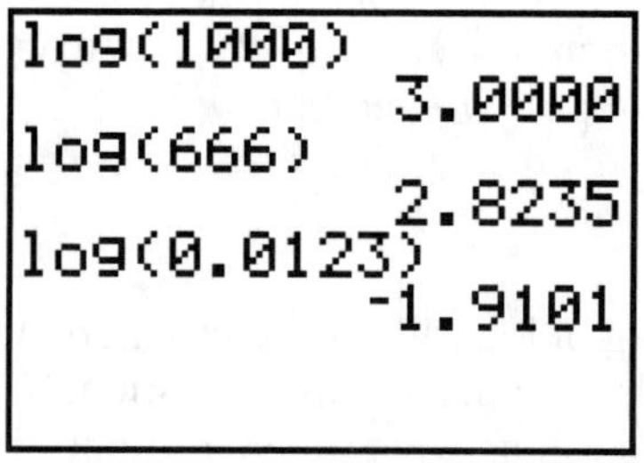

```
log(1000)
            3.0000
log(666)
            2.8235
log(0.0123)
           -1.9101
```

FIGURE 4.2.1 Some typical common logarithms.

Natural Logarithms

In college mathematics, it is customary to use "natural logarithms" instead of common logarithms. These so-called natural logarithms are defined in a similar way, using powers of the base number $e \approx 2.71828$ rather than powers of the base number 10.

> **DEFINITION: Natural Logarithm**
> The natural logarithm of the positive number x is the power to which e must be raised in order to obtain the number x. It is sometimes denoted by $\log_e x$, but more frequently by **ln x** (with **l** for "log" and **n** for "natural"). Thus
>
> $$y = \ln x \quad \text{means that} \quad e^y = x. \qquad (15)$$

Again, it may help to put it in question mode. If we ask what is the natural logarithm of x, we're asking what number to put in the following "question mark" blank:

$$\ln x = ? \quad \text{provided that} \quad e^? = x. \tag{16}$$

Here, however, there are no really easy values to anchor our understanding of natural logarithms.

EXAMPLE 4 | Finding a Natural Logarithm Graphically

Find ln 10, the natural logarithm of 10.

SOLUTION

To find ln 10, we must determine to what power e must be raised in order to obtain 10. Using the e^x key, our calculator gives (as shown in Fig. 4.2.2)

$$e \approx 2.71828, \quad e^2 \approx 7.38906, \quad e^3 \approx 20.08554.$$

```
e^(1)
            2.71828
e^(2)
            7.38906
e^(3)
           20.08554
```

FIGURE 4.2.2 Searching for ln 10.

Thus 10 is greater than e^2 but less than e^3, so we need to raise e to higher than the second power but less than the third power. If our calculator had no **ln** key, we could still solve the equation $e^x = 10$ graphically to find the proper exponent x to use. Figure 4.2.3 shows the graph **Y1 = e^(X)** of the exponential function and the horizontal line **Y2 = 10**. As indicated, the calculator's intersection-finding facility gives $x \approx 2.30259$ for the x-coordinate of the point where $e^x = 10$.

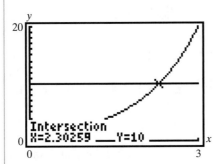

FIGURE 4.2.3 Finding the value ln 10 ≈ 2.30259.

- Thus ln 10 = 2.30259 rounded off accurate to five decimal places.

Figure 4.2.4 shows another graph illustrating Example 4. In addition to showing the point (ln 10, 10) on the graph of $y = e^x$, the graph also shows the general point (ln y, y) where the natural logarithm of y is the x-coordinate of the point at height y on the graph $y = e^x$. Note, however, that this makes sense only if $y > 0$, because the curve $y = e^x$ lies entirely above the x-axis. It follows that

Only *positive* numbers have logarithms.

FIGURE 4.2.4 Finding natural logarithms graphically.

EXAMPLE 5   Finding a Logarithm Using the ln Key

Find each of the following.

a. ln 10
b. ln 100
c. ln 1000

SOLUTION

Instead of solving graphically for natural logarithms as in Example 4, we can simply use our calculator's **ln** key as indicated in Fig. 4.2.5.

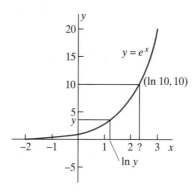

FIGURE 4.2.5 Using the [LN] key to find logarithms.

Thus, ln 10 = 2.3026, ln 100 = 4.6052, and ln 1000 = 6.9078, rounded to four decimal places.

Can you see that the logarithms in Example 5 fit a certain pattern? If we "chopped" each off to one decimal place—ln 10 ≈ 2.3, 100 ≈ 4.6, ln 1000 ≈ 6.9— it would be still more obvious that

$$\ln 100 = 2 \ln 10 \quad \text{and} \quad \ln 1000 = 3 \ln 10.$$

This observation illustrates one of the laws for natural logarithms (analogous to the laws for common logarithms that you may have seen before).

THEOREM 3 Laws of Logarithms

If x and y are any positive real numbers, then

$$\ln xy = \ln x + \ln y, \tag{17}$$

$$\ln \frac{1}{x} = -\ln x, \tag{18}$$

$$\ln \frac{x}{y} = \ln x - \ln y, \tag{19}$$

$$\ln x^y = y \ln x, \tag{20}$$

$$\ln 1 = 0. \tag{21}$$

Some people can learn these laws more readily in words than in symbols:

"The logarithm of a product is the sum of the logarithms of the factors."

"The logarithm of a reciprocal of a number is the negative of the logarithm of that number."

"The logarithm of a quotient is the difference of the logarithm of the numerator and the logarithm of the denominator."

"The logarithm of a power is the exponent times the logarithm of the base."

"The logarithm of one is zero."

Analyze each statement carefully to make certain you see that it says the same thing as the corresponding symbolic law above.

The laws of logarithms in Theorem 3 follow from the laws of exponents in Theorem 2. For instance, consider the last law of exponents, $e^0 = 1$. This law says that the power to which e must be raised to obtain 1 is 0. This is precisely the definition of the natural logarithm of 1. Thus $\ln 1 = 0$, and we have deduced the last law of logarithms from the last law of exponents.

The fact that $\ln 1 = 0$ means that the graph $y = \ln x$ crosses the x-axis when $x = 1$, as indicated in Fig. 4.2.6.

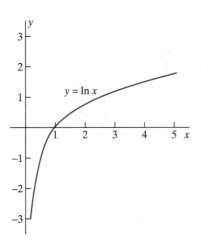

FIGURE 4.2.6 The graph of the natural logarithm function $\ln x$.

Visualizing this graph is a good way to remember that

- the logarithms of numbers *less than* 1 are **negative**, while
- the logarithms of numbers *greater than* 1 are **positive**.

EXAMPLE 6 — Using the Laws of Logarithms

Find each of the following, rounding all values to four decimal places.

a. $\ln 33$

b. $\ln \dfrac{1}{66}$

c. $\ln \dfrac{29}{13}$

d. $\ln 125$

SOLUTION

a. $\ln 33 = \ln(3 \times 11) = \ln 3 + \ln 11 \approx 1.0986 + 2.3979 = 3.4965.$

b. $\ln \dfrac{1}{66} = -\ln 66 \approx -4.1897.$

c. $\ln \dfrac{29}{13} = \ln 29 - \ln 13 \approx 3.3673 - 2.5649 = 0.8024.$

d. $\ln 125 = \ln 5^3 = 3 \ln 5 \approx 3 \times 1.6094 = 4.8283.$

You should verify that in each case the answer matches what you get by using your calculator directly to evaluate the natural logarithm of the given number.

Logarithms and Exponentials as Inverse Functions

The definition of the natural logarithm in Equation (15) says that $e^y = x$ provided that $y = \ln x$. Substitution of the latter equation in the former gives

$$e^{\ln x} = x \tag{22}$$

if the number x is positive (so it *has* a logarithm). That is, if we start with x and apply the logarithm function to get $\ln x$, then use this result as the exponent of e, we get x back. Thus the natural exponential function *undoes* the result of the natural logarithm function. The reverse is true also—the logarithm function undoes the result of the exponential function:

$$\ln(e^x) = x \tag{23}$$

for any number x.

Equations (22) and (23) say that if you start with x and apply both functions (exponential and logarithm) in succession—in either order—then you're back where you started, with the same number x. Figure 4.2.7 illustrates this fact, regarding each

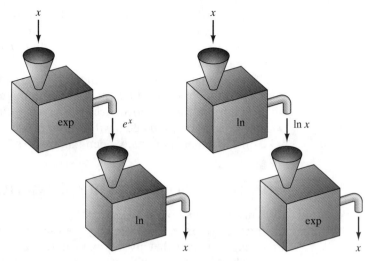

FIGURE 4.2.7 e^x and $\ln x$ as inverse functions.

function as an input-output process. If the output of either function is used as the input to the other, the final output is the same as the original input.

Do symbolic expressions in x tend to make your eyes glaze over? If so, think concretely:

$$e^{\ln 2} = 2, \qquad e^{\ln 7654} = 7654,$$
$$\ln(e^{37}) = 37, \qquad \ln(e^{\pi}) = \pi,$$
$$e^{\ln(\text{CAT})} = \text{CAT}, \qquad \ln(e^{\text{DOG}}) = \text{DOG}.$$

The last line here may seem a bit whimsical, but—assuming that CAT and DOG denote positive numbers—it's just as true as the preceding two lines.

If f and g are two functions that undo each other either way, meaning that

$$f(g(x)) = x \quad \text{and} \quad g(f(x)) = x, \tag{24}$$

then f and g are called ***inverse functions*** of each other. Thus (22) and (23) say that the natural exponential and logarithm functions are inverse functions of each other.

The concept of inverse functions may seem exotic initially, but there are many common and simple inverse-function pairs, such as the linear functions illustrated below.

1. If $f(x) = 2x$ and $g(x) = \dfrac{x}{2}$, then $f(g(x)) = 2\left(\dfrac{x}{2}\right) = x$ and $g(f(x)) = \dfrac{2x}{2} = x$. So the doubling and halving functions are inverses of each other.

2. If $f(x) = 3x + 1$ and $g(x) = \dfrac{x-1}{3}$, then $f(g(x)) = 3\left(\dfrac{x-1}{3}\right) + 1 = x$ and $g(f(x)) = \dfrac{(3x+1) - 1}{3} = x$, and f and g are inverses of each other.

Here you might notice that the constants involved in $g(x)$, the inverse function of $f(x)$, are the same constants (3 and 1) that appear in the original linear function but are used with the inverse operation and in the reverse order. That is, to evaluate $f(x)$ for a particular value of x, we first multiply x by 3 and then add 1 to the result. To evaluate $g(x)$ for a particular value of x, we first subtract 1 and then divide the result by 3.

The "undoing" must be done in the reverse order of the original "doing," and each operation must itself be "undone." You might think of opening a birthday gift that a friend has wrapped nicely for you. If your friend put your gift in a box, wrapped it in paper, and tied it with a ribbon, you must first untie the ribbon, then take off the paper, and finally open the box to see the gift.

Solving Exponential and Logarithmic Equations

The inverse function relations $e^{\ln x} = x$ and $\ln(e^x) = x$ admittedly have an abstract appearance. However, they are used mainly in very practical situations—in solving equations that involve exponentials and logarithms. You may have wondered why we did not solve any exponential equations symbolically in Chapter 3. Perhaps you thought that using the calculator was simply easier, but the real reason was that we did not have the algebraic tools we needed to solve such problems symbolically.

We can use exponential and logarithmic functions to "turn the other one inside out." That is, we take natural logarithms of each side of an exponential equation, and "exponentiate" each side of a logarithmic equation.

EXAMPLE 7 Solving an Exponential Equation Symbolically

Solve the equation $e^{2x-3} = 100$ symbolically.

SOLUTION

If $e^{2x-3} = 100$, then (taking natural logarithms of both sides) we get the equation

$$\ln(e^{2x-3}) = \ln 100.$$

Then the fact that $\ln(e^u) = u$ gives

$$2x - 3 = \ln 100$$

so

$$x = \frac{3 + \ln 100}{2}.$$

While $\frac{3 + \ln 100}{2}$ may appear like a strange solution for the equation, it is just a number, every bit as "normal" as 3 or 7.464. Indeed, this number is the *exact* solution to the equation. The value $x = \frac{3 + \ln 100}{2} \approx 3.8026$, if we round this non-repeating, non-terminating decimal to four decimal places. So 3.8026 is an *approximate* solution to the equation. (Mathematicians love exact solutions, while most other folks are perfectly happy with very good approximate ones.) ●

Similarly, we can use an exponential function to undo a logarithmic function, as in the following example.

EXAMPLE 8 | Solving a Logarithmic Equation Symbolically

Solve the equation $\ln(3x + 4) = 5$ symbolically.

SOLUTION

To turn the left-hand side logarithm "inside out," we exponentiate this equation, that is, take e to each side of the equation:

$$e^{\ln(3x+4)} = e^5.$$

The fact that $e^{\ln u} = u$ now gives

$$3x + 4 = e^5,$$

so

$$x = \frac{e^5 - 4}{3} \approx \frac{148.4132 - 4}{3} \approx 48.1377.$$

As before, $\dfrac{e^5 - 4}{3}$ is the exact solution, while 48.1377 is its four-decimal-place approximation. •

Natural Growth Applications

We can now use exponentials and logarithms to find symbolic solutions of population problems that we previously solved using graphical methods. This will once again give you a choice of methods for solving such problems. In Section 3.3 we saw that a natural growth function of the form

$$P(t) = P_0 \cdot b^{t/N} \qquad (25)$$

describes a population that is multiplied by the factor b every N years. Now we will see that every natural growth function can be written in the standard form

$$P(t) = P_0 e^{rt} \qquad (26)$$

We are then able to describe the population growth in terms of its continuous annual growth rate.

EXAMPLE 9 | Finding the Continuous Annual Growth Rate

Find the continuous annual growth rate of a population

    a. of rabbits that doubles every year;

    b. of people that triples every 25 years.

SOLUTION

a. The fact that the rabbit population doubles in 1 year means that
$$P(1) = P_0 e^{r \cdot 1} = 2P_0.$$
Dividing by P_0, we therefore see that
$$e^r = 2$$
$$\ln(e^r) = \ln 2 \quad \text{(taking natural logs of both sides)}$$
$$r = \ln 2 \approx 0.6931.$$
Thus the population's continuous annual growth rate is *exactly* $r = \ln 2$ and *approximately* $r = 0.6931$.

b. The fact that the human population triples every 25 years means that
$$P(25) = P_0 e^{r \cdot 25} = 3P_0.$$
Dividing by P_0, we therefore see that
$$e^{25r} = 3$$
$$\ln(e^{25r}) = \ln 3 \quad \text{(taking logs of both sides)}$$
$$25r = \ln 3$$
$$r = \frac{\ln 3}{25} \approx 0.0439.$$
Thus this population has a continuous annual growth rate of $r = \frac{\ln 3}{25}$, or approximately 0.0439.

The population of Example 9(b) may be said to grow at the *continuous* annual percentage rate of 4.39%. This use of the word *continuous* is analogous to financial growth with an annual interest rate of 4.39% compounded *continuously*. But we know that this is **not** the same as 4.39% interest compounded annually. For instance, if the population grew by 4.39% each year, then its population after 25 years would be
$$P_0 (1 + 0.0439)^{25} \approx 2.9273 \, P_0,$$
which is somewhat less than triple the original population.

EXAMPLE 10 Using a Continuous Growth Model

In 1993 the world population had reached 5.5 billion and had a continuous annual growth rate of $r = 0.0166$.

a. If this growth continues, when will the world population reach 11 billion?

b. When will it reach 50 billion, which some demographers regard as the maximum population for which the planet can provide food?

SOLUTION

a. The world's population is given (in billions, t years after 1993) by

$$P(t) = 5.5\, e^{0.0166\,t}. \qquad (27)$$

It will be 11 billion when

$$P(t) = 5.5\, e^{0.0166} = 11$$

$$e^{0.0166\,t} = \frac{11}{5.5} = 2$$

$$\ln(e^{0.0166t}) = \ln 2$$

$$0.0166t = \ln 2$$

$$t = \frac{\ln 2}{0.0166} \approx 41.7559.$$

Thus the world population will reach 11 billion in exactly $\dfrac{\ln 2}{0.0166}$ years, or about 42 years (after 1993), that is, in 2035 AD.

b. Using (27) similarly, we find that the world population will be 50 billion when

$$P(t) = 5.5\, e^{0.0166} = 50$$

$$e^{0.0166\,t} = \frac{50}{5.5}$$

$$\ln(e^{0.0166t}) = \ln\frac{50}{5.5}$$

$$t = \frac{\ln(50/5.5)}{0.0166} \approx 132.9684.$$

Thus the world population will reach 50 billion in exactly $\dfrac{\ln(50/5.5)}{0.0166}$ years, or about 133 years, that is, in 2126 AD.

In this section we have used logarithms as a method for solving natural and continuous growth equations symbolically. We can now add this technique to the graphic and numeric methods that we used in Chapter 3.

4.2 Exercises | Building Your Skills

1. Use the laws of exponents to simplify each expression. Write each answer as an integer.

   a. $2^2 \cdot 2^4$ **b.** $3^5 \cdot 3^{-5}$ **c.** $(5^6)^{1/2}$

2. Use the laws of exponents to simplify each expression. Write each answer as an integer.

   a. $10^{11} \cdot 10^{-9}$ **b.** $2^{12^{1/3}}$ **c.** $6^6 \cdot 3^{-5}$

3. Use the laws of exponents to simplify $\sqrt{4e^{6x-8}}$ so that it is in the form $c\, e^{ax+b}$.

4. Use the laws of exponents to simplify $\sqrt[3]{27e^{9x+3}}$ so that it is in the form $c\,e^{ax+b}$.

5. Use the laws of logarithms to express each given natural logarithm in terms of just the three natural logarithms ln 2, ln 3, and ln 5.

   a. ln 9   b. ln 10   c. $\ln\dfrac{8}{27}$

6. Use the laws of logarithms to express each given natural logarithm in terms of just the three natural logarithms ln 2, ln 3, and ln 5.

   a. ln 8   b. ln 15   c. $\ln\dfrac{9}{25}$

7. Which is larger, $2^{(3^4)}$ or $(2^3)^4$?

8. Find the approximate value of ln 25 (correct to four decimal places) by solving the equation $e^x = 25$ graphically.

9. Find the approximate value of ln 0.5 (correct to four decimal places) by solving the equation $e^x = 0.5$ graphically.

In Exercises 10 and 11, find an inverse function $g(x)$ such that $f(g(x)) = g(f(x)) = x$.

10. $f(x) = 7x + 5$

11. $f(x) = \dfrac{x - 3}{4}$

Use natural logarithms and exponentials to find the exact solutions to the equations in Exercises 12–18.

12. $2^x = 100$

13. $5^{-x} = 17$

14. $10^{-2x} = 0.2$

15. $e^{3x-5} = \dfrac{1}{7}$

16. $\ln(3x) = 5$

17. $\ln(10x) + 2 = 0$

18. $\ln(2e^{3x} + 4) = 5$

Applying Your Skills

19. The population of Tokyo, Japan, was 26.6 million in 1975 and 34.5 million in 2000.

    a. Find a natural growth function $P(x) = a \cdot b^x$ that gives the population of Tokyo as a function of years after 1975.
    b. Use the model from part (a) to predict Tokyo's population in the year 2010.
    c. Find the exact value of the year in which the population doubles.
    d. Approximate your answer from part (c), correct to four decimal places.

20. The population of Mumbai, India, was 7.3 million in 1975 and 16.1 million in 2000.

    a. Find a natural growth function $P(x) = a \cdot b^x$ that gives the population of Mumbai as a function of years after 1975.
    b. Use the model from part (a) to predict Mumbai's population in the year 2020.

c. Find the exact value of the year in which the population of Mumbai rises to 20 million.
d. Approximate your answer from part (c), correct to four decimal places.

21. The population of Baltimore, Maryland, was 906 thousand in 1970 and 629 thousand in 2003.
    a. Find a natural growth function $P(x) = a \cdot b^x$ that gives the population of Baltimore as a function of years after 1970.
    b. Use the model from part (a) to predict Baltimore's population in the year 2020.
    c. Find the exact value of the year in which the population of Baltimore falls to 400 thousand.
    d. Approximate your answer from part (c), correct to four decimal places.

22. The population of Washington, DC, was 802.2 thousand in 1950 and 572.1 thousand in 2000.
    a. Find a natural growth function $P(x) = a \cdot b^x$ that gives the population of Washington as a function of years after 1950.
    b. Use the model from part (a) to predict Washington's population in the year 2010.
    c. Find the exact value of the year in which the population of Washington falls to half the initial value.
    d. Approximate your answer from part (c), correct to four decimal places.

23. How long does it take a continuously growing bacteria population to double if it triples in 2.5 hours? Give both the exact value and its four-decimal-place approximation.

24. The number of bacteria in a culture increased fivefold in 8 hours. Assuming continuous growth, how long did it take their number to triple? Give both the exact value and its four-decimal-place approximation.

25. The population of a certain type of fly can be modeled by the function $P(t) = 300e^{0.025t}$, where t is given in days.
    a. How many flies are there initially?
    b. What is the continuous daily growth rate of the population?
    c. How many flies are there after 2 weeks? Give the value to the nearest whole number.
    d. How long will it take for there to be 1000 flies?

26. A colony of 200 bacteria grows at a continuous hourly rate of 1.3%.
    a. Find a function giving the population as a function of t in hours.
    b. How long will it take for the population to double?
    c. When will there be 1000 bacteria? Give both the exact value and its four-decimal-place approximation.
    d. How many bacteria will there be in 1 day? Give the value to the nearest whole number.

27. $A(t) = 500e^{-0.0244t}$ is a function that models the amount (in grams) of radioactive strontium-90 present in a sample after t years.
    a. How much strontium-90 is present initially?
    b. What is the continuous annual decay rate of the sample?
    c. When will only half of the sample remain? Give both the exact value and its four-decimal-place approximation.
    d. How much of the sample will remain after 5 years? Round your answer to four decimal places.

28. A sample of 800 grams of iodine-131 is decaying at a continuous yearly rate of 8.7%.
    a. Find a function giving the amount present as a function of t in years.
    b. How much will be present in 2 years? Round your answer to the nearest whole gram.
    c. How much will be present in 50 years? Round your answer to the nearest whole gram.
    d. How long will it take for there to be only 1 gram remaining? Give both the exact value and its four-decimal-place approximation.

29. The continuous annual rate of decay for carbon-14 is 0.0121%.
    a. Write a function that gives the percentage of carbon-14 remaining after t years.
    b. If a fossil is 6500 years old, how much carbon-14 remains? Round your answer to the nearest whole percent.
    c. If an ancient skull is found to have 31% of its carbon-14 remaining, how old is the skull? Give both the exact value and its four-decimal-place approximation.

30. Most people assume that in a large set of data, the first digits of the numbers will be uniformly distributed—that is, each digit 1 to 9 appears as the first digit in about 1/9 of the numbers. However, in the 1930s, Dr. Frank Benford, a physicist with General Electric, discovered (by looking at the wear pattern on a table of logarithms) that the proportion of numbers having the digit d as the first digit is given by the function $f(d) = \log\left(1 + \frac{1}{d}\right)$. This function is now referred to as Benford's Law, and is used by income tax agencies and insurance companies as a way to detect fraud.

    a. Test Benford's Law by writing down the first digit of each address on a randomly selected page of your local telephone directory. Find the total number of addresses beginning with each digit.
    b. Convert each value into a percentage of addresses beginning with that digit.
    c. A percentage form of Benford's Law is $P(d) = 100 \cdot \log\left(1 + \frac{1}{d}\right)$; this function gives the percentage of a set of data that has each number d (1 through 9) as the first digit. Complete the table to show both the actual percentage of addresses beginning with each digit and the percentage predicted by the function $P(d)$.

DIGIT d	ACTUAL PERCENTAGE OF ADDRESSES HAVING d AS THEIR FIRST DIGIT	PREDICTED PERCENTAGE P OF ADDRESSES HAVING d AS THEIR FIRST DIGIT
1		
2		
3		
4		
5		
6		
7		
8		
9		

    d. Was Benford's Law a good predictor of the actual percentages?

4.3 EXPONENTIAL AND LOGARITHMIC DATA MODELING

In Section 3.4 we analyzed the data in the following table, which shows the U.S. population as recorded during the first six census counts in the decade years 1790, 1800, 1810, 1820, 1830, and 1840.

t	YEAR	U.S. CENSUS POPULATION (MILLIONS)	CHANGE (MILLIONS)
0	1790	3.9	
10	1800	5.3	1.4
20	1810	7.2	1.9
30	1820	9.6	2.4
40	1830	12.9	3.3
50	1840	17.1	4.2

Noting the steadily increasing change figures in the final column, we proceeded to fit the recorded data with the best-fitting *natural growth* model

$$P(t) = 3.9396(1.0300)^t. \tag{1}$$

Now that we know about natural exponentials and logarithms, we can substitute $1.0300 = e^{\ln 1.0300}$ in (1) and write

$$P(t) = 3.9396 \cdot (e^{\ln 1.0300})^t = 3.9396 \cdot e^{(\ln 1.0300)t}.$$

To determine the approximate rate at which this *continuous growth* model is growing, we evaluate $\ln 1.0300 \approx 0.0296$. Hence $P(t) = 3.9396 \cdot e^{0.0296t}$. So from 1790 to 1850, the country was growing at the continuous annual rate of 2.96%.

In general, a natural growth model of the form

$$P(t) = a \cdot b^t \tag{2}$$

is equivalent to the continuous growth model

$$P(t) = a \cdot e^{rt} \tag{3}$$

with continuous growth rate $r = \ln b$. Both the natural growth model and the continuous growth model are referred to as exponential models, since the independent variable t appears as the exponent in the algebraic rule for each function.

We therefore want to revisit here curve fitting for exponential models, but from the viewpoint of the continuous growth model in (3). However, we will see that when we fit data with a graphing calculator, we will need to use first the calculator's **ExpReg** feature to find a natural growth model of the form in (2) and then convert to a continuous growth model of the form in (3) by finding $r = \ln b$. We will refer to such a model as the "best-fitting continuous growth model."

EXAMPLE 1 | Finding a Continuous Growth Model for Data

The data in the following table gives the population of Chula Vista, California, for the indicated years.

YEAR	POPULATION (THOUSANDS)
1970	68
1980	84
1990	135
2003	199

a. Find the best-fitting continuous growth model for these data.
b. Find the average error for this model.
c. Use the model to predict the year in which the population of Chula Vista will reach 250 thousand.

SOLUTION

a. Using the command **ExpReg L1,L2**, we find (as shown in Fig. 4.3.1) that the best-fitting natural growth model is $P(t) = 65.2675(1.0345)^t$, where t is years after 1970.

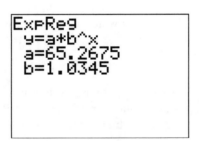

FIGURE 4.3.1 The best-fitting natural growth model.

In order to convert this to a continuous growth model, we find $\ln 1.0345 \approx 0.00339$. So our best-fitting continuous growth model is $P(t) = 65.2675 e^{0.0339t}$.

b. The calculator table in Fig. 4.3.2 gives the predicted values in **L3**, the errors (actual values − predicted values) in **L4**, and the squares of the errors in **L5**.

FIGURE 4.3.2 Calculating SSE.

To find the SSE, we add the values in **L5**, and obtain 107.2270. So the average error is $\sqrt{\dfrac{107.2270}{4}} = 5.1775$ (retaining four decimal places). This means that our predicted population values "miss" the actual population values by an average of a little more than 5 thousand people—a pretty good fit overall.

c. In order to determine the year in which the population grows to 250 thousand, we store $P(t) = 65.2675e^{0.0339t}$ in **Y1** and 250 in **Y2**. Then the **intersect** command gives the screen shown in Fig. 4.3.3, indicating that in 39.6155 years after 1970, the population of Chula Vista is predicted to reach 250 thousand.

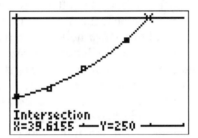

FIGURE 4.3.3 Finding the intersection.

Thus, the population of Chula Vista should reach 250 thousand in the year 2009.

The "Best" Exponential Fit

You may recall that in Section 3.4 we pointed out that the "best-fitting" natural growth model does not always have the smallest possible average error, and we promised to tell you why that is the case. Now that you know about natural logarithms, we can do so.

The calculator's exponential regression technique utilizes the inverse relationship between exponential and logarithmic functions. If the original data set consisting of ordered pairs (x, y) is approximately exponential, then the transformed data set $(x, \ln y)$ is approximately linear. Least squares linear regression then gives the best-fitting (really!) linear model for the transformed data as $\ln y = a + bx$.

If we exponentiate both sides of this linear model equation, we get

$$y = e^{a+bx}.$$

Using the laws of exponents to simplify the right side gives $y = e^a \cdot e^{bx}$. Since a is a constant, e^a is also a constant. Letting $e^a = A$, we have the exponential model $y = A \cdot e^{bx}$ with which we are familiar.

We will continue to refer to the exponential models we obtain using the exponential regression feature of the calculator as the "best-fitting" exponential models—either natural growth or continuous growth, as the situation requires.

Applications of Exponential Models

If $a > 0$ and $r > 0$, then the exponential function

$$P(t) = a \cdot e^{rt}$$

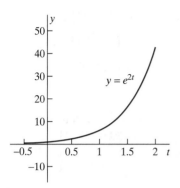

FIGURE 4.3.4 Graph of the exponential growth function $f(t) = e^{2t}$.

with **positive** exponent (for t positive) describes a quantity P that starts with initial value $P(0) = a$ and thereafter *grows* at a steadily *increasing* rate (being multiplied by the factor $b = e^r > 1$ during each unit of time). You should recall from Chapter 1 that an increasing function whose average rate of change is also increasing has a graph that "goes uphill" and curves upward.

Thus, the graph of any such **exponential *growth* function** looks generally like the graph $y = e^{2t}$ (with $a = 1$ and $r = 2$) shown in Fig. 4.3.4 and *rises* with *increasing* steepness (from left to right).

However, if $r < 0$, then the exponential function

$$P(t) = a \cdot e^{rt}$$

with **negative** exponent (for t positive) describes a quantity P that starts with initial value $P(0) = a$ and thereafter *declines* (or *decays*) because the quantity is being multiplied by the factor $b = e^r < 1$ during each unit of time. But what about the average rate of change here? Since the function is decreasing, the average rate of change is negative. Because the size (absolute value) of the changes is getting smaller, the average rate of change is getting closer to 0 or "less negative." So here too the average rate of change is increasing.

We saw in Chapter 1 that a decreasing function whose average rate of change is increasing "goes downhill" and curves upward. The graph of any such **exponential *decay* function** looks generally like the graph $y = 3e^{-t/2}$ (with $a = 3$ and $r = -1/2$) shown in Fig. 4.3.5 and *falls* with *decreasing* steepness (from left to right).

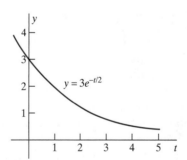

FIGURE 4.3.5 Graph of the exponential decay function $f(t) = e^{-t/2}$.

Just as in Chapter 3, we can discuss the continuous growth and decline of quantities other than populations. The world is full of quantities that appear either to grow at a steadily increasing rate or to decline at a steadily decreasing rate. Any such quantity is fair game for modeling by an exponential function.

In Section 3.4, we discussed **Newton's law of cooling**, which says that the (positive) difference

$$u(t) = T(t) - A \qquad (4)$$

between the temperature A of the surrounding medium and the temperature $T(t)$ of the object at time t is a naturally declining quantity. That is,

$$u(t) = a \cdot b^t \qquad (5)$$

with appropriate values of the positive parameters a and $b < 1$. We can now rewrite this in the (natural) exponential form

$$u(t) = a \cdot e^{rt}, \qquad (6)$$

where $r = \ln b < 0$ (because $b < 1$).

The *heating* of an initially cooler body in a warmer medium at temperature A is similar, except we write [instead of (4)] the difference

$$u(t) = A - T(t) \qquad (7)$$

(the other way round), so that we're still considering a positive quantity. Then **Newton's law of heating** says that this temperature difference $u(t)$ is—just as in the case of cooling—an exponentially decreasing quantity of the same form as in (6). We use either (4) and (7) depending on whether the body is being cooled or heated.

You should not attempt to memorize the fact that (4) corresponds to cooling and (7) corresponds to heating. The easier way to remember "which is which" is this—that in either case $u(t)$ is the **positive** difference between the body's temperature $T(t)$ and the constant temperature A of the surrounding medium.

In the case of a roast being cooked in an oven, for instance, we might measure its rising temperature several times with a thermometer and then attempt to fit the resulting data with an exponential function in order to predict when the roast will be done—that is, hot enough to serve.

EXAMPLE 2 When Is the Roast Ready?

Suppose a 5-pound roast initially at 50°F is placed in a 375°F oven at 5 PM. The temperature of the roast is measured every 15 minutes for the first hour, with the following results.

Time (PM)	5:00	5:15	5:30	5:45	6:00
Temperature (°F)	50	67	82	96	110

When will the roast be 150°F, and hence ready to serve (medium rare)?

SOLUTION

To use Newton's law we must replace the roast temperature T with the difference $u = 375 - T$ that we see in Equation (7), since the oven temperature is $A = 375°F$. Measuring time t in hours after 5:00 PM and subtracting each temperature entry from 375°F, we find that the given data take the form

t	0	0.25	0.5	0.75	1
u	325	308	293	279	265

If we can fit these data with an exponential model $u(t) = a \cdot e^{rt}$, then we should be able to find when the roast will be ready. Since $u(t) = 375 - T(t)$, the temperature of the roast is given by $T(t) = 375 - u(t)$. If we want to serve the roast when it reaches 150°F, then we need to solve the equation $375 - u(t) = 150$.

We begin by entering the list of t-values in **L1** and the list of u-values in **L2**. Then the **STAT CALC** command **ExpReg L1, L2** produces the result shown in Fig. 4.3.6, which describes the best-fitting exponential function

$$u(t) = 324.545 \times 0.816^t.$$

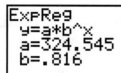

FIGURE 4.3.6 The best-fitting exponential function $u(t) = 324.545 \times 0.816^t = 324.545 e^{-0.203t}$.

Because $\ln 0.816 = -0.203$, this is the same as

$$u(t) = 324.545\, e^{-0.203\, t}.$$

Finally, remembering Equation (7), we subtract $u(t)$ from 375 to obtain the roast's temperature function

$$T(t) = 375 - 324.545\, e^{-0.203\, t}. \qquad (8)$$

It's always important to keep your eye on the roast (ready to pounce when it's ready). Here, literally *no one* cares what is the average error in the optimal exponential model. The only question is when it's ready to eat, that is, when is

$$375 - 324.545\, e^{-0.203\, t} = 150.$$

So we proceed to solve this equation algebraically for t, as follows:

$$324.545\, e^{-0.203\, t} = 375 - 150 = 225$$

$$e^{-0.203\, t} = \frac{225}{324.545}$$

$$-0.203\, t = \ln \frac{225}{324.545} \quad \text{(taking ln of both sides)}$$

$$t = -\frac{1}{0.203} \ln \frac{225}{324.545} \approx 1.805.$$

So it takes 1.805 hours, or about 1 hour and $(0.805)(60) \approx 48$ minutes, for the roast to get done, starting at 5 PM. So we can take it out of the oven and slice it at about 6:48 PM.

If you prefer to solve the equation graphically, store $T(t) = 375 - 324.545\, e^{-0.203\, t}$ in **Y1** and 150 in **Y2**. Then the **intersect** command gives the screen shown in Fig. 4.3.7.

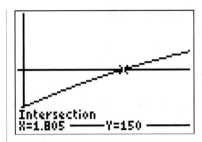

FIGURE 4.3.7 Finding when the roast is done.

The graph indicates that in 1.805 hours after 5 PM—the same bottom line as in our algebraic solution—the roast is ready to eat.

Fitting Logarithmic Models to Data

Figure 4.3.8 shows both the exponential graph $y = e^x$ and the logarithmic graph $y = \ln x$. Both curves rise (from left to right), so both the exponential function and the logarithmic function increase in value as x increases. But the *rates* at which they increase are very different. The exponential graph bends upward and rises ever more steeply, so the exponential function increases more and more *rapidly* as x increases. By contrast, the logarithmic graph bends downward—in the sense that it gets flatter and flatter—so the logarithmic function increases more and more *slowly* as x increases. This contrast often is summarized by saying that

- the exponential function e^x is a rapidly increasing function of x, whereas
- the logarithmic function $\ln x$ is a slowly increasing function of x.

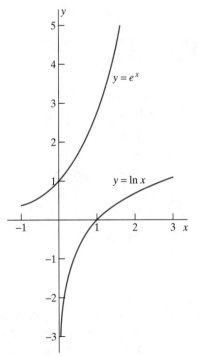

FIGURE 4.3.8 Logarithmic versus exponential growth.

If we have a table of values of either an exponential function or a logarithmic function, we therefore can recognize which it is by looking at the successive differences of the values of the function:

- differences that are getting larger and larger suggest an exponential function, whereas
- differences that are getting smaller and smaller suggest a logarithmic function.

If a logarithmic fit is indicated, we can use our calculator's **LnReg** (**l**ogarithmic **reg**ression) command, as illustrated in the following example.

EXAMPLE 4 — Which Model Is Indicated?

For each table of values given, determine whether an exponential or a logarithmic model is more suited to the data, and use regression to find the indicated model.

a.

x	y = f(x)	CHANGE
1	9.74	
2	18.97	9.23
3	36.95	17.98
4	71.96	35.01
5	140.16	68.20
6	272.99	132.83
7	531.71	258.72

b.

x	y	CHANGE
100	18.82	
200	20.89	2.08
300	22.11	1.22
400	22.97	0.86
500	23.64	0.67
600	24.19	0.55
700	24.65	0.46
800	25.05	0.40
900	25.41	0.35

SOLUTION

a. First, look at the table of values of the function $y = f(x)$. You should notice that the y-values almost double from one to the next, as do the changes recorded in the third column. This is a tip-off that we're looking at a table of values of an exponential function.

Second, apply the calculator **ExpReg** function, which gives the results shown in Fig. 4.3.9.

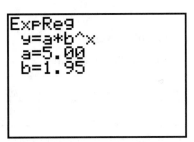

FIGURE 4.3.9 The exponential function of Example 4(a).

Since $\ln 1.95 \approx 0.67$, we find that the function that best fits these data is $y = 5 e^{0.67 x}$. Indeed the y-values in the table are actually the two-decimal-place values of the function $y = 5 e^{2x/3}$, so correct to two decimal places, the calculator got it "spot on."

b. First, look at the table of values of the function $y = f(x)$. You should notice that the y-values are increasing more and more slowly, perhaps as though they're leveling off. In addition, the changes recorded in the third column are decreasing more and more slowly. This is a tip-off that we're looking at a table of values of a logarithmic function.

Second, apply the calculator **LnReg** function (Fig. 4.3.10) to fit the given xy-data with a function of the form $f(x) = a + b \ln x$. Then the command **LnReg L1,L2** gives the result shown in Fig. 4.3.11.

FIGURE 4.3.10 The **LnReg L1, L2** command fits a function of the form $f(x) = a + b \ln x$.

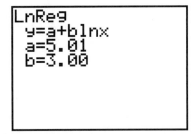

FIGURE 4.3.11 The logarithmic function of Example 4(b).

Thus $y = 5.01 + 3.00 \ln x$. Here the y-values in the table are two-decimal-place values for the function $y = 5 + 3 \ln x$, so once again the calculator found the correct function (with a slight round-off error in the a-coefficient).

EXAMPLE 5 Finding a Logarithmic Model

When your blood pressure is measured, you receive a result of the form "122 over 74" (which lies in the normal range for healthy adults). The larger of the two numbers is your *systolic* blood pressure, recorded in millimeters of mercury. The systolic blood pressure p of a healthy child is lower than that of an adult, and is known to be essentially of the form

$$p = a + b \ln w,$$

where w is the child's weight in pounds.

a. Use the experimental data

w	41	67	78	93	125
p	89	100	103	107	110

to determine the required numerical coefficients a and b.

b. Use your model to find the expected systolic blood pressure of a healthy 55-pound child.

SOLUTION

a. Entering the w-data and p-data in the usual x-list **L1** and y-list **L2** (respectively) and executing the command **LnReg L1,L2,Y1** gives the systolic blood pressure function

$$p = 17.9502 + 19.3775 \ln w$$

(writing w and p in place of x and y). The graph shown in Fig. 4.3.12 (using the window $0 < x, y < 150$) indicates the we have a pretty good fit.

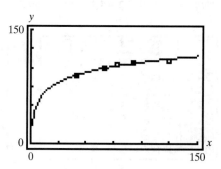

FIGURE 4.3.12 Checking our logarithmic fit with the original data.

b. Substituting $w = 55$ in our function rule gives $p = 17.9502 + 19.38 \ln 55 \approx 95.6023$. Since blood pressures are reported using whole numbers, the expected systolic blood pressure of our healthy 55-pound child is 96.

Perhaps you noticed that in Examples 4 and 5, neither of the tables for the logarithmic models had an input value of 0. There is a very good reason for that—0 is not in the domain of a logarithmic function. This is the case because (as discussed earlier) $\ln 0 = ?$ would mean $e^? = 0$. Since it not possible to replace ? with a real number and get a true sentence, 0 cannot be an output value for an exponential function (of any base whatsoever) and hence 0 cannot be an input value for a logarithmic function. This is important to remember because we have so frequently "reset the clock" in our modeling problems, making the first time input we are given $t = 0$. If your data suggest that a logarithmic function might be indicated, either use the actual input values given or reset the clock so that the first input value represents a positive number.

4.3 Exercises | Building Your Skills

Each of the tables in Exercises 1–6 gives data from a function that is approximately exponential. Use the exponential regression facility of your calculator to find the exponential model of the form $y = a \cdot e^{rx}$ that best fits these points. Find a and r accurate rounded off to three decimal places.

1.

x	0	1	2	3	4
y	5	7	10	15	23

2.

x	1	2	3	4	5
y	35	75	170	385	865

3.

x	1	1.5	2	2.5	3
y	22	23	24	25	27

4.

x	0	1	2	3	4
y	40	29	17	11	7

5.

x	1	3	5	7	9
y	215	40	7	1	0.5

6.

x	0.5	0.75	1	1.25	1.5
y	80	65	55	45	40

Each of the tables in Exercises 7–12 gives data from a function that is approximately logarithmic. Use the logarithmic regression facility of your calculator to find the logarithmic model of the form $y = a + b \ln x$ that best fits these points. Find a and b accurate rounded off to three decimal places.

7.

x	5	10	15	20	25
y	21	28	32	35	37

8.

x	5	10	15	20	25
y	13	10	8	7	6

9.

x	10	20	30	40	50
y	19	24	27	29	30

10.

x	10	20	30	40	50
y	15	10	7	5	4

11.

x	5	15	25	35	45
y	16.0	19.0	20.5	21.5	22.5

12.

x	25	50	75	100	125
y	23	14	9	5	2

Applying Your Skills

In each of Exercises 13–18, the 1970–2000 population census data for a U.S. city is given.

   a. Use a scatter plot of the data to determine whether an exponential model, $P(t) = a \cdot e^{rt}$, or a logarithmic model, $P(t) = a + b \ln t$, is more suited to the data. Then find the optimal function model of the appropriate type; reset the clock so that $t = 0$ in 1960.

   b. Find the average error for your model.

   c. Use your model to predict the city's population in 2010.

13. Akron, Ohio

Year	1970	1980	1990	2000
Population (thousands)	275	237	223	217

14. Charlotte, North Carolina

Year	1970	1980	1990	2000
Population (thousands)	241	315	396	558

15. El Paso, Texas

Year	1970	1980	1990	2000
Population (thousands)	322	425	515	564

16. Miami, Florida

Year	1970	1980	1990	2000
Population (thousands)	335	347	359	362

17. Chesapeake, Virginia

Year	1970	1980	1990	2000
Population (thousands)	90	114	152	199

18. Louisville, Kentucky

Year	1970	1980	1990	2000
Population (thousands)	362	299	269	256

19. The amount $A(t)$ of atmospheric pollutants in a certain mountain valley grows exponentially and was measured (in PU, pollutant units) as follows during the 1990s:

Year	1991	1993	1995	1997	1999
PU	11.7	16.0	21.9	30.0	41.1

   a. Find the best-fitting exponential function of the form $A(t) = A_0 \cdot e^{rt}$.
   b. Suppose it will be dangerous to stay in the valley when the amount of pollutants reaches 100 PU. When will this occur?

20. An accident at a nuclear power plant occurred on February 1, 2000, and left the surrounding area polluted with radioactive material that decays exponentially. The amount still present (in SU, safe units) was measured thereafter at bimonthly intervals, as follows:

Month	April	June	August	October	December
SU	12.8	10.8	9.2	7.8	6.7

   a. Find the best fit of the form $A(t) = A_0 \cdot e^{rt}$ for the amount $A(t)$ of radioactive material (in SU) remaining after t months.
   b. Suppose it will not be safe for people to return to the area until $A = 1$ SU. In what month of what year will this occur?

21. Suppose a roast, initially at 60°F is placed in a 350°F oven at 11 AM. The temperature of the roast is measured every 15 minutes for the first hour, with the following results.

Time	11 AM	11:15 AM	11:30 AM	11:45 AM	12 noon
Temperature (°F)	50	67	82	96	110

   Assuming that $u(t) = 350 - T(t)$ is an exponentially declining function, find the best fit of the form $T(t) = 350 - a \cdot e^{-kt}$. When will the roast be 200°F, and hence ready to serve (medium)?

22. The temperature inside my freezer is $-15°$C and the room temperature in my kitchen is a constant 20°C. At 1 PM one afternoon the power goes off during an ice storm and the freezer's internal temperature $T(t)$ rises as follows for the next few hours:

Time (PM)	1	2	3	4	5
Temperature (°C)	−15.0	−13.3	−11.7	−10.1	−8.7

a. Assuming that $u(t) = 20 - T(t)$ is an exponentially declining function, find the best fit of the form $T(t) = 20 - a \cdot e^{-kt}$.
b. The food in the freezer will begin to thaw once the temperature hits 0°C. When will this occur?

23. During the winter, the amount of heat the body loses depends on both the actual temperature and the speed of the wind. For example, if the actual temperature is 25°F and the wind is blowing at 15 miles per hour, a person feels as cold as she would if the temperature were 2°F with no wind. Meteorologists use the term "wind chill" to describe this phenomenon. The following table relates the wind speed W (in miles per hour) to the wind chill C (in °F) when the actual temperature is 15°F.

W (miles per hour)	5	10	15	35
C (°F)	12	−3	−11	−27

a. Fit these data with a logarithmic function of the form $C = a + b \ln W$.
b. Use your model to predict the wind chill when the wind speed is 45 miles per hour. (Since wind speeds greater than 45 miles per hour have little additional chilling effect, this is, at least theoretically, as cold as you can be when the temperature is 15°F.)

24. The following table gives the estimated life expectancy in the United States as a function of years after 1940.

t (years after 1940)	3	13	43	63
E (years)	63.3	68.8	74.6	77.5

Source: Atlanta Journal-Constitution.

a. Fit these data with a logarithmic function of the form $E(t) = a + b \ln t$.
b. Find $E(20)$ and explain its meaning in this context.
c. In what year does the model predict that the estimated life expectancy will rise to 80 years?

Chapter 4 | Review

In this chapter, you learned about exponential and logarithmic functions and models. After completing the chapter, you should be able to

- Determine whether a relation described numerically, graphically, or symbolically represents an exponential function or a logarithmic function.
- Find the output value of an exponential or logarithmic function for a given input value.
- Find the input value of an exponential or logarithmic function for a given output value.
- Solve an equation or inequality involving an exponential or logarithmic function.
- Find the exponential or logarithmic model that best fits given data.
- Find the continuous rate of growth or decay for an exponential model.

Review Exercises

In Exercises 1–3 the table gives approximate y-values for a function of x. In each case, the function is linear, exponential, or logarithmic. Use the changes in the y-values to determine the type of function.

1.

x	1	2	3	4	5
y	1	1.62	1.99	2.25	2.45

2.

x	1	2	3	4	5
y	1.49	1.98	2.47	2.96	3.45

3.

x	1	2	3	4	5
y	1.14	1.37	2.00	3.73	8.42

In Exercises 4–7, identify each function as exponential, logarithmic, or neither.

4.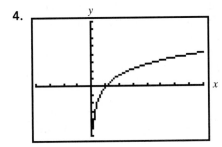

5.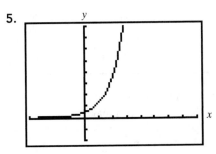

6. $y = 5x^4$

7. $y = 3e^x$

8. An accident at a nuclear power plant has left the surrounding area polluted with a radioactive element that decays naturally. The initial radiation level is 10 times the maximum amount S that is safe, and 100 days later it is still 7 times that amount. What is the continuous annual rate of decay for this element?

9. The value of a 1997 Ford Explorer XLT is given by $V(t) = 26{,}400e^{-0.095t}$, where t represents years after 1997. How much is the Explorer worth in 2002? How long will it take for the value to decline to half of its original value?

10. The following table gives China's trade volume (in billions of U.S. dollars) as a function of year.

t (year)	1980	1985	1995	2004
V (billions of dollars)	38	70	281	1158

Source: Atlanta Journal-Constitution.

a. Find the best-ftting continuous growth exponential model for these data.

b. Use your model to predict the year in which China's trade volume rises to 2000 billion dollars.

11. The following table gives the per capita value of money in circulation in the United States t years after 1990.

t (years)	4	5	6	7
M (thousands)	1428.37	1531.39	1573.15	1664.58

Source: World Almanac and Book of Facts 1998.

a. Find the best-fitting logarithmic function model for these data.

b. Use your model to predict the per capita value of money in circulation in 2015.

12. (Chapter Opener Revisted) In the discussion that introduced this chapter, we looked at a plot that displayed how the center of population (measured to the nearest degree of west longitude) of the United States moved west over the years 1820 to 1940. The plot was based on the data in the following table.

Year	1820	1860	1900	1940
west longitude (degrees)	79	83	86	87

a. Find the best-fitting logarithmic model for these data, with $t = 0$ in 1800.

b. Use your model to predict the west longitude of the center of population in 1950.

c. From 1950 onward, Alaska and Hawaii are included in the calculation of the center. In 1950 the actual center of population was at a longitude of 88 degrees. How does your prediction from part (b) compare to this value?

d. The following table gives the center of the U.S. population for years beginning in 1950.

Year	1950	1970	1980	2000
west longitude (degrees)	88	89	90	91

Make a scatter plot of these data, and use it to determine whether a linear, exponential, or logarithmic model is best suited to fit the data. Then find the best-fitting model of that type.

e. Use your model to predict the year in which the center of population will be at 95 degrees.

INVESTIGATION Interest Rates in the Real World—Buying a Used Car

When you buy a car or a house, your monthly payment is calculated by a method called *amortization*. Amortization is the process of paying off a debt by making a given number of equal payments at specified intervals (usually monthly). These payments include the compound interest. With each payment, the amount of interest declines (as the unpaid balance on the loan declines), while the amount paid toward principal increases.

If equal payments are made monthly, then the payment amount is calculated according to the following formula:

$$\text{payment} = (\text{loan amount}) \times \frac{\text{interest rate}}{12} \times \frac{1}{1 - \left(1 + \frac{\text{interest rate}}{12}\right)^{-12t}},$$

where t is the number of years to repay the loan.

This monthly payment formula looks pretty formidable, but if we use some variables and do a little algebra, it begins to look a bit better. Let P represent the amount borrowed (the *principal*) and m represent the *monthly interest rate* (that is, $m = $ interest rate/12). Then your monthly payment is given by

$$\text{payment} = \frac{P \cdot m}{1 - (1 + m)^{-12t}}.$$

Consider an example. Your rich (and generous) uncle agrees to lend you $3000 at the incredibly low interest rate of 3%, amortized over 2 years. Your monthly interest rate is $0.03/12 = 0.0025$, and your monthly payment is

$$\text{payment} = \frac{(3000)(0.0025)}{1 - 1.0025^{-24}} = 128.94 \quad (\text{dollars}).$$

How do you figure out how much of each payment goes to interest and how much to principal? Each month you must calculate the interest on the current loan balance. If the monthly interest rate is 0.0025 and the initial balance is $3000, then the first month's interest is $7.50. So, of the first payment, only $121.44 is applied toward the principal, leaving a new balance of $2878.56.

Using the monthly interest rate of 0.0025 on this new balance gives a second month's interest of $7.20. So, $121.74 is applied toward principal, leaving a new balance of $2756.82.

If you continue on in this manner, you can construct an *amortization schedule* for your loan. An amortization schedule gives the amount of each payment that goes to interest, the amount that goes to principal, and the new balance after the payment is made. The following table gives the first 4 months of the amortization schedule for this loan.

PAYMENT NO.	INTEREST	PRINCIPAL	NEW BALANCE
			$3000.00
1	$7.50	$121.44	$2878.56
2	$7.20	$121.74	$2756.82
3	$6.89	$122.05	$2634.77
4	$6.59	$122.35	$2512.42

By the time you finish making payments, how much interest do you pay on this loan? If you make 24 payments of $128.94, you repay your uncle $3094.56, so you have paid only $94.56 interest on the loan.

Now suppose that you have found a great deal on the used car of your dreams—a 4-year-old red Mustang convertible, loaded with options, only one-owner, 44,000 miles—for $11,500. You really want to buy this car, and you are hoping that you can afford the payment on a $10,000 loan.

1. If the sales tax in your state is 5%, how much down payment will you need to reduce the loan amount to $10,000?

2. You search the Internet to find a car loan, and find a rate of 7.8%, amortized over 3 years. Construct an amortization table for the first 12 payments on the loan.

3. What is the total cost of the loan repayment? How much interest will you pay on this loan?

4. After more research, you find two different lenders offering the following loans:

   a. 8.16% amortized over 2 years, and
   b. 8.04% amortized over 2½ years.

   If you can afford a car payment of no more than $375.00, can you borrow the money from either of these lenders?

5. Of the loans you can afford, which one is the better deal? What is the total price (including tax and interest) that you will pay for the car?

CHAPTER 5
QUADRATIC FUNCTIONS AND MODELS

5.1 QUADRATIC FUNCTIONS AND GRAPHS

5.2 QUADRATIC HIGHS AND LOWS

5.3 FITTING QUADRATIC MODELS TO DATA

What happens when you toss a ball straight up into the air? Does it travel upward (and then down) with a constant rate of speed? Does it take longer for the ball to reach its highest point than it does to fall back to the point from which it was thrown?

Figure 5.0.1 shows a plot of data obtained in an experiment in which a ball was tossed into the air over a motion detector. The motion detector was attached to a unit that recorded the ball's height (above the motion detector) as a function of time.

What does this graph tell us about the motion of the ball? First, we can see that the ball was thrown upward from a height of about 3 feet and was caught (before it hit the motion detector) at a slightly lower height. Second, the ball was not moving at a constant rate of speed because the distance it traveled from 0.6 second to 0.7 second is much greater than the distance it traveled in the same length of time between 0.9 second to 1.0 second. And, finally, it took (essentially) the same amount of time to travel up as it did to travel down.

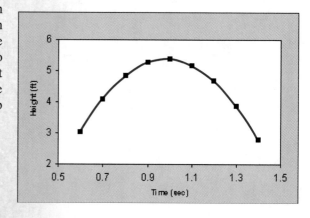

FIGURE 5.0.1 The motion detector measures the height of the ball every tenth of a second.

If we were to construct a mathematical model for the relationship between time and height of the ball, we would surely not choose a linear or an exponential model. It appears that the curve we used to connect our data points is an "upside down" variation of the parabolic graph of the simple quadratic function $f(x) = x^2$. Indeed, the situation of dropping or throwing a ball (or any other object) always yields a quadratic function model of this type.

In this chapter we will study quadratic functions and the relationships for which they provide suitable models.

5.1 QUADRATIC FUNCTIONS AND GRAPHS

Let's take a closer look at a dropping/throwing situation. Suppose a rock is dropped from the top of a tall tower. The following table shows (on the basis of careful measurements) the distance d (in feet) the rock has fallen after t seconds:

t (SEC)	d (FEET)
0	0
1	16
2	64
3	144
4	256
5	400

Note that equal 1-second differences between successive entries in the first column do **not** correspond to equal differences between successive entries in the second column. You might therefore infer that d is not a linear function of t. Indeed, you can verify that d is given as a function of t by the formula

$$d(t) = 16t^2,$$

which involves the *second* (rather than the first) power of the variable t.

> **DEFINITION:** Quadratic Function
>
> A **quadratic function** is a function of the form
>
> $$f(x) = ax^2 + bx + c \qquad (1)$$
>
> with $a \neq 0$, so its formula involves a *square term* as well as linear and constant terms. We call this form the *standard form* of the formula for a quadratic function.

The "squaring function," $f(x) = x^2$, where $a = 1$ and $b = c = 0$ in (1), is the simplest example of a quadratic function. The graph of any quadratic function is a parabola whose shape resembles that of the calculator graph of $f(x) = x^2$ shown in Fig. 5.1.1.

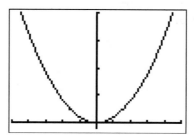

FIGURE 5.1.1 The graph of the squaring function.

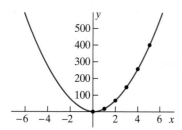

FIGURE 5.1.2 The parabola $y = 16x^2$ and the point $(0, 0)$, $(1, 16)$, $(2, 64)$, $(3, 144)$, $(4, 256)$, and $(5, 400)$.

Indeed, with x and y in place of t and d, the right half of the parabola shown in Fig. 5.1.2 corresponds to the distance function of the falling rock discussed above. The xy-points that correspond to the td-data in the table are marked on the graph.

The parabola $y = -16x^2$ would look similar to the one in Fig. 5.1.2 but would open downward rather than upward. More generally the graph of the equation

$$f(x) = ax^2 \quad (\text{with } a \neq 0) \tag{2}$$

is a parabola with its *vertex* (the point where it appears to change direction) at the origin $(0, 0)$. This parabola opens upward if $a > 0$ and downward if $a < 0$.

The size of the coefficient a in (2) determines the "width" of the parabola. Specifically, the larger $a > 0$ is, the more steeply the curve rises, and hence the narrower the parabola is, as illustrated in Fig. 5.1.3. (If $a < 0$, the larger $|a|$ is, the more steeply the curve falls, and, again, the narrower the parabola is.)

The parabola in Fig. 5.1.4 has its vertex located at the point (h, k) instead of at the origin. In the indicated uv-coordinate system its equation is of the form $v = au^2$, in analogy with Equation (2) but with u and v instead of x and y.

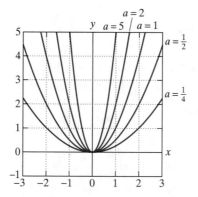

FIGURE 5.1.3 Parabolas with different widths.

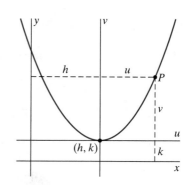

FIGURE 5.1.4 A translated parabola.

It is evident in the figure that the *uv*- and *xy*-coordinates are related by

$$u = x - h, \quad v = y - k.$$

Substitution of these relations in $v = au^2$ then gives the *xy*-equation

$$y - k = a(x - h)^2$$

of this "translated parabola" with vertex (h, k). If we rewrite this equation in "*y*=" form, and then replace *y* with $f(x)$, we obtain the *vertex form* of the formula for a quadratic function:

$$f(x) = a(x - h)^2 + k, \qquad (3)$$

where the coordinates (h, k) of the vertex are easily determined from the function rule.

EXAMPLE 1 Describing the Graph of a Quadratic Function

Describe the graph of the quadratic function

$$f(x) = -2(x + 1)^2 + 3.$$

SOLUTION

Since $a = -2 < 0$, the graph is a parabola that opens downward. Because $|-2| > 1$, the graph is narrower than the graph of the squaring function. In order to determine the vertex, this function rule needs to be written in the vertex form $f(x) = a(x - h)^2 + k$. Since we can replace adding 1 with subtracting -1, the proper vertex form for this function is $f(x) = -2(x - (-1))^2 + 3$. Thus, the vertex of the parabola is the point $(-1, 3)$. Figure 5.1.5 shows the calculator graphs of both the upward-opening squaring function and $f(x)$.

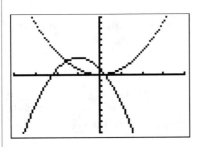

FIGURE 5.1.5 The graphs of the squaring function and $f(x) = -2(x + 1)^2 + 3$.

● Thus we see that we have correctly described the graph of $f(x)$.

What else might we be interested in determining about the graph of a quadratic function $f(x)$? As with the graph of any function, we are usually interested in determining the *y*-intercept (if 0 is in the domain of the function) and any *x*-intercepts.

To find the y-intercept, we merely evaluate $f(0)$. To find any x-intercepts, we must find the x-coordinates of the points on this graph where $f(x) = 0$—that is, we must solve the equation $f(x) = 0$.

The Quadratic Formula

Suppose that our quadratic function $f(x)$ is given in standard form. That is,

$$f(x) = ax^2 + bx + c.$$

Then solving the equation $f(x) = 0$ means solving the quadratic equation

$$ax^2 + bx + c = 0. \tag{4}$$

The **quadratic formula**, which you may have learned in an earlier mathematics course, tells us that this equation has *two* solutions r and s given by

$$r = \frac{-b + \sqrt{b^2 - 4ac}}{2a} \quad \text{and} \quad s = \frac{-b - \sqrt{b^2 - 4ac}}{2a}.$$

This fact—that *both* values $x = r$ and $x = s$ given by these *two* formulas satisfy the given quadratic equation—is what is meant by the quadratic formula written in its usual form

$$x = \frac{-b \pm \sqrt{b^2 - 4ac}}{2a}, \tag{5}$$

which describes both solutions simultaneously.

The key to correct use of the quadratic formula is realizing that a, b, and c in (5) are merely "placeholders," with

- a being the coefficient of the square term,
- b being the coefficient of the first-power term, and
- c being the constant term in the quadratic equation being solved.

EXAMPLE 2   Finding the Intercepts of the Graph of a Quadratic Function

Find the intercepts of the graph of the quadratic function

$$f(x) = -2(x + 1)^2 + 3.$$

SOLUTION

Since any real value of x will give a sensible value for $f(x)$, the domain of f is all real numbers. So $f(0)$ exists, and we can find the y-intercept by evaluating

$$f(0) = -2(0 + 1)^2 + 3 = -2 + 3 = 1.$$

The y-intercept, then, is 1.

To find any x-intercepts, we need to solve the equation $-2(x + 1)^2 + 3 = 0$. In order to do so with the quadratic formula, the quadratic expression must be in standard form, not vertex form. So, expanding the squared binomial and simplifying, we have

$$-2(x^2 + 2x + 1) + 3 = 0,$$

$$-2x^2 - 4x - 2 + 3 = 0,$$
$$-2x^2 - 4x + 1 = 0.$$

Then with $a = -2$, $b = -4$, and $c = 1$, the quadratic formula gives

$$x = \frac{-(-4) \pm \sqrt{(-4)^2 - 4(-2)(1)}}{2(-2)}$$

$$= \frac{4 \pm \sqrt{16 + 8}}{-4}$$

$$= \frac{4 \pm \sqrt{24}}{-4}$$

$$= \frac{4 \pm 2\sqrt{6}}{-4}.$$

So the two solutions of our quadratic equation $f(x) = 0$ are

$$\frac{4 + 2\sqrt{6}}{-4} = -1 - \frac{\sqrt{6}}{2} \approx -2.2 \quad \text{and} \quad \frac{4 - 2\sqrt{6}}{-4} = -1 + \frac{\sqrt{6}}{2} \approx 0.2.$$

Hence the graph $y = f(x)$ intersects the x-axis at the approximate x-values -2.2 and 0.2. If you check the graph of f shown in Fig. 5.1.5, you will see these x-intercepts along with the y-intercept at $x = 1$.

Because the solutions of the quadratic equation $ax^2 + bx + c = 0$ are the x-intercepts of the parabola graph of $f(x) = ax^2 + bx + c$, we see that a parabola can intersect the x-axis

- at *two points*, as in Fig. 5.1.6(a), where the parabola dips beneath the x-axis, or
- at *a single point*, as in Fig. 5.1.6(b), where it just touches the x-axis, or
- at *no points*, as in Fig. 5.1.6(c), where it lies completely above the x-axis.

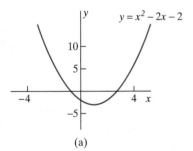
(a)

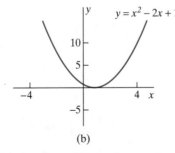
(b)

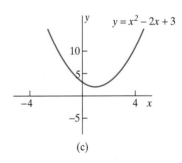
(c)

FIGURE 5.1.6 (a) Two x-intercepts; (b) one x-intercept; (c) no x-intercept.

These three geometric cases correspond to three algebraic possibilities—depending on whether the quantity $b^2 - 4ac$ under the radical in (5) is

- *positive*, so there are two different solutions of the quadratic equation;
- *zero*, so the equation has only one solution;
- *negative*, so the equation has no real solutions at all.

(If $b^2 - 4ac = 0$, then whether we add or subtract the radical, we get the same thing. If $b^2 - 4ac$ is negative, then the radical represents the square root of a negative number, and thus is an imaginary number, which means the equation has no real solutions.)

The **factor theorem** of elementary algebra says that *if* the quadratic equation $ax^2 + bx + c = 0$ has solutions r and s, *then* the quadratic function $f(x) = ax^2 + bx + c$ factors as

$$f(x) = a(x - r)(x - s). \tag{6}$$

We call this the *factored form* of a quadratic function. Here the x-intercepts r and s of the parabola are easily determined from the function rule.

EXAMPLE 3 — Finding the Factored Form of a Quadratic Function

Find the factored form of $f(x) = x^2 - 10x - 119$.

SOLUTION

You may have factored quadratic expressions in a previous course; if you recognize the factors here, good for you! But (just for the practice) we'll assume you don't. We will use the quadratic formula to find the solutions of the equation $x^2 - 10x - 119 = 0$, and then use the factor theorem to write the factored form.

With $a = 1$, $b = -10$, and $c = -119$, the quadratic formula gives the two solutions

$$r, s = \frac{-(-10) \pm \sqrt{(-10)^2 - 4(1)(-119)}}{2(1)}$$

$$= \frac{10 \pm \sqrt{100 + 476}}{2}$$

$$= \frac{10 \pm \sqrt{576}}{2}$$

$$= \frac{10 \pm 24}{2}$$

$$= \frac{34}{2}, \frac{-14}{2} \quad \text{(taking the two signs separately)}$$

$$r, s = 17, -7$$

We write r, s here to emphasize that we're calculating two different solutions simultaneously. In plain language, the quadratic equation $x^2 - 10x - 119 = 0$ has the two solutions $x = 17$ and $x = -7$.

We all make algebraic mistakes from time to time. (The authors of this textbook have made more algebraic mistakes than any student reader has ever thought about making, but we hope to have found them before the book was printed.)

Therefore, you should always check your algebraic solutions. Substitute them into the equation to verify that they really do satisfy it:

$$(17)^2 - 10(17) - 119 = 289 - 170 - 119 = 0$$
$$(-7)^2 - 10(-7) - 119 = 49 + 70 - 119 = 0$$

So the factors of f are $(x - 17)$ and $(x - (-7)) = (x + 7)$, and the quadratic function $f(x) = x^2 - 10x - 119$ can be written in factored form as $f(x) = (x - 17)(x + 7)$. ●

Most of the quadratic equations we'll see in applications will have two real solutions that can be found either

- *algebraically*—by using the quadratic formula,
- *graphically*—by locating the intersection points of a parabola and either the *x*-axis or some other horizontal line, or
- *numerically*—by using a table of function values.

We have done a great deal of just plain (boring?) algebra here so that you will once again have a choice of methods when you solve application problems involving quadratic functions. Many students are familiar with the algebraic methods shown; others prefer to use the graphical or numerical methods we used extensively in previous chapters. The choice is yours, but you should practice whichever method you like best until you can use it easily and reliably.

Quadratic Population Models

In Section 2.4 we analyzed the 1950–1990 population of Charlotte, and found that it is reasonably "well fitted" by a linear model of the form

$$P(t) = P_0 + at, \qquad (7)$$

where P_0 is the initial population (at time $t = 0$) and a is the annual change in the population. But a linear model is appropriate only when the rate of change is *constant* (or at least approximately so). The following table displays the 1950–1990 census data for Austin, Texas:

YEAR	t (YEARS AFTER 1950)	POPULATION (THOUSANDS)	CHANGE DURING DECADE (THOUSANDS)
1950	0	132	55
1960	10	187	67
1970	20	254	92
1980	30	346	120
1990	40	466	—

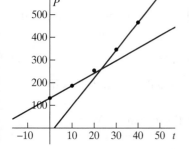

FIGURE 5.1.7 No single straight line appears to fit the Austin population data.

The final column shows the population change during the decade until the next census year. Observe that the population was *not* increasing at a constant rate. Indeed, it appears that the rate of change *itself* was increasing throughout the 1950–1990 period. Figure 5.1.7 shows two lines, one through

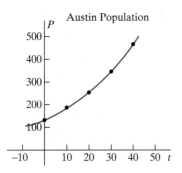

FIGURE 5.1.8 The Austin population data are better fitted by the parabola $P(t) = 132 + 3.84t + 0.113t^2$.

the points (0, 132) and (10, 187) and the second through the points (30, 346) and (40, 466).

Neither of these lines appears to fit the Austin population data points well. The first line—though it passes through the 1950 and 1960 data points—"rides below" the last three data points. Though the second line contains the 1980 and 1990 data points, it lies below the first three data points. It appears that no single straight line fits well all five of Austin's 1950–1990 population data points.

Perhaps a close inspection of Fig. 5.1.7 will suggest to you that any smooth curve drawn through all five Austin data points must "bend upward" like a parabola does. Indeed, Fig. 5.1.8 shows how well the quadratic function

$$P(t) = 132 + 3.84t + 0.113t^2 \qquad (8)$$

appears to fit the actual 1950–1990 population growth of Austin. (In Section 5.3 we will talk about how to find such a quadratic model, but here we take it as given.)

Equation (8) is an example of the general **quadratic population model**

$$P(t) = P_0 + bt + at^2. \qquad (9)$$

If the coefficients are all positive, this model describes a population that—roughly speaking—grows more rapidly as t increases than any linear function. Note that (9) differs from the linear model (7) only in the inclusion of a t^2-term.

Previously we wrote the typical quadratic function in the standard form

$$f(x) = ax^2 + bx + c \qquad (10)$$

and observed that, if the coefficient of the square-term in (9) is positive, then the graph $y = f(x)$ is a parabola opening upward. The function $P(t)$ in (9) is a quadratic function of t, while the function $f(x)$ in (10) is a quadratic function of x. The coefficients P_0, a, and b in (9)—and the coefficients a, b, and c in (10)—denote numerical constants, and these constants can be denoted by any letters we choose. We denote the constant term in the quadratic population model by P_0 simply because substitution of $t = 0$ yields $P(0) = P_0$.

EXAMPLE 4 | Using a Quadratic Population Model

The population of Austin t years after January 1, 1950, is described (in thousands) by the quadratic model

$$P(t) = 132 + 3.84t + 0.113t^2. \qquad (11)$$

a. What is the y-intercept of the graph of this function? What does it represent in terms of the population of Austin?

b. Are there any x-intercepts of the graph of this function? What do x-intercepts represent in terms of the population of Austin?

c. According to this model, what is the predicted population of Austin in 2010?

d. In what month of what calendar year does the population of Austin reach 400 thousand?

SOLUTION

a. The *y*-intercept of the graph is the value $P(0) = 132 + 3.84(0) + 0.113(0)^2 = 132$. This value represents the population of Austin when $t = 0$, that is, in 1950.

b. The *x*-intercepts of the graph are values of t for which $P(t) = 0$. Let's use the quadratic formula to find any such values algebraically, and then verify our result by looking at a graph of the function.

Writing $P(t)$ in standard form as $P(t) = 0.113t^2 + 3.84t + 132$, we see that $a = 0.113$, $b = 3.84$, and $c = 132$. Thus,

$$t = \frac{-3.84 \pm \sqrt{3.84^2 - 4(0.113)(132)}}{2(0.113)}$$

The expression under the radical gives us pause here. If we calculate $4(0.113)(132)$, we get 59.664. Since 3.84^2 must be somewhat less than $4^2 = 16$, $b^2 - 4ac$ is negative, and there are no *x*-intercepts of this graph. Figure 5.1.9 shows the graph of $P(t)$ in the window $-100 \le x \le 100$, $-100 \le y \le 600$, and confirms this conclusion.

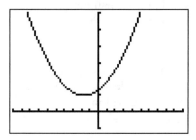

FIGURE 5.1.9 The graph of the Austin population function.

If there *were* *x*-intercepts, they would represent the years after 1950 in which the population of Austin was 0 thousand people (unlikely, barring some disaster).

c. Here we are asked for the population when $t = 60$, that is, the output for an input of 60. So $P(60) = 132 + 3.84(60) + 0.113(60)^2 = 769.2$. Thus, the predicted population of Austin in 2010 is 769.2 thousand or 769,200 people.

d. We need to find the value of t when $P(t) = 400$, that is, so that

$$132 + 3.84x + 0.113x^2 = 400 \tag{12}$$

We could rewrite this equation in standard form and use the quadratic formula to solve it. However, the arithmetic involved is likely to be quite messy (even with the calculator). So we will use a graphical solution instead, as we have done in previous chapters.

With *x* in place of *t*, Fig. 5.1.10 shows the calculator graphs of the parabola **Y1 = 132+3.84X+0.113X^2** and the horizontal line **Y2 = 400** (in the same window as before).

One of the intersection points in Fig. 5.1.10 has a negative *x*-coordinate and the other has a positive *x*-coordinate. Hence Equation (12) has both a negative solution and a positive solution. However, because $t = 0$ represents January 1, 1950, the negative solution $t < 0$ would be a long time in the past rather than in the future. Thus we are interested in only the positive solution of the equation.

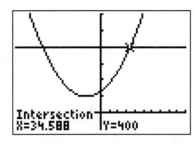

FIGURE 5.1.10 Finding when the population of Austin is 400 thousand.

Figure 5.1.10 indicates that we have already used the graphing calculator's intersection-finding feature to locate the intersection point (34.588, 400) on the right. Thus, the positive solution is

$$t = 34.588 \text{ years}$$
$$= 34 \text{ years} + (0.588 \times 12) \text{ months}$$
$$= 34 \text{ years} + 7.056 \text{ months}.$$

According to our function model, Austin's population reached 400 thousand early in August 1984.

While we have used a population example here to illustrate a quadratic function model, such models apply to many other situations as well. In the exercises for this section, you will investigate similar models in which we have "reset the clock" to make your calculations—graphical, numerical, or symbolic—easier.

Quadratic Models for Projectile Motion

Just solving quadratic equations by themselves—though sometimes necessary for practice—can seem like pretty dry stuff. But such equations appear in real-world problems.

For instance, suppose an object like a ball or a bullet is thrown or fired at time $t = 0$ from a point h_0 feet above the ground. The object (ball, book, bullet whatever) is referred to as a "projectile," and so its movement is called "projectile motion."

If the projectile's initial velocity is v_0 feet per second (ft/sec), then (neglecting air resistance) its height h in feet above the ground after t seconds is given by the formula

$$h(t) = -16t^2 + v_0 t + h_0, \qquad (13)$$

as illustrated in Fig. 5.1.11.

The initial velocity v_0 is positive if the object is thrown or shot upward and negative if the object is thrown or shot downward. We take $h_0 = 0$ if the object's initial position is on the ground. (Notice that here, as with our population functions, the subscript 0 indicates an initial value.)

Everyone knows that what goes up must come back down. If the values of h_0 and v_0 are given, then Equation (13) is all we need in order to find out how long the ball stays in the air before returning to the ground.

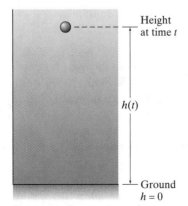

FIGURE 5.1.11 The height function for an object moving vertically.

EXAMPLE 5 Describing an Arrow's Height

Suppose an arrow is shot straight upward with an initial velocity of 160 ft/sec from a point on the ground beside a building 256 feet high.

a. How high is the arrow after 5 seconds?
b. When does it pass the top of the building on the way up? On the way back down?
c. How long is the arrow in the air before it returns to the ground?

SOLUTION

a. With an initial velocity of 160 ft/sec and an initial height of 0, the arrow's height in feet t seconds after it is shot is given by

$$h(t) = -16t^2 + 160t. \tag{14}$$

Then substitution of $t = 5$ in (14) gives

$$h(5) = -16(5)^2 + 160(5) = 400 \text{ feet}$$

for the arrow's height after 5 seconds.

b. If we set $h(t) = 256$ in (14) we find that the arrow is 256 feet high when

$$-16t^2 + 160t = 256.$$

We could use the quadratic formula, but this equation is reasonably easy to solve by factoring:

$$-16t^2 + 160t - 256 = 0$$
$$-16(t^2 - 10t + 16) = 0$$
$$-16(t - 2)(t - 8) = 0.$$

(Even if factoring like this is not easy for you, it should be easy to multiply out and verify that the indicated factorization is correct.) The two solutions $t = 2$ and $t = 8$ are the times when the arrow is 256 feet high. So it passes the top of the building on the way up after 2 seconds and on the way back down 8 seconds after it was shot upward.

Figure. 5.1.12 shows a graphic solution for the time the arrow passes the top of the building on the way back down.

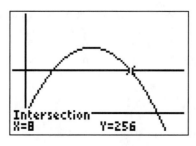

FIGURE 5.1.12 When the ball passes the top of the building on the way down.

c. Note that the arrow's height is again zero when the arrow returns to ground level. Therefore, the time t at which it hits the ground satisfies the equation

$$h(t) = -16t^2 + 160t = -16t(t - 10) = 0,$$

or when $t = 0$ or when $t = 10$. But $t = 0$ is when it was originally shot upward, so the arrow remains in the air for 10 seconds.

5.1 Exercises | Building Your Skills

In each of Exercises 1–4, a brief table of values of a function $f(x)$ is given. Use the function's average rate of change to determine whether it is a linear or quadratic function. Verify your answer by constructing a scatter plot of the data.

1.

x	-3	-1	2	5	7
$f(x)$	11	-3	-9	3	21

2.

x	-4	-2	1	2	5
$f(x)$	13	7	-2	-5	-14

3.

x	-5	-1	1	3	4
$f(x)$	-13	-5	-1	3	5

4.

x	-3	-1	0	2	4
$f(x)$	-35	5	10	-10	-70

In Exercises 5–8, determine the value of the coefficient a so that the parabola $y = ax^2$ passes through the point P with given coordinates.

5. $P(3, 18)$

6. $P(-2, 24)$

7. $P(4, -64)$

8. $P(-5, -250)$

In Exercises 9–14, match the given quadratic function $f(x)$ with its graph among those shown in Figs. 5.1.13–5.1.18. Don't use a graphing calculator. Instead, use the information that can be easily determined from the function rule, such as the vertex, the x- and y-intercepts, and whether the parabola opens upward or downward.

9. $f(x) = x^2 + 1$

10. $f(x) = 4 - x^2$

11. $f(x) = -x(x + 2)$

12. $f(x) = x(x - 2)$

13. $f(x) = (x - 1)^2 - 4$

14. $f(x) = -(x + 1)^2 + 4$

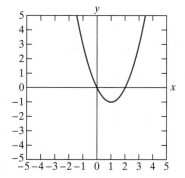
FIGURE 5.1.13

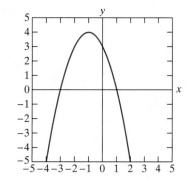
FIGURE 5.1.14

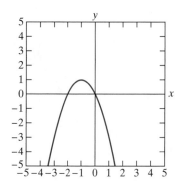
FIGURE 5.1.15

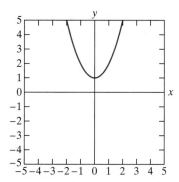
FIGURE 5.1.16

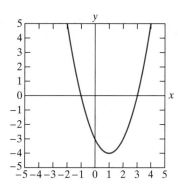
FIGURE 5.1.17

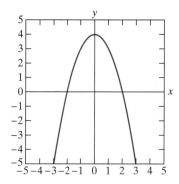
FIGURE 5.1.18

15. Determine a value of c such that the equation $x^2 - 10x + c = 0$ has

  a. two solutions;
  b. exactly one solution;
  c. no solutions.

Suggestion: Use your graphing calculator. Start by constructing graphs with different values of c.

Applying Your Skills

In Exercises 16 to 19, the given quadratic population model predicts the population (in thousands) of a city t years after January 1, 1997. Find how long it takes this city's population to reach the given level P_1. Give your answer in the form m years and n days. (Assume exactly 365 days per year.)

16. $P(t) = 0.25t^2 + 5t + 100; \quad P_1 = 200$

17. $P(t) = 0.3t^2 + 6t + 150; \quad P_1 = 250$

18. $P(t) = 0.7t^2 + 12t + 200$; $P_1 = 350$

19. $P(t) = 1.5t^2 + 21t + 300$; $P_1 = 500$

20. The population of Montgomery, Alabama, from 1970 through 2003 can be modeled approximately by the quadratic function $P(t) = 135.18 + 4.37t - 0.07t^2$, where P is measured in thousands and t is years after 1970.

    a. According to the model, what was the population of Montgomery in 1970?
    b. When does the model predict that the population of Montgomery will fall to half its 1970 population?
    c. What is Montgomery's predicted population in 2010?

21. The population of Omaha, Nebraska, from 1970 through 2003 can be modeled approximately by the quadratic function $P(t) = 344.86 - 4.49t + 0.20t^2$, where P is measured in thousands and t is years after 1970.

    a. According to the model, what was the population of Omaha in 1970?
    b. When does the model predict that the population of Omaha will be double its 1970 population?
    c. What is Omaha's predicted population in 2010?

22. The typical distance (in feet) that a car travels after the brakes are applied is given by the quadratic function $D(s) = 0.05s^2 + s$, where s is the car's speed in miles per hour.

    a. Find any intercepts of the graph of this function, and explain their meaning in terms of the situation.
    b. If a car is traveling at 60 miles per hour, how far does the car travel after the brakes are applied?
    c. If a car travels 120 feet after the brakes are applied, how fast was it traveling?

23. According to data reported by the *Statistical Abstract 2005*, the percentage of U.S. households leasing a vehicle for personal use in the years 1992–2001 can be modeled by the quadratic function $L(t) = -0.061t^2 + 0.903t + 2.76$, where t is years after 1992.

    a. According to the model, what is the percentage of households leasing a vehicle for personal use in 1998? How does your value compare to the actual value of 6.4%?
    b. According to the model, in what year(s) does 5% of U.S. households lease a car for personal use?
    c. Use the model to determine when the percentage of households leasing a car for personal use is predicted to fall below 2%.

24. According to data reported by the *Atlanta Journal-Constitution*, the Hispanic unemployment rate in the United States from 2000 through 2005 can be modeled by the quadratic function $U(t) = -0.284x^2 + 1.503x + 5.596$, where U represents the percentage unemployed and t is years after 2000.

    a. According to the model, what is the percentage of unemployed in 2002? How does your value compare to the actual value of 7.5%?
    b. According to the model, what is the percentage of unemployed in 2010? Do you think this is a good estimate of Hispanic unemployment in 2010?
    c. Use the model to determine when the Hispanic unemployment rate is predicted to fall below 3%.

25. A powerful crossbow fires an arrow upward from a tree stand. The height of the arrow (in feet) is given by the quadratic function $h(t) = -16t^2 + 240t + 10$, where t is given in seconds.

   a. What is the initial velocity of the arrow?
   b. How high is the tree stand?
   c. How long does the arrow stay in the air?

26. A ball is thrown straight upward from the ground beside a 100-foot tree. Its initial velocity is $v_0 = 96$ ft/sec.

   a. When does it pass the top of the tree on the way up?
   b. When does it pass the top of the tree on the way down?
   c. How long does the ball stay in the air?

27. A ball is thrown or dropped from the top of the Empire State Building, which is 960 feet tall.

   a. If it is simply dropped (with an initial velocity of 0 ft/sec), how long does it take to hit the ground?
   b. If it is thrown downward at 50 ft/sec, how long does it take to hit the ground?
   c. If it is thrown upward at 50 ft/sec, how long does it take to hit the ground?

28. In each part of Exercise 25, when does the ball pass the 50th floor (480 feet above the ground) on its way down?

29. Cezar drops a stone into a well, and it hits the water at the bottom after 3 seconds. How deep is the well?

30. Sydney drops a rock into a well in which the water surface is 300 feet below ground level. How long does it take the rock to hit the water at the bottom?

31. This problem is a personal challenge. Suppose the population of Your City t years after January 1, 1998 is given (in thousands) by

$$P(t) = 100 + kt + \left[\frac{k}{10}\right]^2 t^2,$$

where k is the largest digit in your student ID number. On what calendar day of what year will the population of Your City reach 200 thousand? For the purpose of this problem you can ignore leap years and assume that every year has 365 days, but you will need to take into account the differing number of days in a month.

5.2 QUADRATIC HIGHS AND LOWS

"Buy low, sell high" is the goal of those who invest in the stock market. But how does an investor know when the highs and lows will occur? Both people and functions have highs and lows, and in either case it can be important to know when and where they occur.

By looking at the graph of a function $y = f(x)$ on an interval $a \leq x \leq b$ of x-values we can usually spot its highest or lowest point visually, but determining the function value there may require a bit of guesswork. In the case of the quadratic functions we study in this chapter, the solution is fairly straightforward.

The graph of a quadratic function is a parabola; the parabola opens upward if the coefficient of the square term is positive, and it opens downward if the coefficient of the square term is negative. If the parabola opens upward, then it has a lowest point but no highest point. If the parabola opens downward, then it has a highest point but no lowest point. In either case this "extreme point" is the vertex of the parabola.

In the last section we looked at quadratic functions in various forms—standard, vertex, and factored. If we are fortunate enough to have the quadratic function in vertex form, the task of finding a maximum or minimum value for the function (as the case may be) is an easy one.

EXAMPLE 1   Finding Maximum or Minimum Function Values

Find the maximum or minimum value for each of the following quadratic functions and the x-value at which it occurs:

a. $f(x) = -2(x - 4)^2 + 1$

b. $g(x) = 0.5(x + 1)^2 + 7$

SOLUTION

a. The graph of $f(x) = -2(x - 4)^2 + 1$ is a parabola that opens downward, so it has a maximum value but no minimum value. Since the function is in vertex form, we can see that the vertex of the parabola is $(4, 1)$. Thus, the maximum value of the function is 1; this maximum y-value occurs when $x = 4$.

b. The graph of $g(x) = 0.5(x + 1)^2 + 7$ is an upward-opening parabola, so it has a minimum value but no maximum value. Here our function is not quite in vertex form; vertex form requires a subtraction inside the binomial and an addition on the outside. But when we rewrite the function as $g(x) = 0.5(x - (-1))^2 + 7$, we see that the vertex of the parabola is $(-1, 7)$. Therefore, the minimum value of $g(x)$ is 7, and this value occurs when $x = -1$.

If the function is not in vertex form, but its coefficients are "nice" (integers, for example), then the task of locating high and low points is only a bit more difficult. If the quadratic function is given in standard form

$$f(x) = ax^2 + bx + c, \qquad (1)$$

then the vertex of its parabola graph has the x-coordinate $-\dfrac{b}{2a}$. But the vertex of the parabola is its high or low point. Therefore, the maximum or minimum value of the quadratic function (depending on whether a is positive or negative) is simply $f\left(-\dfrac{b}{2a}\right)$.

Thus to find the maximum or minimum value of a standard-form quadratic function given as in (1), you need only remember that the x-coordinate of its graph is $-\dfrac{b}{2a}$. But if you were to forget this, you could work it out by completing the square to put the quadratic function in vertex form:

$$f(x) = ax^2 + bx + c$$
$$= a\left(x^2 + \frac{b}{a}x\right) + c$$
$$= a\left(x^2 + 2\frac{b}{2a}x + \frac{b^2}{4a^2}\right) + c - \frac{b^2}{4a}$$
$$= a\left(x + \frac{b}{2a}\right)^2 + \left(c - \frac{b^2}{4a}\right)$$
$$f(x) = a\left(x - \left(-\frac{b}{2a}\right)\right)^2 + \left(c - \frac{b^2}{4a}\right).$$

You can now see why the vertex x-coordinate is $-b/2a$. (Can't you?) And you would surely rather just remember this rather than derive it every time. (Wouldn't you?)

EXAMPLE 2 — Finding a Minimum Function Value

Find the minimum value of $f(x) = x^2 - 6x + 7$.

SOLUTION

Since $f(x)$ is (essentially) in standard form, with $a = 1$, $b = -6$, and $c = 7$, then the x-coordinate $-b/2a$ of the vertex is

$$x = -\frac{(-6)}{2(1)} = -(-3) = 3.$$

This is not the minimum value of the function, but rather it is the x-value at which the minimum occurs. The minimum function value is then $f(3) = 3^2 - 6(3) + 7 = -2$.

The result found in Example 2 can be stated in a couple of different ways:
- The *lowest point* on the parabola $y = x^2 - 6x + 7$ is the point $(3, -2)$.
- The *minimum value* of the function $f(x) = x^2 - 6x + 7$ is the number $f(3) = -2$.

Thus we distinguish between the minimum value of the function $f(x)$ and the lowest point on the graph $y = f(x)$. The minimum value of the function is the y-coordinate of the low point, and the x-coordinate of the low point is the input that gives the minimum value as the corresponding output.

Example 2 provides a symbolic method that will always locate the vertex of a parabola. However, when applications have quadratic function models, the coefficients are seldom so easy to work with. In such cases, we may prefer to use a graphical method, as illustrated in Example 3, using the same function as Example 2.

EXAMPLE 3   Locating the Vertex of a Parabola

Use a graphical approach to locate the vertex of the graph of $f(x) = x^2 - 6x + 7$.

SOLUTION

Figure 5.2.1 shows the graph $y = x^2 - 6x + 7$ plotted in the standard window $-10 \leq x, y \leq 10$. The parabola opens upward from the low point that's visible. So it's just a matter of zooming in to see exactly where this low point lies. At first glance, it looks to be near the point $(3, -2)$. Figure 5.2.2 shows the graph in the window $0 \leq x \leq 6, -3 \leq y \leq 3$, and now it looks like it really is *exactly* the point $(3, -2)$.

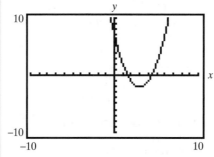

FIGURE 5.2.1 The parabola $y = x^2 - 6x + 7$ in the window $-10 \leq x, y \leq 10$.

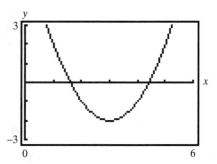

FIGURE 5.2.2 The parabola in the window $0 \leq x \leq 6, -3 \leq y \leq 3$, where it looks like the low point is $(3, -2)$.

If we didn't already know the exact answer, how could we be sure the low point is not $(3.002, -2.001)$ instead? Such a tiny difference from $(3, -2)$ might not be visible on a graph. Fortunately, the **CALC** menu on our calculator offers **minimum** and **maximum** functions that give us a better approximation than we can get visually.

With **Y1=x²−6x+7**, we select **minimum** from the **CALC** menu. We are prompted to first enter a *left bound*, then a *right bound*, and finally a *guess*. We move the cursor to the left of the apparent low point and press **ENTER** to record our left bound (Fig. 5.2.3), then to the right of the low point (and hit **ENTER**) to record our right bound, then back to the low point itself (or near it) for our guess. When we hit **ENTER** once more, the calculator screen in Fig. 5.2.4 results.

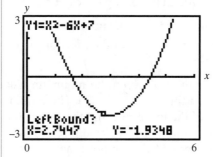

FIGURE 5.2.3 Entering a left bound.

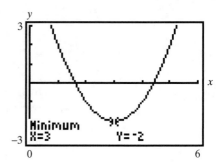

FIGURE 5.2.4 The low point on the graph has coordinates $x = 3$ and $y = -2$.

● We see that $(3, -2)$ is, indeed, the best guess for the low point on the parabola.

When, as in Example 3, your calculator reports integer values for the coordinates of your maximum or minimum point, it is a reasonable bet that you have found the exact answer. If your calculator reports coordinates like $x = 2.9999993$ and $y = -2.0000001$, you may have found approximate answers. To be on the safe side, however, you should consider the answers found by the **minimum** and **maximum** functions to be very good approximations but not necessarily exact values.

EXAMPLE 4 Describing the Motion of a Ball

Suppose a baseball is thrown upward with velocity 45 miles per hour (mph) from the top of a tower 40 feet tall.

a. How long does it take the ball to reach its maximum height?
b. How high in the air does the ball go before starting back downward?
c. How long is the ball in the air before it hits the ground?

SOLUTION

The initial velocity expressed in miles per hour probably means more to most people than one given in feet per second. However, our function rule from Section 5.1,

$$h(t) = -16t^2 + v_0 t + h_0, \tag{2}$$

only applies if we express all quantities in feet and seconds. For this purpose we use the handy conversion factor

$$60 \text{ mph} = 88 \text{ ft/sec},$$

which (for whatever reason) is sometimes mentioned in driver's license exam study booklets. We then see that our baseball's initial velocity in feet per second is

$$v_0 = 45 \text{ mph} \times \frac{88 \text{ ft/sec}}{60 \text{ mph}} = \frac{45 \times 88}{60} \frac{\text{ft}}{\text{sec}} = 66 \frac{\text{ft}}{\text{sec}}.$$

Then substitution of $v_0 = 66$ (ft/sec) and $h_0 = 40$ (ft) in (2) yields the ball's height function

$$h(t) = -16t^2 + 66t + 40. \tag{2}$$

In principle, this function tells everything there is to know about the ball's ascent and descent. In particular, it's the key to answering questions (a), (b), and (c).

a. Figure 5.2.5 shows the result we get when we define **Y1 = −16X^2 + 66X + 40** and use the calculator's **CALC maximum** facility to locate

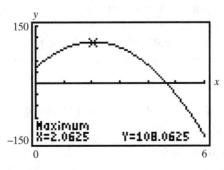

FIGURE 5.2.5 Graph of the baseball's height function

the visible high point. The x-coordinate of the high point marked on the graph gives the approximate time $t = 2.0625$ seconds that it takes the ball to reach its maximum height.

b. The y-coordinate of the high point in Fig. 5.2.5 indicates that the ball's approximate maximum height is 108.0625 feet (above the ground, not above the top of the tower from which it was thrown).

c. To find how long the ball is in the air (another way of asking when it hits the ground), we need to find when $-16t^2 + 66t + 40 = 0$. There are several methods for doing this:

- algebraically, using the quadratic formula,
- numerically, zooming in on a table of values of **Y1**,
- graphically, using the **CALC intersect** command (after entering **Y2 = 0)**, or
- graphically, using the **CALC zero** command.

The **CALC zero** command finds an x-intercept and requires the same kind of entries as the **CALC minimum** or **maximum**—a left bound, a right bound, and a guess. Figure 5.2.6 shows that $y = 0$ when $x = 4.6613$.

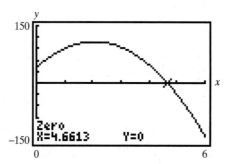

FIGURE 5.2.6 The baseball's impact with the ground.

Thus the ball is in the air for a total of approximately $t = 4.6613$ seconds.

Frequently we need to know the maximum or minimum value attained by a given function $f(x)$ for values of x in a *given interval* $a \leq x \leq b$ (rather than over its whole domain). As illustrated in the following example, each of these extreme values must occur *either*

- at an **interior point** x of the interval where $a < x < b$, or
- at one of the interval's two **endpoints** $x = a$ and $x = b$.

If we graph the function on the given interval, then we can

- spot visually an extreme value that occurs at an endpoint of the interval, and can
- use the calculator's maximum or minimum function to find an extreme value that occurs at an interior point.

EXAMPLE 5 Finding Maximum and Minimum Values on an Interval

Find the maximum and minimum values of the function $f(x) = 5 + 12x - 4x^2$ on the interval $0 \leq x \leq 4$.

SOLUTION

Figure 5.2.7 shows the graph $y = 5 + 12x - 4x^2$, a parabola that opens downward, plotted on the given interval $0 \leq x \leq 4$. There is therefore no lowest point on the whole parabola, but we are concerned only with the portion that corresponds to the interval $0 \leq x \leq 4$.

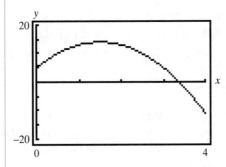

FIGURE 5.2.7 The graph $y = 5 + 12x - 4x^2$ with low point at the right-hand endpoint and high point in the interior of the interval $0 \leq x \leq 4$.

We see that minimum value of the function $f(x)$ occurs at the right-hand endpoint $x = 4$ of the interval. Thus the minimum is the value

$$f(4) = 5 + 12(4) - 4(4)^2 = -11.$$

It is equally obvious in Fig. 5.2.7 that the maximum value of the function occurs at an interior point of the interval, the vertex of the parabola. The **CALC maximum** screen in Fig. 5.2.8 indicates that this maximum value is $f(1.5) = 14$.

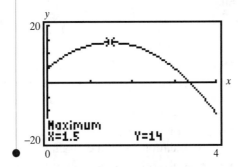

FIGURE 5.2.8 The high point on the graph has coordinates $x = 1.5$ and $y = 14$.

Example 5 illustrates the **Graphical Maximum-Minimum Method** for solving maximum-minimum problems:

To find the maximum value of the function $f(x)$ for values of x in the interval $a \leq x \leq b$, we carry out the following steps.

- Construct the graph $y = f(x)$ on the interval $a \leq x \leq b$.
- Find the highest point (x_m, y_m) on the graph.
- Then $y_m = f(x_m)$ is the maximum value of the function.

With a graphing calculator this is easily done using the **CALC maximum** command. To find the minimum value of $f(x)$, proceed similarly, except use the **CALC minimum** facility to find the lowest point on the graph.

We can use this method with purely algebraic problems or with function models for any type of application. In Example 6, we apply the method to a quadratic population model.

EXAMPLE 6 Finding the Maximum and Minimum Population

A city has an initially declining population, given (in thousands) t years after January 1, 1999, by the quadratic model

$$P(t) = 100 - 2.4t + 0.2t^2.$$

a. What is the minimum population the city reaches before rebounding?
b. When does it regain its initial population?
c. What is the city's maximum predicted population in the years 1999–2014?

SOLUTION

a. Figure 5.2.9 shows the result when we define **Y1 = 100 − 2.4X + 0.2X^2** (with the calculator's **X** instead of t as the independent variable and **Y** instead of P as the dependent variable) and plot our city's population function in the window $0 \le x \le 20, 75 \le y \le 125$.

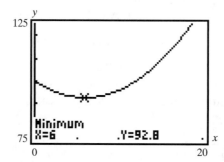

FIGURE 5.2.9 Graph of the population function $P(t) = 100 - 2.4t + 0.2t^2$ for $0 \le t \le 20, 75 \le P \le 125$.

The plot confirms what we know about such a quadratic function—the population declines at first, reaches a minimum value, then "rebounds," and thereafter increases. The **CALC minimum** command locates the visible low point as (6, 92.8). Thus, after $t = 6$ years, the city reaches its minimum population of 92.8 thousand people. Starting on January 1, 1999, this minimum population would be attained on January 1, 2005.

b. Here we need to solve the especially simple (why?) quadratic equation

$$100 - 2.4t + 0.2t^2 = 100.$$

Regardless of the method we use (factoring, the quadratic formula, graphical intersection, or a table of function values), when we solve this equation we obtain $t = 12$. (Of course, you should verify this with *your* method of choice.) Thus the city regains its original population of 100 thousand people at the beginning of the year 2011.

c. Finally, we are interested in the maximum population in the interval $0 \le t \le 15$. As indicated in the Graphical Maximum-Minimum Method, we first graph $P(t)$ on the given interval. Since the function is already stored in **Y1**, we just need to change our x-window to $0 \le x \le 15$ to represent the calendar years from 1999 to 2014. Figure 5.2.10 shows that the maximum function value on this interval occurs at the right endpoint, that is, when $t = 15$.

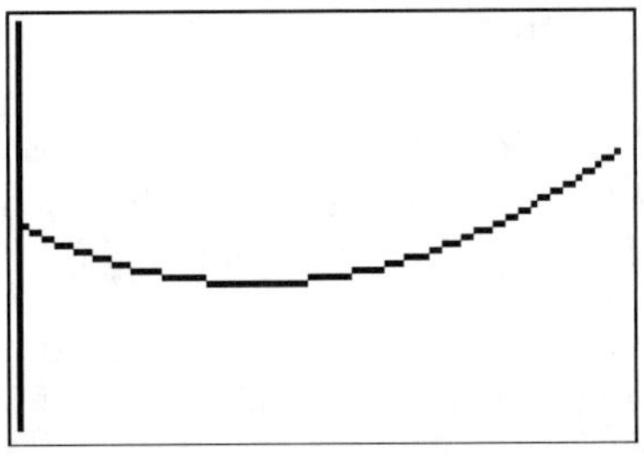

FIGURE 5.2.10 The maximum function value on the interval occurs at the right endpoint.

The maximum population in this time interval is then $P(15) = 100 - 2.4\,t + 0.2\,t^2 = 109$ thousand people, occurring on January 1, 2014.

5.2 Exercises | Building Your Skills

In Exercises 1–10, find the maximum and/or minimum value(s) of the function $f(x)$ for the indicated values of x.

1. $f(x) = x^2 - 4x + 3$, $\quad x$ in the domain of f
2. $f(x) = x^2 - 4x + 3$, $\quad -2 \leq x \leq 1$
3. $f(x) = x^2 - 4x + 3$, $\quad 2 \leq x \leq 5$
4. $f(x) = x^2 - 4x + 3$, $\quad -1 \leq x \leq 4$
5. $f(x) = 5 + 2x - x^2$, $\quad x$ in the domain of f
6. $f(x) = 5 + 2x - x^2$, $\quad -2 \leq x \leq 1$
7. $f(x) = 5 + 2x - x^2$, $\quad 2 \leq x \leq 5$
8. $f(x) = 5 + 2x - x^2$, $\quad -3 \leq x \leq 4$
9. $f(x) = 2x^2 - 10x + 9$, $\quad 0 \leq x \leq 5$
10. $f(x) = 10 - 15x - 3x^2$, $\quad -10 \leq x \leq 5$

In Exercises 11 and 12, the initially declining population (in thousands) of a city t years after January 1, 1999, is given by the specified quadratic population model. Find

  a. the minimum population reached by the city,
  b. when it regains its initial population, and
  c. when it reaches double its initial population.

11. $P(t) = 200 - 5.2t + 0.47t^2$
12. $P(t) = 350 - 11t + 1.16t^2$

Applying Your Skills

13. A ball is thrown straight upward with a velocity of 50 feet per second from the top of building 100 feet tall.

    a. When does the ball reach its maximum height?
    b. What is the maximum height reached by the ball?
    c. How long does the ball remains in the air before hitting the ground?

14. A ball is thrown straight upward with a velocity of 75 feet per second from a tower 125 feet tall.
    a. When does the ball reach its maximum height?
    b. What is the maximum height reached by the ball?
    c. How long does the ball remains in the air before hitting the ground?

15. A powerful crossbow fires a bolt straight upward from a deck 10 feet off the ground and with an initial velocity of 320 feet per second.
    a. How high does the bolt go?
    b. What is the bolt's maximum height during its first 5 seconds in the air?

16. A bullet is fired straight upward from the ground from a powerful rifle with a muzzle velocity of 640 feet per second. Express the bullet's maximum height in miles (with 1 mile = 5280 feet).

17. The population of Omaha, Nebraska, from 1970 through 2003 can be modeled approximately by the quadratic function $P(t) = 344.86 - 4.49t + 0.20t^2$, where P is measured in thousands and t is years after January 1, 1970.
    a. According to the model, when was the population of Omaha smallest?
    b. What was the minimum population of Omaha?
    c. In what month and year did Omaha regain its January 1, 1970 population?

18. The population of Montgomery, Alabama, from 1970 through 2003 can be modeled approximately by the quadratic function $P(t) = 135.18 + 4.37t - 0.07t^2$, where P is measured in thousands and t is years after January 1, 1970.
    a. According to the model, what was the maximum population of Montgomery over this time period?
    b. When did the maximum population occur?
    c. In what month and year did the population of Montgomery fall to its January 1, 1970 level?

19. According to data reported by the *Statistical Abstract 2005*, the percentage of U.S. households leasing a vehicle for personal use in the years 1992–2001 can be modeled by the quadratic function $V(t) = -0.061t^2 + 0.903t + 2.76$, where t is years after 1992.
    a. According to the model, in what year was the percentage of households leasing a vehicle for personal use highest?
    b. What was the maximum percentage of households leasing a vehicle for personal use?
    c. Use your model to predict the percentage of households leasing a vehicle for personal use in 2010. Is this a reliable prediction? Why or why not?

20. According to data reported by the *Statistical Abstract 2005*, the percentage of private airports in the United States with lighted runways from 1980 through 2000 can be modeled by the quadratic function $L(t) = 0.0437t^2 - 1.2483t + 14.9057$, where t is years after 1980.
    a. What percentage of private airports had lighted runways in 1988?
    b. When was the percentage of private airports with lighted runways smallest?
    c. Use the model to predict the year in which the percentage of private airports with lighted runways returns to its 1980 level.

21. Based on data reported by *USA Today*, the quadratic function $B(t) = 0.175t^2 - 0.675t + 3.025$ provides an approximate model for U.S. imports of Venezuelan crude oil (in millions of barrels per day), with $t = 0$ in 2001.

a. Over what interval of years were U.S. imports of Venezuelan crude oil decreasing?
b. When were U.S. imports of Venezuelan crude oil at a minimum?
c. Use the model to predict the year in which U.S. imports of Venezuelan crude oil will rise to 5 million barrels a day.

22. Based on data reported by the *Atlanta Journal-Constitution*, the quadratic function $C(t) = 2.461t^2 - 79.586t + 745.412$ provides an approximate model for yearly cotton production (in thousands of bales) in Georgia for the years 1960–1999 (with $t = 0$ in 1960).
    a. Over what interval of years was yearly cotton production increasing?
    b. What was the minimum cotton production? In what year did it occur?
    c. What was the maximum cotton production? In what year did it occur?

In Exercises 23 and 24 a projectile is fired at a 45° angle from the ground with the vertical and horizontal components of its velocity both being v_0 ft/sec. Its height y feet above the ground after it has traveled x feet horizontally is then given by $y = x - 16(x/v_0)^2$.

23. If the projectile is a rifle bullet fired with $v_0 = 250$ ft/sec, find
    a. the maximum height reached by the bullet; and
    b. how far the bullet has traveled horizontally when it hits the ground.

24. If the projectile is a ball thrown with $v_0 = 64$ ft/sec, find
    a. the maximum height reached by the ball; and
    b. how far the ball has traveled horizontally when it hits the ground.

5.3 FITTING QUADRATIC MODELS TO DATA

The data in the following table shows the population of Austin, Texas, as recorded in the decade census years of 1950–1990.

YEAR	POPULATION (THOUSANDS)	CHANGE (THOUSANDS)
1950	132	
1960	187	55
1970	254	67
1980	346	92
1990	466	120

The change figures in the final column of the table show that the population of Austin was changing at a steadily increasing rate. Thus the population of Austin does *not* exhibit the constant rate of change that is characteristic of a linear model. Indeed, we indicated in Section 5.1 that these data can be fitted well by a quadratic model of the form

$$P(t) = P_0 + bt + at^2. \tag{1}$$

In this section we discuss the fitting of data by quadratic models. Much of this discussion will parallel our discussions in Chapters 2–4, in which we fitted linear, exponential, and logarithmic models to data.

People frequently use spreadsheets to record and analyze data. Figure 5.3.1 shows a spreadsheet that we prepared to analyze the Austin population data. It is set up so that, when the population census figures are entered in the Population column and the desired numerical coefficients P_0, a, and b in (1) are entered in the first three shaded cells, the resulting predicted populations are automatically calculated and displayed in the $P(t)$ column, the errors are shown in the next two columns, and the average error is shown in the final shaded cell.

FIGURE 5.3.1 Spreadsheet modeling the population of Austin, Texas.

	A	B	C	D	E	F	G
1	The Population of Austin, Texas					austin.xls	
2							
3			P(t) = P0 + a t + bt^2				
4							
5		P0 =	140	a =	4	b =	0.1
6							
7		Year	t	Pop.	P(t)	Error E	E-squared
8		1950	0	132	140	-8	64.0
9		1960	10	187	190	-3	9.0
10		1970	20	254	260	-6	36.0
11		1980	30	346	350	-4	16.0
12		1990	40	466	460	6	36.0
13						161.0	5.675
14						SSE	Ave Error

We chose the indicated values $P_0 = 140$, $a = 4$, and $b = 0.1$ by a process of trial and error. We actually started with the values $P_0 = 130$ (close to Austin's actual initial population in the table) and the outright guesses $a = 5$ and $b = 0.1$. Then we adjusted slightly the values of P_0 and a so as to decrease the calculated average error. In this way we discovered that the specific quadratic model

$$P(t) = 140 + 4t + 0.1t^2 \tag{2}$$

fits the 1950–1990 Austin census data with an average error of 5.675 thousand people.

t	ACTUAL P	$P(t)$	ERROR E	E^2
0	132	140	-8	64
10	187	190	-3	9
20	254	260	-6	36
30	346	350	-4	16
40	466	460	6	36

The meaning of the term "average error" here is the same as before. The foregoing table was copied from the spreadsheet of Fig. 5.3.1 and lists in its P column the *actual* 1950–1990 census populations P_1, P_2, P_3, P_4, and P_5 at the successive times t_1, t_2, t_3, t_4, and t_5. The $P(t)$ column lists the *predicted* populations $P(t_1)$, $P(t_2), P(t_3), P(t_4)$, and $P(t_5)$ calculated using the quadratic model in (2). The E column of the table lists the successive errors (actual $-$ predicted) E_1, E_2, E_3, E_4, and E_5. Finally, the last column of the table lists the squares of these errors.

For the Austin data and the quadratic model $P(t) = 140 + 4t + 0.1t^2$, we find that the sum of the squares of the errors is given by

$$SSE = (-8)^2 + (-3)^2 + (-6)^2 + (-4)^2 + (6)^2$$
$$= 64 + 9 + 36 + 16 + 36 = 161.$$

Thus, the average error is $\sqrt{\dfrac{161}{5}} \approx 5.675$, and we have verified with our own calculations the average error displayed in the spreadsheet of Fig. 5.3.1.

The Best Fit—The Least Possible Average Error

Of course, the "optimal" quadratic model is the one that best fits the actual data. As we have seen previously, the better of two models is the one giving the lesser average error.

EXAMPLE 1   Which Is the Better Model?

Which of the two quadratic models

$$P(t) = 140 + 4t + 0.1t^2 \quad \text{and} \quad P(t) = 125 + 4.5t + 0.1t^2 \quad (3)$$

fits the Austin census data best?

SOLUTION

We saw that the average error of the first quadratic model in (3) is 5.675 thousand people. Use your calculator to verify that the second of these quadratic models yields the figures shown in the following table:

t	ACTUAL P	P(t)	ERROR E	E^2
0	132	125	7	49
10	187	180	7	49
20	254	255	−1	1
30	346	350	−4	16
40	466	465	1	1

Consequently, the second model in (3) has the sum of the squares of the errors

$$SSE = 49 + 49 + 1 + 16 + 1 = 116$$

and the resulting average error

$$\text{average error} = \sqrt{\dfrac{116}{5}} \approx 4.817.$$

Thus the average error in the second model is only 4.817 thousand people as compared with the average error of 5.675 in the first model. Thus the quadratic model $P(t) = 125 + 4.5t + 0.1t^2$ fits the Austin population data better than does the first model $P(t) = 140 + 4t + 0.1t^2$.

For best fitting, the question is this: What choice of the numerical parameters P_0, a, and b in the quadratic model $P(t) = P_0 + at + bt^2$ will minimize the average error? That is, what values of these three coefficients will give the quadratic model with the least possible average error? This optimal quadratic model will be the one that we say "best fits" the given data.

Finding the best-fitting quadratic model is easily accomplished using the regression capability of the calculator. Figure 5.3.2 shows the **STAT EDIT** menu displaying the Austin data in table form. Item **5: QuadReg** in the **STAT CALC** menu is the calculator's quadratic regression facility for finding the parabola that best fits the selected data points. Using the command **QuadReg L1,L2,Y1** results in the display shown in Fig. 5.3.3 that results when this command is entered.

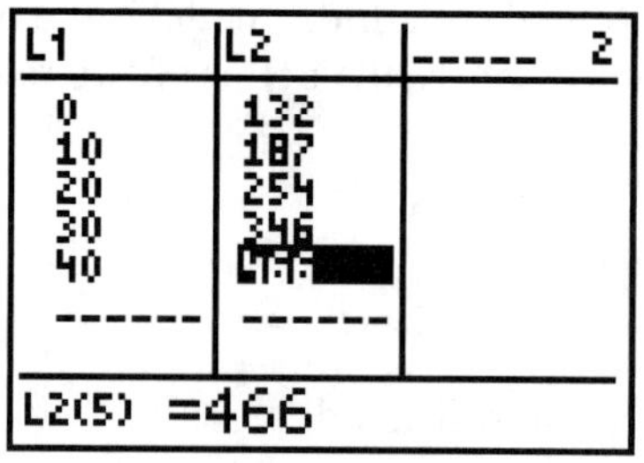

FIGURE 5.3.2 TI-83 **STAT EDIT** menu showing the Austin population data in table form.

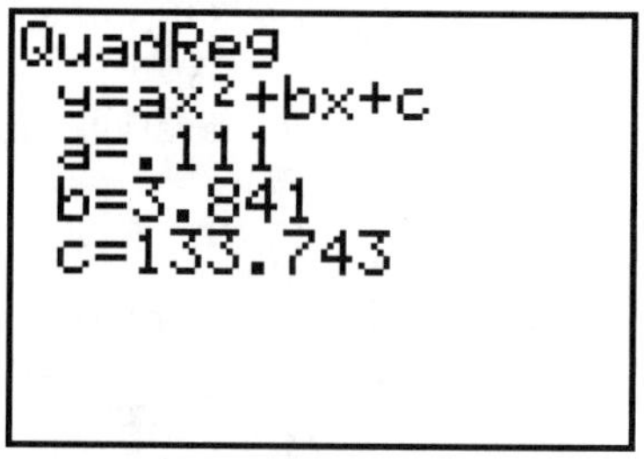

FIGURE 5.3.3 The best quadratic polynomial fit to our data points.

In the calculator's x-y notation, the displayed results say that the quadratic polynomial best fitting the given data is

$$y = 0.111x^2 + 3.841x + 133.743 \tag{4}$$

The **QuadReg L1, L2, Y1** command automatically enters Equation (4) in the **Y=** menu, ready for plotting. With **Plot 1** turned **On**, **GRAPH** then gives the plot of the best-fitting parabola (quadratic polynomial graph) shown in Fig. 5.3.4, along with the original census data points.

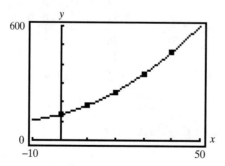

FIGURE 5.3.4 The optimal quadratic fit.

In terms of calendar year as the independent t-variable and population as the dependent P-variable, Equation (4) says that the quadratic model that best fits the 1950–1990 Austin census data is given by

$$P(t) = 133.743 + 3.841t + 0.111t^2. \tag{5}$$

EXAMPLE 2   Finding SSE and Average Error

Find the SSE and the average error for the best-fitting quadratic model for the Austin census data.

SOLUTION

The following table compares the original census data with the population figures predicted by this quadratic model:

t	ACTUAL P	P(t)	ERROR E	E^2
0	132	133.743	−1.743	3.038
10	187	183.253	3.747	14.040
20	254	254.963	−0.963	0.927
30	346	348.873	−2.873	8.254
40	466	464.983	1.017	1.034

[You should use a calculator to calculate the indicated values of $P(t)$ and to verify the other values shown in this table.] As usual, the final column shows the squares of the errors. Hence the SSE associated with the optimal quadratic model in (5) is

$$3.038 + 14.040 + 0.927 + 8.254 + 1.034 = 27.293.$$

Therefore the average error is given by

$$\text{average error} = \sqrt{\frac{27.293}{5}} \approx 2.336.$$

Thus the quadratic model $P(t) = 133.743 + 3.841t + 0.111t^2$ predicts 1950–1990 census year populations that differ (on average) by 2,336 people from those actually recorded for Austin. Note that this minimal average error of 2.336 thousand is less than the average errors for the quadratic models in Example 1.

Applications of Quadratic Modeling

Thus far, we have discussed a best-fitting quadratic model of population growth. But the world is full of other apparently quadratic data waiting to be modeled.

EXAMPLE 3   Finding a Best-Fitting Quadratic Model

From 1820 through 1950, the median age of the U.S. population was increasing; data from selected years is shown in the accompanying table.

Year t	1820	1840	1860	1870	1880	1890
Median Age (years) M	16.7	17.8	19.4	20.2	20.9	22.0

a. Use a scatter plot of the data to verify that a quadratic function is a suitable model for these data.

b. Find the quadratic function that best fits these data.

SOLUTION

a. We can see from the table that the median age is increasing, but not at a constant rate. (The average rate of change over the first 20 years is 0.055 years per year, while over the last 10 years it is 0.11 years per year.) Thus median age is not a linear function of year.

If we "reset" our starting point so that $t = 0$ represents the year 1820 and view our data in the window **Xmin=−10, Xmax=100, Ymin=15, Ymax=25** (Fig. 5.3.5), it appears to have a slight "upward bend." Hence, we believe that a quadratic function might fit these data better than a linear function would.

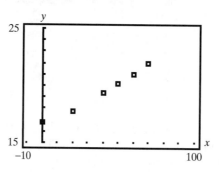

FIGURE 5.3.5 The median age data of Example 6.

b. Using the command **QuadReg L1,L2,Y1,** we find that the quadratic function that best fits these data is given by

$$y = 0.00031x^2 + 0.054x + 16.68,$$

where y represents median age in years and x represents years after 1820. So the quadratic function we are looking for (in terms of t and M) is

$$M(t) = 0.00031t^2 + 0.054t + 16.68,$$

where t denotes years after 1820.

As always, once we have a given function model, we can use the model to make predictions about input or output values for which we do not have data. Example 4 illustrates these procedures.

EXAMPLE 4 Using a Best-Fitting Quadratic Model

Use the best-fitting quadratic model from Example 3 to

a. predict the median age of the population in 1920, and
b. find the year in which the median age of the population was 27 years.

SOLUTION

a. Since t denotes years after 1820, the year 1920 is $t = 100$. So the predicted median age in 1920 is given by

$$M(100) = 0.00031(100)^2 + 0.054(100) + 16.68 = 25.18.$$

Thus the median age of the U.S. population in 1920 was approximately 25.18 years.

b. Here we need to solve the equation $M(t) = 27$, that is,

$$0.00031t^2 + 0.054t + 16.68 = 27.$$

As always, we have a choice of methods with which to solve this equation.

Graphical Method Enter the left-hand side of the equation ($0.00031t^2 + 0.054t + 16.68$) as **Y1** and the right-hand side 27 as **Y2**. Since **Ymax** is set at 25, we must reset our window to include $y = 27$; an **Xmax** of 100 may not be big enough either. Let **Xmax=130** (since we originally said that the median age was increasing from 1820 to 1950) and **Ymax=30**. Then use the calculator's intersection-finding command to find the intersection point (Fig. 5.3.6) of (115.08, 27). Thus, the median population is predicted to reach 27 years of age 115.08 years after 1820, or early in the year 1935.

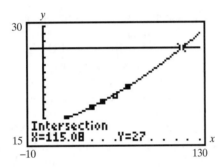

FIGURE 5.3.6 Finding the intersection point.

Numerical Method With $0.00031t^2 + 0.054t + 16.68$ in **Y1**, we look in the table and "zoom in" on $y = 27$. Since our last data point is (70, 22), we start the table at $x = 70$, and increment by steps of $\Delta x = 10$ (Fig. 5.3.7).

We see that the median age reaches 27 years somewhere between 110 and 120 years, so we reset the table to start at 110 and increment by steps of $\Delta x = 1$ (Fig. 5.3.8).

X	Y1
70.000	21.979
80.000	22.984
90.000	24.051
100.00	25.180
110.00	26.371
120.00	27.624
130.00	28.939

X=110

FIGURE 5.3.7 Table of median ages, with 10-year increments.

X	Y1
110.00	26.371
111.00	26.494
112.00	26.617
113.00	26.740
114.00	26.865
115.00	26.990
116.00	27.115

X=115

FIGURE 5.3.8 Table of median ages, with 1-year incements.

Here we find that at $x = 115$, $y = 26.99$, so we reset the table to start at 115. Since 26.99 is so close to 27, we now increment by steps of $\Delta x = 0.01$. Thus we again find (Fig. 5.3.9) that 115.08 years after 1820, the median age reaches 27 years.

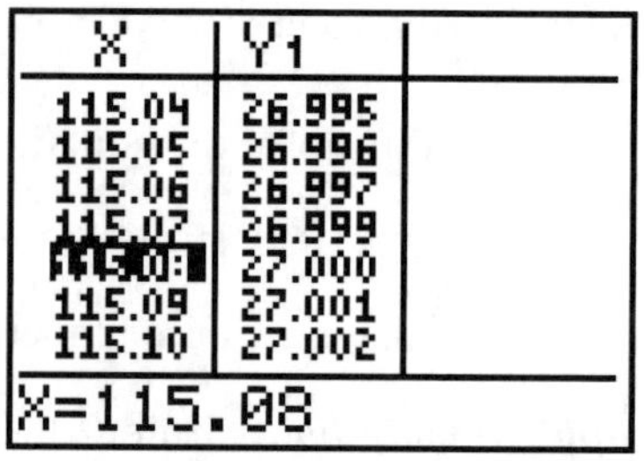

FIGURE 5.3.9 Table of median ages, with one-hundredth-year increments, after scrolling down to the target value.

Symbolic Method If we transform the quadratic equation $0.00031t^2 + 0.054t + 16.68 = 27$ into standard form, we get $0.00031t^2 + 0.054t - 0.32 = 0$. We can then use the quadratic formula to obtain the two solutions

$$t = \frac{-0.054 + \sqrt{0.0157128}}{0.00062} \quad \text{or} \quad t = \frac{-0.054 - \sqrt{0.0157128}}{0.00062}.$$

Approximating these solutions gives $t = 115.08$ years (as before) and $t = -289.28$ years. Since the latter, negative solution corresponds to approximately 289 years *before* 1820, it does not make sense in the context of this problem, and we discard it.

Thus, no matter which method we use, we arrive at the same conclusion—that the median age is predicted to be 27 years early in the year 1935.

Note that this is an example in which hindsight is available as a check. Because we are using selected data to model population trends in the past, we can check the reasonableness of our model using any additional data available to us. As it happens, the median age of the population in 1920 was 25.3 years, quite close to our prediction of 25.18 years. Furthermore, the median age of the U.S. population was 26.5 years in 1930 and 29.0 years in 1940, so our prediction of early in 1935 for a median age of 27 years looks quite reasonable.

The only reason to use a mathematical model to predict a data value that we know is to check its reasonableness to predict values that we don't know. As always, we must be careful about predicting too far beyond our known data. Since recent data are readily available, we would not use the model from Example 3 to predict the current median age of the U.S. population. We have used the example here to illustrate the principle of modeling data with a quadratic function.

The following discussion revisits the cigarette consumption–lung cancer data explored in Chapter 2 and displays another pitfall that may appear when creating mathematical models based on real-world data.

The per capita consumption of cigarettes in 1930 and the lung cancer death rate (deaths per million males) for 1950 in the four Scandinavian countries were as follows:

COUNTRY	CIGARETTE CONSUMPTION x	DEATH RATE y
Norway	250	95
Sweden	300	120
Denmark	350	165
Finland	1100	350

It is hard to ignore the fact that higher cigarette consumption appears to be correlated with higher lung cancer death rates. In Section 2.4 we found that the optimal linear model fitting these data is

$$y = 0.28x + 40.75. \qquad (6)$$

This straight-line fit to the data is shown in Fig. 5.3.10. We note that this straight line appears to go through the second and fourth data points but does not fit the first and third points so well. Perhaps a parabola that "levels off" a bit as it approaches the fourth point would fit all of the points better.

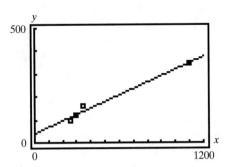

FIGURE 5.3.10 Straight line $y = 0.28x + 40.75$ fitting the cigarette data.

We therefore explore a *quadratic* model fitting these same data. After entering the foregoing x- and y-data in the lists **L1** and **L2** and executing the command **QuadReg L1, L2, Y1**, we find that the best-fitting quadratic function is given by

$$y = -0.000521x^2 + 1.01x - 127.99. \qquad (7)$$

With an appropriate window and the same **STAT PLOT** settings as before, we get the plot of this parabola and the table data points that is shown in Fig. 5.3.11.

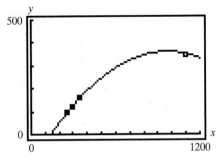

FIGURE 5.3.11 The parabola best fitting the cigarette consumption versus lung cancer deaths data.

FIGURE 5.3.12 Table of values of the optimal quadratic model $y = -0.000521x^2 + 1.01x - 127.99$.

Figure 5.3.12 shows a calculator table of values of the quadratic function in (7). In the following table we have added these values to our original data:

CIGARETTE CONSUMPTION x	ACTUAL DEATH RATE y	PREDICTED DEATH RATE $y(x)$	ERROR E	SQUARED ERROR E^2
250	95	91.95	3.05	9.30
300	120	128.12	−8.12	65.93
350	165	161.69	3.31	10.96
1100	350	352.60	−2.60	6.76

The final column shows the squares of the individual errors. Hence the average error of our quadratic model (10) is

$$\text{average error} = \sqrt{\frac{9.30 + 65.93 + 10.96 + 6.76}{4}} = 4.82.$$

Thus the average error in the quadratic model's predictions (for the four Scandinavian countries) is only 4.82 lung cancer deaths per million males.

In Example 6 of Section 2.4 we found that the average error in the linear model (6) was 15.50. Thus it appears that the optimal quadratic model fits the data considerably better than does the optimal linear model.

But does it really? When we look at the graph in Fig. 5.3.11 it appears that the rate of cancer deaths peaks *before* we reach the fourth point consumption of $x = 1100$. Indeed, by tracing the graph as indicated in Fig. 5.3.13, we see that the quadratic model appears to predict a maximum of about 361.5 cancer deaths per million at a cigarette consumption level of about $x = 969$. Does it sound reasonable that you are worse off if you smoke 968 cigarettes, but every additional cigarette smoked thereafter improves your chances? Surely not!

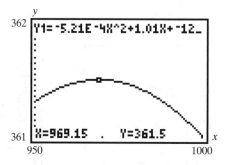

FIGURE 5.3.13 Consumption $x = 969$ appears to yield the maximum predicted death rate of $y = 361.5$.

So the important lesson is this: There's more to the validity of a mathematical model than its average error. If we wanted to predict the effect of cigarette consumption at higher levels than $x = 1100$, we might well rely more on the linear model—despite its greater average error for the known data points—because at least it predicts a further increase in death rates for greater consumption rates. Whenever you are using a mathematical model to make predictions, it is critical that you examine the results in terms of the context and evaluate them based on your common sense.

5.3 Exercises | Building Your Skills

In Exercises 1–6, both a quadratic polynomial $p(x) = ax^2 + bx + c$ and several (x, y) data points are given. Find the average error with which the indicated polynomial approximates these points. If you wish, you may use the table facility of your calculator to compute an appropriate table of values of $p(x)$.

1. $p(x) = 2x^2 - 5$

x	-2	0	1	3
y	2	-4	-2	12

2. $p(x) = 3x^2 - 7x$

x	-3	-1	4	5
y	45	13	17	43

3. $p(x) = 5 - 3x + x^2$

x	-1	2	4	6
y	11	2	6	25

4. $p(x) = 17 - 3x + 2x^2$

x	-2	1	3	5
y	35	15	30	50

5. $p(x) = 25 + 4x - 3x^2$

x	-2	0	1	3	5
y	6	15	24	8	-40

6. $p(x) = 100 + 5x - x^2$

x	-4	-2	0	2	6
y	60	80	110	110	90

7–12. Find the quadratic model $q(x) = Ax^2 + Bx + C$ that best fits the data in Exercises 1–6 (respectively), as well as the average error of this optimal quadratic model.

Applying Your Skills

In each of Exercises 13–16, the 1950–1990 population census data for a U.S. city is given.

    a. Find the quadratic model $P(t) = P_0 + at + bt^2$ (with $t = 0$ in 1950) that best fits the census data.
    b. Construct a table showing the actual and predicted populations (and errors) for the decade years 1950-1990.
    c. Use your quadratic model to predict the year 2000 population of the city.

13. Mesa, Arizona:

t (year)	1950	1960	1970	1980	1990
P (thousands)	17	34	63	152	288

14. St. Petersburg, Florida:

t (year)	1950	1960	1970	1980	1990
P (thousands)	96	181	216	239	240

15. Arlington, Texas:

t (year)	1950	1960	1970	1980	1990
P (thousands)	8	45	90	160	262

16. Corpus Christi, Texas:

t (year)	1950	1960	1970	1980	1990
P (thousands)	108	168	205	232	257

17. The annual production of tobacco (in millions of pounds) in the United States for selected years is given in the following table:

t (year)	1988	1990	1992	1994	1996
P (millions of pounds)	1370	1626	1722	1583	1517

Source: World Almanac and Book of Facts 1998.

    a. Find the quadratic function that best fits these data.
    b. Find the average error for this model.
    c. Use the model to find the year in which tobacco production falls to 1400 million pounds.

18. The number of U.S. bank failures (banks closed or assisted) in the years from 1935 to 1940 is given in the following table:

t (year)	1935	1936	1937	1938	1939	1940
B (banks)	32	72	84	81	72	48

Source: World Almanac and Book of Facts 1998.

a. Find the quadratic function that best fits these data.
b. Find the average error for this model.
c. Use the model to find the number of bank failures in 1945.

19. The following table gives the per capita value (to the nearest dollar) of money in circulation in the United States in selected years:

t (year)	1985	1990	1994	1996	1997
P (dollars)	779	1029	1428	1573	1665

Source: World Almanac and Book of Facts 1998.

a. Find the quadratic function that best fits these data.
b. Use the model to predict the year in which the per capita value of money is $650.
c. Use the model to predict the year in which the per capita value of money is $1000.
d. Which of your answers (b) or (c) is likely to be more accurate? Why?

20. The data in the following table give the number of convictions and cases closed as a result of the FBI's anti–child pornography initiative:

t (year)	2000	2001	2002	2003	2004	2005
C (convictions/cases closed)	476	557	646	714	881	994

Source: USA Today.

a. Find the quadratic function that best fits these data.
b. Use the model to predict the year in which the number of convictions and cases closed reaches 1500.
c. Use the model to predict the number of convictions and cases closed in 2010.

21. The data in the following table give the total number of Waffle House Restaurants in the United States in selected years:

t (year)	1956	1960	1970	1980	1990
R (restaurants)	1	48	401	672	1228

Source: Waffle House, Inc.

a. Find the quadratic function that best fits these data.
b. Use the model to predict the number of Waffle House restaurants in 2010.
c. According to the model, in what year will the number of Waffle House restaurants rise to 2000?

22. The data in the following table give the amount of money (in millions of dollars) appropriated by Congress for alternative medicine programs for the given fiscal years. (Money is allocated to the Office of Alternative Medicine and the National Center for Complementary and Alternative Medicine.)

t (year)	1992	1994	1996	1998	2000	2002
M (millions of dollars)	2	3.4	7.7	19.5	68.7	104.6

Source: Atlanta Journal-Constitution.

a. Find the quadratic function that best fits these data.
b. Find the average error for this model.
c. Use the model to predict the year in which the funding for alternative medicine programs reaches $150 million.

23. The following table gives per capita consumption of milk (in gallons) in the United States for the selected years.

t (year)	1980	1985	1990	1995	1999	2000
M (gallons)	27.6	26.7	25.7	23.9	22.9	22.5

Source: The 2006 Statistical Abstract.

a. Find the quadratic function that best fits these data.
b. Find the average error for this model.
c. Use the model to predict the per capita milk consumption in the United States in the year 2008.

Chapter 5 | Review

In this chapter, you learned about quadratic functions and models. After completing the chapter, you should be able to

- determine whether a relation described numerically, graphically, or symbolically represents a quadratic function;
- find the domain and range of a quadratic function;
- find the output value of a quadratic function for a given input value;
- find the input value(s) of a quadratic function for a given output value;
- solve an equation or inequality involving a quadratic function;
- find the maximum or minimum value of a quadratic function;
- find a quadratic function that models given quadratic data; and
- find the best-fitting quadratic model for data that are approximately quadratic.

Review Exercises

In Exercises 1–7, determine whether, based on the table, graph, or formula, y is a quadratic function of x.

1.
x	2	4	5	6	8	11
y	−2	−8	−12.5	−18	−32	−60.5

2.
x	−2	−1	1	3	4	5
y	9	2	0	−26	−63	−124

3.
x	1	2	3	4	5	6
y	300	150	75	37.5	18.75	9.375

4.

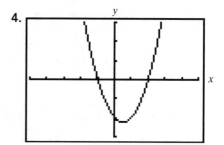

5.

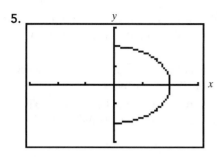

6. $x^2 + y - 4 = 0$

7. $y = 3(x + 2)(x - 7)$

In Exercises 8–10, find the domain and range of each function.

8. The height h (in feet) of a math book dropped from a second floor window as a function of the time t (in seconds), according to the following table:

t (seconds)	0	0.2	0.4	0.6	0.8	1.0
h (feet)	16	15.36	13.44	10.24	5.76	0

9.

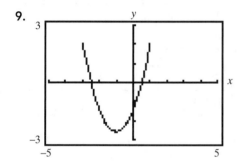

10. $f(x) = -3(x - 2)^2 + 4$

11. In Exercise 10 of the Chapter 2 Review, we found that the relationship between price and demand (number of items sold) for Pomelia's decorative lawn sprinkler business was given by $p(x) = -1.25x + 85$, where x was the number of items sold and p was the price per item. Since **revenue** (the money resulting from sales) is found by the product

(number of items) × (price per item) = $x \times p(x)$,

we see that the revenue is a quadratic function of the number of items sold.

a. Find a quadratic function $R(x)$ giving revenue as a function of x, the number of sprinklers sold per month.

b. How many sprinklers must be sold in order to maximize monthly revenue?

c. What is the maximum monthly revenue?

12. Maria throws a ball straight upward with an initial velocity of 48 feet per second from the top of a building 160 feet high.

a. When does the ball pass the top of the building on its way down?

b. How long does the ball remain in the air?

13. The population data from the U.S. census for Stockton, California, is given in the following table:

t (years)	1950	1960	1970	1980	1990
M (thousands)	71	86	110	148	211

a. Find the quadratic population model that best fits these data.

b. Use your model to predict the year in which the population of Stockton reaches 250 thousand.

14. The following table displays the percentage P of the U.S. population that was foreign-born in the year y:

x	1930	1940	1950	1960	1970	1980	1990	1996
$P(\%)$	11.6	8.8	6.9	5.4	4.8	6.2	7.9	9.3

Source: World Almanac and Book of Facts 1998.

a. Find the quadratic model that best fits these data.

b. Use your model to predict the percentage of the population that was foreign-born in 2000.

c. According to your model, what is the minimum percentage of the U.S. population that was foreign-born from 1930 to 1996?

d. According to your model, in what year did the minimum percentage occur?

15. (Chapter Opener Revisited) In the discussion that introduced this chapter, we looked at the graph of the height function of a ball tossed straight up into the air. The following table gives the data from which the graph was constructed; t represents time in seconds after the ball is thrown and H is the height of the ball in feet:

t (seconds)	0.6	0.7	0.8	0.9	1.0	1.1	1.2	1.3	1.4
H (feet)	3.0213	4.0944	4.8326	5.2575	5.3764	5.1675	4.6634	3.8603	2.7728

a. Find the quadratic function that best fits these data, rounding all values to four decimal places.

b. Recall that we use the formula $h(t) = -16t^2 + v_0 t + h_0$ to give a projectile's height h in feet above the ground after t seconds if the projectile's initial velocity is v_0 feet per second and we ignore the effect of air resistance. How does the coefficient of the square term in your model compare to the coefficient in the formula $h(t)$? Why are they different?

c. If the coefficient of the linear term in your model represents the ball's initial velocity, at what rate was the ball thrown upward?

d. Does the constant term in your model make sense in this situation? Why or why not?

e. If the ball had not been caught at an approximate height of 2.8 feet, at what time would it have hit the ground?

INVESTIGATION Quadratic Models in Baseball

In this chapter, you have learned about quadratic models for a dropped or thrown ball. Does a batted ball behave in the same manner? In this activity, you will investigate what happens when a ball is batted under different conditions.

You may have observed, either from watching or playing baseball, that when a ball is hit in the air it appears to travel in a parabolic path. A fly ball to the warning track looks like a "wide" parabola, while a pop-up to the third baseman looks more like a narrow one. If we ignore air resistance and any wind that may be blowing, a batted baseball does indeed follow a parabolic path.

In the discussion that follows, we will also assume that the ball is hit on the "sweet spot," that place on the bat that minimizes the loss of energy of the pitched ball contacting the bat. The distance that the ball travels depends on both the angle at which it is batted and the velocity of the bat.

If a ball is batted at an angle of 35°, the distance that the ball travels as a function of the bat speed is given approximately by

$$D(v) = 0.029v^2 + 0.002v - 0.068,$$

where v is the bat speed in miles per hour and D is distance traveled in feet.

1. Find the distance a batted ball will travel if the ball is batted with a velocity of 90 miles per hour.
2. With what velocity must you hit the ball in order for it to travel 400 feet?

Each of the following tables gives data for distance traveled in feet as a function of the angle in degrees at which the ball is hit for various bat velocities. (An average bat velocity is about 100 miles per hour.) For each of the tables, (a) find the quadratic function of best fit, and (b) use this quadratic model to determine the maximum distance the ball can be hit and the angle at which it must be hit to achieve the maximum distance.

3. Velocity = 90 mph

Angle (A)	30	40	55	70	75	80
Distance (D)	219	249	238	163	127	87

4. Velocity = 100 mph

Angle (A)	35	40	55	65	70	75
Distance (D)	294	308	294	239	201	156

5. Velocity = 110 mph

Angle (A)	30	40	55	70	75	80
Distance (D)	327	372	355	243	189	129

6. If you were to advise an aspiring player on how to increase the distance of his hits, what would you suggest?

In constructing these models, we have considered the force of gravity acting on the ball but not the effect of air resistance. The air causes a "drag force" on the ball, which slows it down. Interestingly enough, this drag force is typically a quadratic function of the velocity of the ball—still another example of a quadratic function. Another factor in the drag force function is the density of the air through which the ball is traveling. The denser the air, the greater is the drag force. This explains why balls travel farther in the "thin" air of Denver, the home of the Colorado Rockies.

In its web page on information and facts about Coors Field, the Colorado Rockies provide the chart in Fig. 5.R.4, which shows the effect of altitude on the distance a ball travels. (To see the chart, and find out about Coors Field, go to http://colorado.rockies.mlb.com/NASApp/mlb/col/ballpark/history.jsp.) The graph in the chart curves upward, suggesting that a quadratic function might provide a good model.

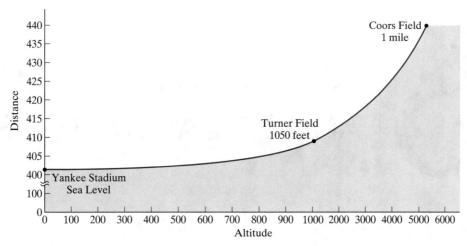

FIGURE 5.R.4 Distance a batted baseball travels as a function of altitude above sea level.
Source: Colorado Rockies Online.

7. The following table summarizes the data given in the graph; find the quadratic function of best fit for these data.

STADIUM	ALTITUDE IN FEET (x)	DISTANCE IN FEET (y)
Yankee Stadium	0	400
Turner Field	1050	408
Coors Field	5280	440

8. Use your quadratic model to predict the distance a ball would travel in Wrigley Field, where the altitude is approximately 600 feet.

9. Find the linear function of best fit for the altitude–distance data, and use it to predict the distance the ball would travel in Wrigley Field.

10. Your predictions for the distance the ball would travel in Wrigley Field are essentially the same regardless of which model you use. This seems to indicate that distance is *not* a quadratic function of altitude. Look carefully at the models from numbers (7) and (9). What indicates that the quadratic model is virtually the same as the linear model?

11. In order to verify your result from (10), make a plot of the data and the quadratic function. Does your graph look like the Rockies' graph?

12. Look carefully at the Rockies' graph, and explain why their graph is misleading.

References: Michigan Electronic Library and colorado.rockies.mlb.com.

CHAPTER 6
POLYNOMIAL MODELS AND LINEAR SYSTEMS

6.1 SOLVING POLYNOMIAL EQUATIONS

6.2 SOLVING PAIRS OF LINEAR EQUATIONS — LOTS OF WAYS!

6.3 LINEAR SYSTEMS OF EQUATIONS

6.4 POLYNOMIAL DATA MODELING

"**B**ring me your tired, your poor, Your huddled masses yearning to breathe free, The wretched refuse of your teeming shore, Send these, the homeless, tempest-tossed, to me, I lift my lamp beside the golden door!" The Statue of Liberty, with its inscription by Emma Lazarus, has long been a symbol of freedom and welcome to people from around the world. Unless you are a Native American, either you or your ancestors were born somewhere else. The United States has grown and prospered due to the efforts of those from other lands. Some periods of our history saw many newcomers to our shores; in other times there were fewer. The chart in Fig. 6.0.1 shows the percentage of the U. S. population that was born abroad in the indicated years of the twentieth century.

The dots on the graph represent the actual percentages of individuals in the United States born abroad, while the curve is the graph of a third-degree polynomial function that very closely approximates these percentages.

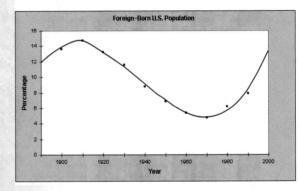

FIGURE 6.0.1 Percentage of U.S. population born abroad.

We see from this chart that the percentage of individuals born abroad rose from 1900 to 1910, then fell until 1970, and has since been rising. Over the time interval in question, the maximum percentage of individuals born abroad occurred in 1910, while the minimum occurred in 1970. Notice also that, at some point near 1930, the graph appears to change from curving downward to curving upward.

This graph does not look like any of the models we have studied so far; rather, with its up-and-down wiggles or bends, it has the distinct characteristics of a higher-degree polynomial graph. In this chapter we will study such polynomial functions, along with systems of linear equations and their real-world applications.

6.1 SOLVING POLYNOMIAL EQUATIONS

In Chapters 2 and 5, we solved problems that involved linear and quadratic equations. We found that a linear (first-degree polynomial) equation $ax + b = 0$ has one real number solution, while a quadratic (second-degree) equation $ax^2 + bx + c = 0$ has at most two (real number) solutions. Furthermore, we noted that the solutions to such a quadratic equation are given by the quadratic formula

$$x = \frac{-b \pm \sqrt{b^2 - 4ac}}{2a}.$$

However, a **cubic** (third-degree) polynomial equation of the form

$$ax^3 + bx^2 + cx + d = 0 \tag{1}$$

can have as many as **three** real number solutions. And a **quartic** (fourth-degree) equation of the form

$$ax^4 + bx^3 + cx^2 + dx + e = 0 \tag{2}$$

can have as many as **four** solutions.

Extremely complicated formulas for the solution of cubic and quartic equations are known. For instance, one of the three solutions of (1) is given by

$$x = \frac{-b + \sqrt[3]{R} + (b^2 - 3ac)/\sqrt[3]{R}}{3a},$$

where

$$R = \frac{1}{2}\left(-2b^3 + 9abc - 27a^2d + \sqrt{4(3ac - b^2)^3 + (9abc - 2b^3 - 27a^2d)^2}\right),$$

and the other two solutions can be found using similar formulas. The formulas for the solutions of the quartic equation (2) are still more complicated! Nowadays, these formulas are used only by calculators and computers, and it's a safe bet that no human being knows them all by memory. (Certainly, no one *should* use his or her precious memory cells that way!)

So don't let the appearance of the cubic formula above intimidate you. Indeed, we quote it only to make you feel grateful for your graphing calculator. Recall the **principle of graphical solution**: The **real solutions** of the equation $f(x) = 0$ are simply the **x-intercepts** of the graph $y = f(x)$. A glance at the graph can therefore reveal the number and approximate locations of the solutions of the equation. We can then use graphical techniques to approximate each solution more accurately.

Solving Cubic Equations

As illustrated in Fig. 6.1.1, the graph of a *cubic* polynomial typically (though not always) has *two* bends or "wiggles" as—assuming the "leading coefficient" a in (1) is positive—it traverses the xy-plane from southwest (lower left) to northeast (upper right).

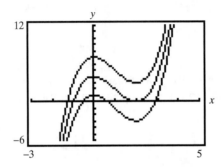

FIGURE 6.1.1
$y = x^3 - 3x^2 + 7$ (top graph);
$y = x^3 - 3x^2 + 4$ (middle graph);
$y = x^3 - 3x^2 + 1$ (bottom graph).

If your window is big enough to show both these bends, then you should be able to see how many solutions the corresponding cubic equation has. We see in Fig. 6.1.1 that

- the equation $x^3 - 3x^2 + 7 = 0$ has just one solution;
- the equation $x^3 - 3x^2 + 4 = 0$ has two solutions; and
- the equation $x^3 - 3x^2 + 1 = 0$ has three solutions.

A higher-degree polynomial equation can have complex or imaginary solutions in the same way that the quadratic equation $x^2 - 4x + 5 = 0$ has the quadratic formula solutions $x = 2 \pm \sqrt{-1}$ involving the imaginary number $\sqrt{-1}$. Such solutions are not found by graphing. So, here, by *solutions* we mean the "real solutions" of an equation.

EXAMPLE 1 Verifying Apparent Solutions

Verify that the equation

$$x^3 - 3x^2 + 4 = 0 \qquad (3)$$

has the two solutions $x = 2$ and $x = -1$.

SOLUTION

Looking at the middle graph in Fig. 6.1.1, we see that -1 and 2 appear to be the x-intercepts of the graph of $f(x) = x^3 - 3x^2 + 4$. But the graph would probably look the same to the naked eye if the two solutions were, say, $x = -1.0002$ and $x = 1.9997$.

In Fig. 6.1.2 we've applied our calculator's **CALC** zero-finding facility to locate the right-hand solution.

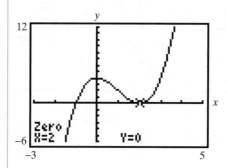

FIGURE 6.1.2 $y = x^3 - 3x^2 + 4$. It appears that $x = 2$ is a solution.

The calculator seems to say it's exactly 2, but this still doesn't prove it—if it were actually 2.00000003, then (displaying four decimal places) the calculator would round it off to 2. The only way to nail it down is to substitute $x = 2$ in (3) and see what happens:

$$(2)^3 - 3(2)^2 + 4 = 8 - 12 + 4 = 0$$

So $x = 2$ does, indeed, check out as an exact solution—substitution in the right-hand side of Equation (3) yields exactly 0. Similarly, substituting $x = -1$ in (3) gives

$$(-1)^3 - 3(-1)^2 + 4 = -1 - 3 + 4 = 0,$$

● and -1 is also an exact solution.

EXAMPLE 2 Finding Approximate Solutions

Find approximate values for the real solutions of the equation

$$x^3 - 3x^2 + 1 = 0. \tag{4}$$

SOLUTION

Looking at the bottom graph in Fig. 6.1.1, we see that the solutions of the equation appear to be approximately $-1/2$, $1/2$, and 3 (as closely as the eye can see). Are these solutions actually such simple fractions as this? The answer here is no! In Fig. 6.1.3 we have used our calculator's **CALC** zero-finding facility to see that the solution between 0 and 1 actually is $x \approx 0.6527$ (rounded off accurate to four decimal places).

If we proceed in the same fashion, we find that the other two solutions are $x \approx -0.5321$ and $x \approx 2.8794$. (You should verify this for yourself.)

Figure 6.1.4 shows that the calculator's **solve** function—entered from the **CATALOG** menu—can be used to find the real solutions of a cubic equation essentially "all at once" (after having defined **Y1 = X^3 − 3X² + 1** in the **Y=** menu). This command solves the equation **Y1 = 0** for **X**, with a specific starting "guess."

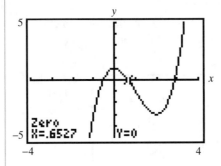

FIGURE 6.1.3 $y = x^3 - 3x^2 + 1$ and the approximate solution $x = 0.6527$.

FIGURE 6.1.4 Using **solve** to final all three solutions of $Y_1 = x^3 - 3x^2 + 1 = 0$.

Here the three starting guesses $-0.5, 0.5$, and 3 (obtained from our "eyeball" approximation) returned the three approximate solutions $-0.5321, 0.6527$, and 2.8794, respectively.

EXAMPLE 3 — Verifying Fractional Solutions

Use a graph to determine likely fractional solutions for the cubic equation

$$24x^3 + 2x^2 - 227x + 110 = 0. \tag{5}$$

SOLUTION

Figure 6.1.5 shows the graph of $y = 24x^3 + 2x^2 - 227x + 110$ plotted in the window $-5 \le x \le 5, -500 \le y \le 500$, indicating that the cubic equation has three real solutions. Using **solve** as in Fig. 6.1.6 suggests that these three solutions have the *approximate* values $x \approx -3.3333$, $x \approx 0.5000$, and $x \approx 2.7500$.

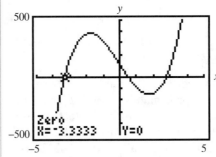

FIGURE 6.1.5 $y = 24x^3 + 2x^2 - 227x + 110$ and the approximate solution $x = -3.3333$.

FIGURE 6.1.6 Using **solve** to find all three solutions of $Y_1 = 24x^3 + 2x^2 - 227x + 110 = 0$.

This leads us to believe—but does not prove—that the given equation has the *exact* fractional solutions $x = -3\frac{1}{3}$, $x = \frac{1}{2}$, and $x = 2\frac{3}{4}$.

To verify that $x = -3\frac{1}{3} = -\frac{10}{3}$ is an exact solution, we substitute $-\frac{10}{3}$ in (5) and obtain

$$24\left(-\frac{10}{3}\right)^3 + 2\left(-\frac{10}{3}\right)^2 - 227\left(-\frac{10}{3}\right) + 110$$

$$= -\frac{24000}{27} + \frac{200}{9} + \frac{2270}{3} + 110$$

$$= \frac{-8000 + 200 + 6810 + 990}{9} = \frac{0}{9} = 0.$$

This is horribly messy, but in fact it proves that $x = -3\frac{1}{3}$ does satisfy Equation (5) exactly. You should verify similarly that $x = \frac{1}{2}$ and $x = 2\frac{3}{4} = \frac{11}{4}$ are also exact solutions. (Only when we have done so do we actually *know* that $x = -3\frac{1}{3}$, $x = \frac{1}{2}$, and $x = 2\frac{3}{4}$ are exact rather than merely approximate solutions of the cubic equation.)

An Application

Suppose you are a product designer for a consulting firm. Your client is a packaging manufacturer that has acquired cheaply a large surplus of rectangular cardboard sheets of various sizes. This cardboard will be used to make open-topped popcorn trays for use in movie theaters. Each tray will be constructed by cutting equal squares out of the corners of a cardboard sheet and then folding up the remaining flaps to form a box (Fig. 6.1.7). First an x by x square is to be cut from each corner of the original p by q cardboard rectangle. Then the four flaps are to be folded up to form the vertical sides of the tray.

Your specific task is to investigate the way in which the volume V of the resulting tray depends upon the edge length x of the corner squares that are cut

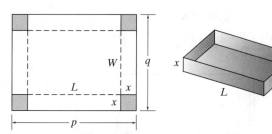

FIGURE 6.1.7 The popcorn tray construction.

out. Looking at the right-hand sketch in Fig. 6.1.7, we see that resulting tray has height x, so its volume is given by

$$V = LWx.$$

Looking at the left-hand sketch in Fig. 6.1.7, we see that

$$L = p - 2x \quad \text{and} \quad W = q - 2x, \tag{6}$$

so it follows that

$$V = x(p - 2x)(q - 2x). \tag{7}$$

This formula tells precisely how the volume V of the box depends on the corner notch size x.

You can use the formula in (7) to determine how to construct a box of a given desired volume if you know the length p and the width q of the cardboard sheet you are using. The next example illustrates how to determine the dimensions of a popcorn tray that will hold a half-liter (500 cubic centimeters) of popcorn.

EXAMPLE 4 Constructing a Popcorn Tray

Starting with a 20 cm by 30 cm sheet of cardboard, we construct a popcorn tray as indicated in Fig. 6.1.7. If its volume is precisely 500 cm³, what are its dimensions?

SOLUTION

To answer this question, we substitute $p = 30$ and $q = 20$ in (7). It follows that the volume V of the tray is described in terms of the size x of the corner notch by the function

$$V(x) = x(30 - 2x)(20 - 2x). \tag{8}$$

(You should verify this volume formula yourself, starting with your own sketch like Fig. 6.1.7, with the cardboard sheet labeled 30×20 rather than $p \times q$.) Then our task is to determine what x should be in order that

$$x(30 - 2x)(20 - 2x) = 500. \tag{9}$$

Without actually expanding the left-hand side, you should be able to see that a *cubic* would result.

If we transpose the right-hand-side number 500 and graph **Y₁= X(30−2X)(20−2X)−500** in the window $0 \leq x \leq 20$, $-1000 \leq y \leq 1000$, we get the cubic graph shown in Fig. 6.1.8.

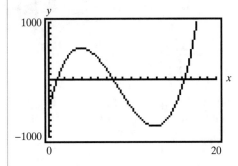

FIGURE 6.1.8 The graph **Y₁=X(30−2X)(20−2X)−500**, $0 \leq x \leq 20$, $-1000 \leq y \leq 1000$.

The tick marks on the horizontal axis represent single units, so we see three solutions—very roughly, near $x = 1$, $x = 8$, and $x = 16$. But the last one **cannot** be used to make an actual tray from our 30×20 cardboard sheet because we cannot cut a 16-cm square off of each end of a piece of cardboard of this size.

Consequently, we actually need to find only the two smaller solutions that are visible in Fig. 6.1.8. Again using the calculator's **solve** function, we find the real solutions $x \approx 0.9903$ and $x \approx 7.7748$ (Fig. 6.1.9).

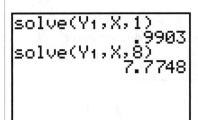

FIGURE 6.1.9 Using **solve** to find the two smaller zeros of $Y_1 = X(30-2X)(20-2X) - 500$.

Thus we have the two possible corner notch sizes $x \approx 0.9903$ cm and $x \approx 7.7748$ cm. Substituting these two possibilities into (6), we get

$$L = 30 - 2(0.9903) = 28.0194 \text{ and}$$
$$W = 20 - 2(0.9903) = 18.0194 \text{ with } x = 0.9903$$

and

$$L = 30 - 2(7.7748) = 14.4504 \text{ and}$$
$$W = 20 - 2(7.7748) = 4.4504 \text{ with } x = 7.7748.$$

Therefore (rounding off to two decimal places), we have the dimensions

28.02 cm by 18.02 cm by 0.99 cm of one possible tray, and
14.45 cm by 4.45 cm by 7.775 cm of another one.

Hence it apparently is possible to make *two* different "trays" of entirely different shapes from the same-size cardboard sheets, each having the desired volume of 500 cm^3. It is important to note that the cubic equation in (9) has three mathematical solutions, but our actual physical question has only two answers.

You should always check your answer to a problem by substituting it in the original equation(s) to make certain they're all satisfied. Here it's a question of whether the dimensions we found actually yield boxes with volume 500 cm^3. The calculator multiplications

$$28.0194 \times 18.0194 \times 0.9903 = 499.9953, \text{ and}$$
$$14.4504 \times 4.4504 \times 7.7748 = 499.9979$$

● corroborate our results; some "round-off" inaccuracy is to be expected in computations when we retain only a few decimal places. But any mistake in our solution would likely have produced a clear-cut disparity in this final check.

Example 4 is fairly typical of applied questions that have polynomial equations like (9) as mathematical models. Not every mathematical solution of the equation necessarily provides a real answer to the applied question. Each mathematical solution must be "checked out"—to see that it actually answers physically the question that was originally posed. In Example 4, each of the two mathematical solutions $x \approx 0.9903$ cm and $x \approx 7.7748$ cm of (9) yields positive tray dimensions with product 500 cm^3, but you should check that the third solution $x \approx 16.2349$ would give negative (hence physically impossible) dimensions if substituted in (6).

Solving Quartic Equations

Figure 6.1.10 illustrates the fact that the graph of a *quartic* polynomial $ax^4 + bx^3 + cx^2 + dx + e$ has

- *three* or fewer bends and
- *four* or fewer real-number zeros.

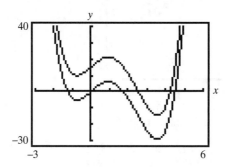

FIGURE 6.1.10
$y = x^4 - 5x^3 + 10x + 15$ (top graph);
$y = x^4 - 5x^3 + 10x$ (bottom graph).

If the leading coefficient a is positive, then the graph rises to the northwest on the left and to the northeast on the right. If there are three bends—the largest possible number—and if your window is big enough to show all of these bends, then you should be able to see how many solutions the corresponding quartic equation has. Simply count the number of intersections of the graph with the x-axis. Thus we see in Fig. 6.1.10 that

- the equation $x^4 - 5x^3 + 10x = 0$ has four solutions; but
- the equation $x^4 - 5x^3 + 10x + 15 = 0$ has just two solutions.

Once such a graph is plotted on our calculator screen, we can locate its solutions graphically by zooming in and using our calculator's zero-finding facility. Alternatively, we can simply apply the **solve** function.

EXAMPLE 5 | Finding Approximate Solutions

Find approximate solutions of the quartic equation $x^4 - 5x^3 + 10x + 15 = 0$.

SOLUTION

Looking at Fig. 6.1.10, we see that the upper graph $y = x^4 - 5x^3 + 10x + 15$ has one x-intercept between 2 and 3 and the other a bit greater than 4. Using the calculator's **solve** function with initial guesses 2 and 4, we see in Fig. 6.1.11 (with **Y1 = X^4 − 5X^3 + 10X + 15**) that the *two* real solutions of the equation

$$x^4 - 5x^3 + 10x + 15 = 0$$

are $x \approx 2.5441$ and $x \approx 4.2516$ (rounded off accurate to four decimal places).
In a similar fashion you can find all *four* real solutions of the equation

$$x^4 - 5x^3 + 10x = 0.$$

SECTION 6.1 Solving Polynomial Equations

```
solve(Y₁,X,2)
              2.5441
solve(Y₁,X,4)
              4.2516
```

FIGURE 6.1.11 The two real solutions of the equation $x^4 - 5x^3 + 10x + 15 = 0$.

● Can you guess how to alter the constant term in these equations to obtain a quartic equation with *no* real solutions?

EXAMPLE 6 | Determining the Number of Solutions

For each quartic equation, use a graph to determine the number of real solutions, and then verify your answer algebraically.

a. $x^4 - 2x^3 + 11x^2 - 20x + 10 = (x^2 + 10)(x - 1)^2 = 0$

b. $x^4 - 10x^3 + 21x^2 + 40x - 100 = (x^2 - 4)(x - 5)^2 = 0$

SOLUTION

a. From the graph in Fig. 6.1.12, we see that $x^4 - 2x^3 + 11x^2 - 20x + 10 = 0$ has exactly *one* real solution.

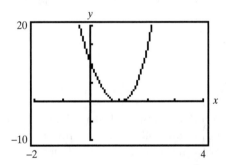

FIGURE 6.1.12 The graph $y = x^4 - 2x^3 + 11x^2 - 20x + 10$.

Solving the factored form of the equation, we find that since

$$(x^2 + 10)(x - 1)^2 = 0, \text{ then}$$

$$x^2 + 10 = 0 \text{ or } (x - 1)^2 = 0.$$

Since $x^2 + 10 = 0$ has no real solutions, and $(x - 1)^2 = 0$ has only one solution, the quartic equation $x^4 - 2x^3 + 11x^2 - 20x + 10 = (x^2 + 10)(x - 1)^2 = 0$ has only one solution $x = 1$.

b. The graph in Fig. 6.1.13 shows that $x^4 - 10x^3 + 21x^2 + 40x - 100 = 0$ has exactly *three* real solutions.

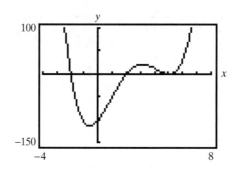

FIGURE 6.1.13 The graph $y = x^4 - 10x^3 + 21x^2 + 40x - 100$ for $-4 \leq x \leq 8$.

Again working with the factored form of the equation, we have

$$(x^2 - 4)(x - 5)^2 = 0$$

$$x^2 - 4 = 0 \quad \text{or} \quad (x - 5)^2 = 0$$

$$(x + 2)(x - 2) = 0 \quad \text{or} \quad (x - 5)^2 = 0.$$

Thus, we can see that the solutions to the quartic equations are the three real numbers $-2, 2,$ and 5.

Examples 5 and 6 illustrate all the possibilities for a quartic equation. It can have no (real) solutions, exactly one solution, two (distinct) solutions, three solutions, or four solutions, but it cannot have more than four solutions. Similarly, it cannot have more than three bends. Based on these facts, you ordinarily should be able to tell whether you've graphed a given quartic polynomial in a sufficiently large window to see its essential features. If so, you should be able to see all its x-intercepts, and can therefore proceed to solve for its real zeros.

The next example presents a problem that appears, at first glance, to have nothing to do with polynomials. You may be surprised to discover that its solution requires solving a quartic equation, which can be accomplished by the techniques we have just illustrated.

EXAMPLE 7   Where Does the Ladder Go?

Figure 6.1.14 shows a 12-foot ladder that leans across a 5-foot fence and just touches a tall wall standing 3 feet behind the fence. How far is the foot of the ladder from the bottom of the fence?

SOLUTION

Observe that the figure contains three *similar* (that is, equal-angled) right triangles. The unknown dimension we seek is the base x of one of the two smaller right triangles. The unknown height of the other small right triangle is labeled y. Recall from geometry the equality of base/height ratios,

$$\frac{b}{h} = \frac{B}{H}$$

in similar right triangles—as in Fig. 6.1.15, where we see a small right triangle with base b and height h and a similar but larger triangle with base B and height H.

FIGURE 6.1.14 The leaning ladder of Example 8.

Applied to the two smaller right triangles in Fig. 6.1.14, with bases 3 and x and with heights y and 5, this fact gives the equation

$$\frac{3}{y} = \frac{x}{5}. \tag{10}$$

The large right triangle in the figure has base $x + 3$, height $y + 5$, and hypotenuse 12. Hence the Pythagorean formula for this triangle is

$$(x + 3)^2 + (y + 5)^2 = 144. \tag{11}$$

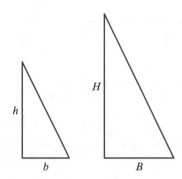

FIGURE 6.1.15 Similar right triangles with bases b, B and heights h, H.

Equations (10) and (11) are two equations that the two unknowns x and y must satisfy. The usual first approach to solving two simultaneous equations is to solve one equation for one unknown and substitute the result in the other equation. If we solve (10) for $y = 15/x$ and substitute the result in (11), we get the equation

$$(x + 3)^2 + \left(\frac{15}{x} + 5\right)^2 = 144. \tag{12}$$

Expansion of the two binomials here gives

$$(x^2 + 6x + 9) + \left(\frac{225}{x^2} + \frac{150}{x} + 25\right) = 144. \tag{13}$$

Finally, we multiply by x^2 (to clear fractions),

$$(x^4 + 6x^3 + 9x^2) + (225 + 150x + 25x^2) = 144x^2$$

and collect like terms to obtain (at last) the promised quartic equation

$$x^4 + 6x^3 - 110x^2 + 150x + 225 = 0 \tag{14}$$

that we need to solve for x.

Figure 6.1.16 shows the graph $y = x^4 + 6x^3 - 110x^2 + 150x + 225$ in the standard window $-10 \leq x \leq 10$.

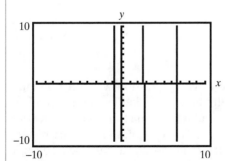

FIGURE 6.1.16 The graph $y = x^4 + 6x^3 - 110x^2 + 150x + 225$ for $-10 \leq x, y \leq 10$.

It appears to show three x-intercepts and hence three solutions. However, the steepness of the graph pieces we see indicates the need for a much wider y-view, which in turn indicates the need for a larger scale on y.

In Fig. 6.1.17 we have increased the y-range to $-1000 \leq y \leq 1000$, but we still see only three solutions rather than the *four* solutions our quartic ought to have. Because we see the graph rising to the northeast as it should but not to the northwest, we evidently need to look further to the west—in the negative x-direction—to find the fourth solution. So, finally, the graph in Fig. 6.1.18 for $-20 \leq x \leq 10$ shows all four solutions.

252 CHAPTER 6 Polynomial Models and Linear Systems

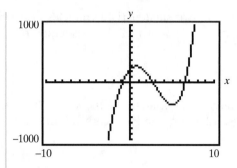

FIGURE 6.1.17 The graph
$y = x^4 + 6x^3 - 110x^2 + 150x + 225$ for
$-10 \leq x \leq 10, -1000 \leq y \leq 1000$.

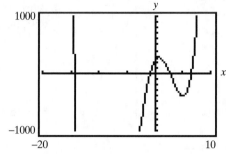

FIGURE 6.1.18 The graph
$y = x^4 + 6x^3 - 110x^2 + 150x + 225$
for $-20 \leq x \leq 10, -1000 \leq y \leq 1000$.

The ticks on the *x*-axis in Fig. 6.1.18 mark *x*-intervals of length 5, so it appears that the four solutions of Eq. (14) are, very roughly, near $-14, -1, 3$, and 6. Using these rough approximations as initial guesses, we find the negative solutions shown in Fig. 6.1.19 and the positive solutions shown in 6.1.20. Thus our quartic equation has

- two negative solutions $x \approx -14.3301$ and $x \approx -0.8921$, and
- two positive solutions $x \approx 2.6972$ and $x \approx 6.5251$.

```
solve(Y₁,X,-14)
           -14.3301
solve(Y₁,X,-1)
             -.8921
```

```
solve(Y₁,X,3)
             2.6972
solve(Y₁,X,6)
             6.5251
```

FIGURE 6.1.19 The two negative solutions of Equation (14).

FIGURE 6.1.20 The two positive solutions of Equation (14).

But the original question was not how many solutions Equation (14) has, but where the foot of the ladder should be placed in order to just graze the top of the fence and lean against the wall as shown in Fig. 6.1.14. Our algebraic derivation of Eq. (14) shows that—if *x* is the distance from the foot of the fence to the foot of the ladder—then *x* must be a solution of the equation, and therefore must be one of the four numbers $-14.3301, -0.8921, 2.6972$, and 6.5251. However, the physical dimension *x* cannot be negative, so only the two positive solutions "qualify." This leaves the two positive solutions 2.6972 and 6.5251 as possibilities.

But which of these two values of *x* is the **answer** to the original question? Precisely where should the foot of the ladder be placed so that it will lean across the fence and just touch the tall wall behind it?

You might get a ruler to serve as a ladder, make a diagram, and try different placements until you're convinced that this problem has two correct answers!

Thus, using Equation (10), $y = 15/x$, to find the corresponding y-values, we find that

- the smaller positive value $x \approx 2.6972$ ft gives a *high* position of the ladder with $y \approx 15/2.6972 \approx 5.5613$ ft, and
- the larger positive value $x \approx 6.5251$ ft gives a *low* position of the ladder with $y \approx 15/6.5251 \approx 2.2988$ ft.

These two ladder positions are shown in Fig. 6.1.21.

FIGURE 6.1.21 The two possible positions of the ladder.

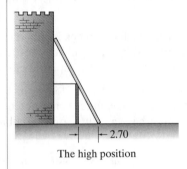
The high position

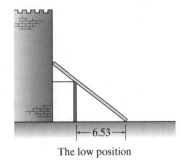
The low position

A final check to assure that both these two xy-pairs "work" is to verify that they satisfy not only Equation (10) but also Equation (11), $(x + 3)^2 + (y + 5)^2 = 144$. Rounding the answers to four decimal places, we find that

$$(2.6972 + 3)^2 + (5.5613 + 5)^2 \approx 143.9991 \approx 144$$

and

$$(6.5251 + 3)^2 + (2.2988 + 5)^2 \approx 144.0000 \approx 144.$$

Thus we do indeed have **two** physically possible solutions.

6.1 Exercises | Building Your Skills

The (exact) solutions of the cubic and quartic equations in Exercises 1–10 are all integers or simple fractions. Discover these exact solutions using the approach of Example 3. That is, first find approximate solutions by graphing, and then verify each solution by substitution in the given equation.

1. $x^3 - 2x^2 - 5x + 6 = 0$
2. $x^3 - 4x^2 - 11x + 30 = 0$
3. $x^4 - 2x^3 - 13x^2 + 14x + 24 = 0$
4. $x^4 - 37x^2 - 24x + 180 = 0$
5. $6x^3 - 7x^2 - 43x + 30 = 0$
6. $4x^4 - 8x^3 - 39x^2 + 43x + 70 = 0$
7. $12x^4 + 13x^3 - 188x^2 - 167x + 330 = 0$

8. $40x^3 + 94x^2 - 261x - 468 = 0$
9. $144x^3 - 522x^2 - 757x + 1365 = 0$
10. $120x^4 + 118x^3 - 1859x^2 - 318x + 819 = 0$

Applying Your Skills

11. A 2 inch × 4 inch × 7 inch iron block is plated with brass so that each of its three dimensions is increased by the same amount x. The volume of the plated block is 50% greater than the volume of the unplated block. What are its dimensions?

12. A 3 inch × 5 inch × 9 inch iron block is plated with brass so that each of its three dimensions is increased by the same amount x. The volume of the plated block is exactly twice the volume of the unplated block. What are its dimensions?

13. A square piece of cardboard with side 45 inches is to be made into an open-topped tray by cutting out square corner notches and turning up the remaining flaps. If the tray must have a volume of 4500 cubic inches, what size squares must be cut out?

14. A square piece of cardboard with side 55 inches is to be made into an open-topped tray by cutting out square corner notches and turning up the remaining flaps. If the tray must have a volume of 6500 cubic inches, what size squares must be cut out?

15. A 17-foot ladder that leans across a 6-foot fence and just touches a tall wall located 4 feet behind the fence. How far is the foot of the ladder from the bottom of the fence?

16. A 25-foot ladder that leans across an 8-foot fence and just touches a tall wall located 5 feet behind the fence. How far is the foot of the ladder from the bottom of the fence?

17. A 72-foot tree stands 18 feet from a 10-foot wall. The tree is to be cut at a height of x feet. Then it will be bent—as indicated in Fig. 6.1.22—so that it leans across the wall and its tip just touches the ground on the other side. It can be shown that the height x of the tree cut must satisfy the cubic equation.

$$4x^3 - 215x^2 + 3280x - 14400 = 0.$$

This equation has three real solutions. However, only two of these three solutions of the equation give actual ways to cut the tree as indicated in the figure. Express these two in feet and inches, rounded off accurate to the nearest inch.

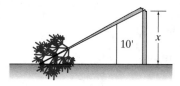

FIGURE 6.1.22 The bent tree of Exercises 17 and 18.

18. An 80-foot tree stands 24 feet from a 10-foot wall. The tree is to be cut at a height of x feet. Then it will be bent (as in Exercise 17) so that it leans across the wall and its tip just touches the ground on the other side. The height x of the tree cut must satisfy the cubic equation

$$5x^3 - 282x^2 + 4500x - 20000 = 0.$$

Find all ways that the tree can be cut, expressing each answer correct to the nearest inch.

Exercises 19–21 refer to a floating cork ball that has a radius of $a = 1$ foot, as illustrated in Fig. 6.1.23. The density of the ball is d times the density of water.

SECTION 6.2   Solving Pairs of Linear Equations—Lots of Ways!   **255**

According to Archimedes, the volume V of the submerged "spherical segment" of the ball is given by

$$V = \frac{\pi x}{6}(3r^2 + x^2) = \frac{4}{3}\pi d a^3. \qquad (12)$$

The unknowns r and x also satisfy the Pythagorean formula

$$r^2 + (a - x)^2 = a^2 \qquad (13)$$

for the right triangle in the figure. In Exercises 19–21,

a. first eliminate r in (12) and (13) with $a = 1$ and the given value of d to obtain a cubic equation in the depth x to which the ball sinks in the water; write your answer in the form $ax^3 + bx^2 + cx + d = 0$ with integer coefficients $a, b, c,$ and d;

b. then find the actual depth x to which the ball sinks in the water, rounded off accurate to the nearest thousandth of an inch.

19. $d = 1/4$

20. $d = 1/5$

21. $d = 2/3$

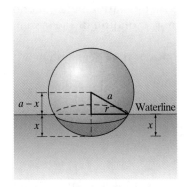

FIGURE 6.1.23 The floating cork ball of Exercises 19–21.

6.2 SOLVING PAIRS OF LINEAR EQUATIONS—LOTS OF WAYS!

In Section 6.1 we talked only about solving *single* equations. Now we talk about solving two equations simultaneously, but both of them must be *linear*. Thus, we consider a pair of linear equations of the form

$$\begin{aligned} ax + by &= p \\ cx + dy &= q, \end{aligned} \qquad (1)$$

where x and y are the unknowns and the coefficients $a, b,$ and p in the first equation and $c, d,$ and q in the second equation are given constants. A solution of this system is simply a pair (x, y) of numbers that satisfy both of the equations.

EXAMPLE 1   Verifying a Solution

Show that the numbers $x = 2, y = -1$ constitute a solution of the linear system

$$\begin{aligned} 2x - y &= 5 \\ x + 2y &= 0. \end{aligned} \qquad (2)$$

SOLUTION

Substituting $x = 2$ and $y = -1$ into each equation yields

$$2 \cdot (2) - (-1) = 5 \quad \text{and} \quad (2) + 2 \cdot (-1) = 0.$$

● So the ordered pair $(2, -1)$ is a solution of the system in (2).

By contrast, the values $x = 3$, $y = 1$ satisfy the first equation in (2) but do not satisfy the second one. Thus the pair $(3, 1)$ is not a solution of the system in (2).

Some linear systems have no solutions at all. Consider the linear system

$$x + y = 1$$
$$2x + 2y = 3. \qquad (3)$$

If $x + y = 1$, then

$$2x + 2y = 2(x + y) = 2(1) = 2 \neq 3.$$

Thus, if the first equation in (3) is satisfied, then $2x + 2y \neq 3$, so the second equation is not satisfied. Hence, the two equations cannot be satisfied simultaneously.

Graphical Solutions

The graph of each of the equations in (1) is a straight line in the plane—because by transposing the x-term and then dividing by the coefficient of y, either can be written in the form $y = mx + b$. Hence the two equations together define two straight lines L_1 and L_2. The numbers x_0 and y_0 then constitute a solution of the system provided that the point (x_0, y_0) lies on both lines.

From a geometric viewpoint, there are three ways that two lines in the xy-plane can look:

- The two lines intersect at a single point (as in Fig. 6.2.1).
- The two lines are parallel and therefore non-intersecting lines (as in Fig. 6.2.2).
- The two lines coincide—they actually are the same line (see Fig. 6.2.3).

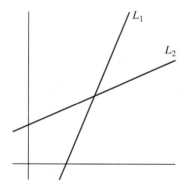

FIGURE 6.2.1 Two intersecting lines—exactly one solution.

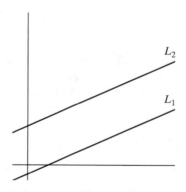

FIGURE 6.2.2 Two parallel lines—no solution.

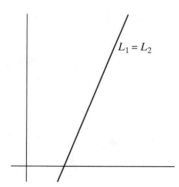

FIGURE 6.2.3 Two coincident lines—infinitely many solutions.

It follows that there are just three possibilities for the solution of a linear pair of equations:

- exactly one solution;
- no solution; or
- infinitely many solutions.

EXAMPLE 2 | Infinitely Many Solutions

Verify that the equations $2x + 2y = 6$ and $4x + 4y = 12$ are algebraic representations of the same line, showing that the linear system

$$2x + 2y = 6$$
$$4x + 4y = 12 \qquad (4)$$

has infinitely many solutions.

SOLUTION

The equations $2x + 2y = 6$ and $4x + 4y = 12$ are both equivalent to the simplified form $x + y = 3$. Hence *every* point (x, y) on this line—for instance, $(1, 2)$ or $(3, 0)$ or $(7, -4)$—provides a solution of the system in (4). So the solution set for this system contains infinitely many ordered pairs—all ordered pairs for which the sum of the coordinates is 3. ●

In the typical applied problem we are interested mainly in the case of exactly one solution. One way to find this single solution is to graph both equations and locate the point of intersection of their straight-line graphs. You are already familiar with this technique because we used it in Chapter 2 when we wanted to determine when two linear functions (perhaps describing population) had the same output value.

EXAMPLE 3 | Solving a Linear System Graphically.

Use a graphical technique to solve the linear system

$$2x - y = 2$$
$$4x + 2y = 10. \qquad (5)$$

SOLUTION

In order to use a graphical approach, we must first have our equations written in $y =$ form. So first we solve each equation for y in terms of x:

$$y = 2x - 2$$
$$y = 5 - 2x. \qquad (6)$$

We can now graph **Y1 = 2X−2** and **Y2 = 5−2X**, with the result shown in Fig. 6.2.4. We have already used our calculator's intersection-finding facility to locate the point of intersection $(1.75, 1.5)$.

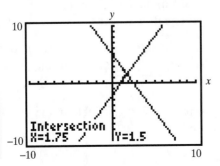

FIGURE 6.2.4 Solving the equations $2x - y = 2, 4x + 2y = 10$ of Example 3.

We naturally wonder whether this means that $x = 1.75$, $y = 1.5$ is the exact solution of the original system in (5) or only an approximate one. We should always verify an apparent graphical solution by direct substitution in the original system:

$$2(1.75) - (1.5) = 3.5 - 1.5 = 2 \text{ and}$$
$$4(1.75) + 2(1.5) = 7 + 3 = 10.$$

Thus, both equations in (5) are satisfied, so the ordered pair (1.75, 1.5) is indeed the exact solution. ●

We can also solve two linear equations numerically. Figure 6.2.5 shows a tabulation of the functions **Y1 = 2X−2** and **Y2 = 5−2X** in (6), and we see again that **Y1** and **Y2** have the same value $y = 1.5$ when $x = 1.75$, thereby showing again that (1.75, 1.5) is the solution to the system.

X	Y1	Y2
1.6000	1.2000	1.8000
1.6500	1.3000	1.7000
1.7000	1.4000	1.6000
1.7500	1.5000	1.5000
1.8000	1.6000	1.4000
1.8500	1.7000	1.3000
1.9000	1.8000	1.2000

X=1.75

FIGURE 6.2.5 Two intersecting lines—exactly one solution. Solving the same two equations numerically.

Algebraic Solutions

In elementary algebra we learn the symbolic **method of elimination** for solution of pairs of linear equations. The method is implemented like this: First we add an appropriate constant multiple of the first equation to the second equation. The idea is to choose the constant multiple that serves to eliminate the variable x from the second equation. We can then solve readily the resulting equation for the remaining variable y. Finally, now knowing y, we find the value of x by substitution of this value of y into the original first equation.

Let's consider again the same system

$$\begin{aligned} 2x - y &= 2 \\ 4x + 2y &= 10. \end{aligned} \tag{7}$$

as in Example 3. We note that the coefficient of x in the second equation is twice the coefficient of x in the first equation. Hence we can eliminate x by subtracting 2 times the first equation from the second equation:

$$\begin{array}{rl} 4x + 2y = 10 & \text{(second equation)} \\ -(4x - 2y = 4) & -\text{(2 times the first equation)} \\ \hline 4y = 6 & \text{(result)} \end{array}$$

This gives $y = 6/4 = 3/2 = 1.5$. Substitution of this value of y into the first equation in (7) gives

$$2x - (1.5) = 2,$$

so
$$2x = 2 + 1.5 = 3.5.$$

Hence $x = \dfrac{3.5}{2} = 1.75$, so we have found algebraically the same solution (1.75, 1.5) that we discovered graphically (and then verified numerically) in Example 3.

You may recall that it really doesn't matter which variable you eliminate initially; it is generally best to eliminate the one that seems easiest to "get rid of." This algebraic method is not as simple to apply if the coefficient of x (or y) in one equation is not an integer multiple of the coefficient of x (or y) in the other equation. The next example illustrates a "sure-fire" elimination technique that always works to eliminate x from a linear pair of equations. We multiply the first equation by the coefficient of x in the second equation, and we multiply the second equation by the coefficient of x in the first equation. Then we subtract one of the resulting equations from the other. This gives the desired equation involving only y.

EXAMPLE 4 Solving a Linear System Algebraically

Use algebraic techniques to solve the linear system

$$7x + 6y = 202$$
$$11x + 9y = 245. \qquad (8)$$

SOLUTION

This system looks pretty formidable. But let's multiply the first equation by 11 and the second one by 7. This gives the pair

$$77x + 66y = 11 \times 202 = 2222$$
$$77x + 63y = 7 \times 245 = 1715 \qquad (9)$$

in which the coefficients of x are the same. Hence subtraction of the second equation in (9) from the first one gives

$$3y = 2222 - 1715 = 507,$$

so

$$y = \frac{507}{3} = 169.$$

Finally, substitution of this value of y into the original first equation gives

$$7x + 6(169) = 202,$$
$$7x = 202 - 6(169) = -812$$
$$x = \frac{-812}{7} = -116$$

Thus we have found the solution $(-116, 169)$ of the system in (8).

The algebra and arithmetic here are rather messy, but it may have been easier in the long run than trying to find an appropriate window to solve the system graphically. Once again, the choice of method is yours. We present all the various options so that you can choose a method that works well for you.

More Symbolic Solutions

We can derive a formula solution by applying the method of Example 4 to the general pair

$$ax + by = p$$
$$cx + dy = q \tag{10}$$

of linear equations in the two unknowns x and y. Multiplication of the first equation by c and multiplication of the second equation by a yields the new pair

$$acx + bcy = pc$$
$$acx + ady = qa.$$

Then subtraction of the first equation here from the second one gives

$$(ad - bc)y = qa - pc,$$

so

$$y = \frac{qa - pc}{ad - bc}. \tag{11}$$

Instead of substituting this value of y into (10) and solving for x, it's simpler to start afresh and multiply the first equation in (10) by d and the second one by b:

$$adx + bdy = pd$$
$$bcx + bdy = qb.$$

Subtraction of the second equation here from the first one yields

$$(ad - bc)x = pd - qb,$$

so

$$x = \frac{pd - qb}{ad - bc}. \tag{12}$$

One can then solve any linear pair of equations by plugging the coefficients $a, b, c, d, p,$ and q into the formulas in (11) and (12).

EXAMPLE 5 Using Formulas to Solve a Linear System

Use formulas (11) and (12) to solve the linear system

$$2x - y = 2$$
$$4x + 2y = 10. \tag{13}$$

SOLUTION

Matching the coefficients in our system with the coefficients in (10), we find that

$$a = 2, \quad b = -1, \quad p = 2,$$
$$c = 4, \quad d = 2, \quad q = 10. \tag{14}$$

[Observe that, if the literal symbols and the equal signs are removed in (13) and in (14), then the same array of six numbers remains.] Then the formulas in (11) and (12) give

$$y = \frac{10 \cdot 2 - 2 \cdot 4}{2 \cdot 2 - (-1) \cdot 4} = \frac{12}{8} = \frac{3}{2} \quad \text{and} \quad x = \frac{2 \cdot 2 - 10 \cdot (-1)}{2 \cdot 2 - (-1) \cdot 4} = \frac{14}{8} = \frac{7}{4},$$

the same solution (1.75, 1.5) found previously. [Notice that we calculated y first and then x, but the ordered pair must still be given in (x, y) format.]

Determinants

Although it's virtually automatic, the method of solution illustrated in Example 5 surely seems useless as a practical matter because you could never remember formulas (11) and (12) when you need them. Or could you?

Observe that the fractions giving x and y in (11) and (12) have the same denominator $ad - bc$. This is the value of the 2 by 2 (or 2×2) **determinant** defined by

$$\Delta = \begin{vmatrix} a & b \\ c & d \end{vmatrix} = ad - bc, \tag{15}$$

where Δ is the upper-case Greek letter delta.

Note that the value $ad - bc$ is the product of the (left-to-right) downward diagonal elements a and d, minus the product of the upward diagonal elements c and b. For instance,

$$\begin{vmatrix} 5 & 3 \\ 21 & 17 \end{vmatrix} = (5)(17) - (21)(3) = 85 - 63 = 22.$$

You should teach your "fingers to do the walking"—from upper left to lower right, then from lower left to upper right.

The particular determinant in (15) is called the **coefficient determinant** of the linear system

$$ax + by = p$$
$$cx + dy = q \tag{16}$$

because the numbers a, b, c, d appear as the same "array"—both in the determinant [on the left in (15)] and as the coefficients of the unknowns on the left in (16).

Now let's look at the numerators in (11) and (12). The numerator in the expression for x is

$$\begin{vmatrix} p & b \\ q & d \end{vmatrix} = pd - qb,$$

while the numerator in the expression for y is

$$\begin{vmatrix} a & p \\ c & q \end{vmatrix} = qa - pc.$$

Therefore, Equations (11) and (12) say that

$$x = \frac{\begin{vmatrix} p & b \\ q & d \end{vmatrix}}{\begin{vmatrix} a & b \\ c & d \end{vmatrix}} \quad \text{and} \quad y = \frac{\begin{vmatrix} a & p \\ c & q \end{vmatrix}}{\begin{vmatrix} a & b \\ c & d \end{vmatrix}} \quad (17)$$

Finally, the point is that—unlike Equations (11) and (12)—these formulas for x and y are easy to remember! But how, and why? Well, it's because the numerator determinant for x is obtained from the coefficient determinant upon replacing the coefficients $\begin{bmatrix} a \\ c \end{bmatrix}$ of x in (16) by the constants $\begin{bmatrix} p \\ q \end{bmatrix}$. Similarly, the numerator determinant for y is obtained from the coefficient determinant upon replacing the coefficients $\begin{bmatrix} b \\ d \end{bmatrix}$ of y by the same constants $\begin{bmatrix} p \\ q \end{bmatrix}$. In short, when setting up the numerator determinant to solve for either unknown x or y, *its* coefficients are replaced by the constants on the other side in the pair of equations.

It is important to remember that in order to divide by it, the coefficient determinant appearing in the denominators in (17) must be *nonzero*. If $\Delta = 0$, then it turns out that the system has *either*

- infinitely many different solutions (because the two equations have the same straight line graph), *or*
- no solution at all (because the two equations have parallel nonintersecting straight-line graphs).

However, we need not be concerned here with these exceptional cases.

EXAMPLE 6 Using Determinants to Solve a Linear System

Use determinants to solve once more the system

$$2x - y = 2$$
$$4x + 2y = 10.$$

SOLUTION

We start by calculating the system's coefficient determinant

$$\begin{vmatrix} 2 & -1 \\ 4 & 2 \end{vmatrix} = (2)(2) - (-1)(4) = 4 - (-4) = 8$$

that appears in the denominators for both x and y. Then

$$x = \frac{\begin{vmatrix} 2 & -1 \\ 10 & 2 \end{vmatrix}}{8} = \frac{(2)(2) - (10)(-1)}{8} = \frac{4 + 10}{8} = \frac{14}{8} = \frac{7}{4}$$

$$y = \frac{\begin{vmatrix} 2 & 2 \\ 4 & 10 \end{vmatrix}}{8} = \frac{(2)(10) - (4)(2)}{8} = \frac{20 - 8}{8} = \frac{12}{8} = \frac{3}{2}.$$

As we expect, we once again get the same solution $(1.75, 1.5)$ to our system of linear equations.

EXAMPLE 7　Another Determinant Solution

Solve again the system

$$7x + 6y = 202$$
$$11x + 9y = 245.$$

SOLUTION

Once again we start by calculating the coefficient determinant

$$\begin{vmatrix} 7 & 6 \\ 11 & 9 \end{vmatrix} = (7)(9) - (11)(6) = 63 - 66 = -3.$$

that appears in both denominators in (17). Then

$$x = \frac{\begin{vmatrix} 202 & 6 \\ 245 & 9 \end{vmatrix}}{-3} = \frac{(202)(9) - (245)(6)}{-3} = \frac{1818 - 1470}{-3} = \frac{348}{-3} = -116$$

$$y = \frac{\begin{vmatrix} 7 & 202 \\ 11 & 245 \end{vmatrix}}{-3} = \frac{(7)(245) - (11)(202)}{-3} = \frac{1715 - 2222}{-3} = \frac{-507}{-3} = 169.$$

The bottom line is this: Whereas it may seem that the solution of a pair of linear equations by elimination involves some ingenuity, the solution using determinants is a (nearly) fail-safe matter. Just substitute in the determinant formulas (17), and complete the calculation.

Using Matrices

A **matrix** is simply a rectangular array—that is, a rectangular pattern—of numbers. For instance, the determinant formulas in (17) suggest that the solutions of the linear system

$$ax + by = p$$
$$cx + dy = q$$

are determined by its matrix

$$\mathbf{A} = \begin{bmatrix} a & b \\ c & d \end{bmatrix} \tag{18}$$

of left-hand side coefficients (of the unknowns) and its matrix

$$\mathbf{B} = \begin{bmatrix} p \\ q \end{bmatrix} \tag{19}$$

of right-hand side constants. The letters **A** and **B** simply represent the indicated arrays in (18) and (19).

The **coefficient matrix A** in (18) is called a 2 × 2 (or 2 by 2) matrix, for obvious reasons—it has the two *horizontal* **rows**

$$[a \quad b] \text{ and}$$
$$[c \quad d],$$

as well as the two *vertical* **columns**

$$\begin{bmatrix} a \\ c \end{bmatrix} \text{ and } \begin{bmatrix} b \\ d \end{bmatrix}.$$

The **constant matrix B** in (19) is a 2 × 1 matrix because it has the two rows [a] and [b] but just one vertical column.

In terms of its coefficient and constant matrices, the linear system in (16) can be abbreviated by writing

$$\begin{bmatrix} a & b \\ c & d \end{bmatrix} \begin{bmatrix} x \\ y \end{bmatrix} = \begin{bmatrix} p \\ q \end{bmatrix},$$

or simply

$$\mathbf{A} \begin{bmatrix} x \\ y \end{bmatrix} = \mathbf{B}, \tag{20}$$

where

$$\begin{bmatrix} x \\ y \end{bmatrix}$$

is a 2 × 1 *unknown matrix* consisting of the two unknown numbers x and y.

In some ways, matrices can be used symbolically like other algebraic symbols. For instance, think of the fact that the solution of the single equation

$$Ax = B \tag{21}$$

(where A, B, and x represent ordinary numbers) can be written in the form

$$x = \frac{B}{A} = A^{-1}B. \tag{22}$$

There is an obvious analogy between (20) and (21)—in each we see a coefficient number or matrix, an unknown number or matrix, and a constant number or matrix. In analogy with (22) we can symbolically write

$$\begin{bmatrix} x \\ y \end{bmatrix} = \mathbf{A}^{-1}\mathbf{B} \tag{23}$$

to represent the solution matrix for the original linear system in (16).

The amazing thing is that an algebra of matrices has been devised so that (23) actually makes sense! That is, the "inverse matrix" $\mathbf{A}^{-1}$ can be calculated as a 2 × 2 matrix, and then the product of the matrices $\mathbf{A}^{-1}$ and $\mathbf{B}$ can be calculated, so that the result is the 2 × 1 **solution matrix** $\begin{bmatrix} x \\ y \end{bmatrix}$ consisting of the two unknown values x and y. Indeed, our graphing calculator can do this!

EXAMPLE 8   Solving a Linear System Using Matrices

Use matrices to solve (yet again) the linear system

$$7x + 6y = 202$$
$$11x + 9y = 245.$$

SOLUTION

This linear system has its coefficient and constant matrices given by

$$\mathbf{A} = \begin{bmatrix} 7 & 6 \\ 11 & 9 \end{bmatrix} \quad \text{and} \quad \mathbf{B} = \begin{bmatrix} 202 \\ 245 \end{bmatrix}.$$

As shown in Fig. 6.2.6, we enter these matrices in the calculator using the **MATRIX EDIT** menu, where available matrices are denoted by **[A]**, **[B]**, **[C]**, and so forth. (Menus regarding matrices are labeled as "**MATRIX**" in the TI-84 and "**MATRX**" in the TI-83.) If we select matrix **[A]** and enter 2 × 2, we see a 2 by 2 array ready to receive our entries (Fig. 6.2.7).

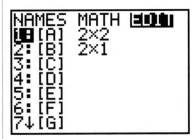

FIGURE 6.2.6 The **MATRX EDIT** menu.

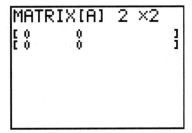

FIGURE 6.2.7 The matrix **[A]** before entry of the coefficients in our system.

We can now enter the coefficients in our linear system one at a time, first those on the first row, then those on the second row (Fig. 6.2.8). Next we enter similarly our system's 2 × 1 constant matrix **[B]** (Fig. 6.2.9).

FIGURE 6.2.8 The matrix **[A]** after entry of the coefficients in our system.

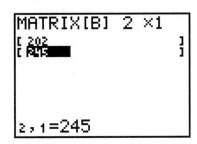

FIGURE 6.2.9 Our system's constant matrix **[B]**.

Now we're ready! We need only enter **[A]**$^{-1}$***[B]** (as in Eq. (23))—using the **MATRIX NAMES** menu—to calculate the solution (Fig. 6.2.10).

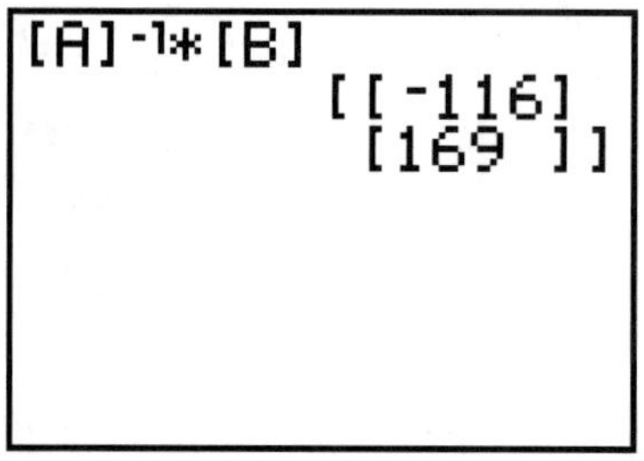

```
[A]-1*[B]
         [[-116]
          [169 ]]
```

FIGURE 6.2.10 Calculating the solution matrix $\begin{bmatrix} x \\ y \end{bmatrix}$.

The result

$$\begin{bmatrix} x \\ y \end{bmatrix} = \begin{bmatrix} -116 \\ 169 \end{bmatrix}$$

tells us that the solution of our system is given by $x = -116$, $y = 169$. (No surprise, since we have solved this system twice already.)

To solve any other linear system of two equations in two unknowns, we need only re-enter the elements of the coefficient and constant matrices **[A]** and **[B]** for our new system, then re-calculate the solution matrix **[A]**$^{-1}$***[B]**.

Applications

Algebra textbooks usually contain several types of "word problems"—such as coin problems, mixture problems, and the like—that may seem unrelated but actually fit a single pattern. Such a problem typically deals with the unknown numbers x and y of two different items, such as widgets and wodgets. The total number N of widgets and wodgets is given, so we know that

$$x + y = N.$$

We also know that the value of each widget is c, and the value of each wodget is d. Finally, the total value V of the all the widgets and wodgets is given, so we also know that

$$cx + dy = V.$$

We can therefore find the number x of widgets and the number y of wodgets by solving the linear system

$$x + y = N$$
$$cx + dy = V. \qquad (24)$$

These equations provide a common mathematical model for a wide range of "value problems." Except for inflation, the following example is like a coin problem—in which one typically is asked to find the numbers of nickels and dimes in a bag of coins.

EXAMPLE 9 Solving a Value Problem

You are walking down the street minding your own business when you spot a thick envelope lying on the sidewalk. It turns out to contain fives and twenties, a total of 61 bills with a total value of $875. How many of each type of bill is there?

SOLUTION

If there are x fives and y twenties, then

$$x + y = 61 \quad \text{(total number)}$$
$$5x + 20y = 875 \quad \text{(total value)} \tag{25}$$

because each of the x fives is worth 5 dollars, and each of the y twenties is worth 20 dollars. Using the determinant method in Equation (17), we find that

$$x = \frac{\begin{vmatrix} 61 & 1 \\ 875 & 20 \end{vmatrix}}{\begin{vmatrix} 1 & 1 \\ 5 & 20 \end{vmatrix}} = \frac{1220 - 875}{20 - 5} = \frac{345}{15} = 23$$

and

$$y = \frac{\begin{vmatrix} 1 & 61 \\ 5 & 875 \end{vmatrix}}{\begin{vmatrix} 1 & 1 \\ 5 & 20 \end{vmatrix}} = \frac{875 - 305}{20 - 5} = \frac{570}{15} = 38.$$

● Thus our lucky envelope contained 23 fives and 38 twenties.

The denominator determinant in this example is 15, however many fives and twenties there are. Can you see that the total value of an envelope full of fives and twenties must therefore be an integer multiple of $15? For instance, an envelope could contain $990 (a multiple of 15) in fives and twenties, but it could not contain $1000 (not a multiple of 15) in fives and twenties.

The following example is a mixture problem. We can regard it as a value problem modeled by Equations (24) by regarding the water in an alcohol–water solution as worthless. Then the value of the solution is measured by the amount of alcohol it contains.

EXAMPLE 10 Solving a Mixture Problem

Suppose large quantities of both a 7% alcohol solution and a 17% alcohol solution are available. How many gallons of each must be mixed in order to get 42 gallons of a 13% alcohol solution?

SOLUTION

Suppose we mix x gallons of the 7% solution with y gallons of the 17% solution. We must have

$$x + y = 42$$

in order to get 42 gallons of the mixed 13% solution. Now

$$\begin{pmatrix} \text{alcohol in the} \\ \text{7\% solution} \end{pmatrix} + \begin{pmatrix} \text{alcohol in the} \\ \text{17\% solution} \end{pmatrix} = \begin{pmatrix} \text{alcohol in the} \\ \text{13\% solution} \end{pmatrix},$$

that is,

$$(7\% \text{ of } x) + (17\% \text{ of } y) = (13\% \text{ of } 42),$$

so

$$0.07x + 0.17y = 0.13 \times 42 = 5.46.$$

This last equation is of the form of the second equation in (24), with $c = 0.07$, $d = 0.17$, and $V = 5.46$. Thus we can find x and y by solving the two equations

$$\begin{aligned} x + y &= 42 \\ 0.07x + 0.17y &= 5.46. \end{aligned} \quad (26)$$

Using the determinant method in Equation (17), we find that

$$x = \frac{\begin{vmatrix} 42 & 1 \\ 5.46 & 0.17 \end{vmatrix}}{\begin{vmatrix} 1 & 1 \\ 0.07 & 0.17 \end{vmatrix}} = \frac{42 \times 0.17 - 5.46 \times 1}{0.17 - 0.07} = \frac{1.68}{0.1} = 16.8$$

and

$$y = \frac{\begin{vmatrix} 1 & 42 \\ 0.07 & 5.46 \end{vmatrix}}{\begin{vmatrix} 1 & 1 \\ 0.07 & 0.17 \end{vmatrix}} = \frac{1 \times 5.46 - 0.07 \times 42}{0.17 - 0.07} = \frac{2.52}{0.1} = 25.2.$$

Thus we should mix 16.8 gallons of a 7% solution with 25.2 gallons of a 17% solution in order to get 42 gallons of a 13% solution. ●

SUMMARY

In this section you have seen five different methods of solving two linear equations in two unknowns:

- graphical—as in Example 3;
- elimination—as in Example 4;
- symbolic formulas—as in Example 5;
- determinants—as in Examples 6 and 7; and
- matrices—as in Example 8.

Each of these methods yields its own insights, but in applications like Examples 9 and 10 the determinant method is likely to yield the quickest and most routine calculation. The matrix method is used mainly for larger linear systems (as in Section 6.3).

6.2 Exercises

Building Your Skills

Solve the linear systems in Exercises 1–10 using two or more of the methods listed in the Summary for this section. Each x- and y-value is an integer or a simple fraction whose denominator is not greater than 10. Verify that the values you obtain actually satisfy the given equations exactly.

1. $7x + 4y = 26$
   $5x + 3y = 19$

2. $5x + 11y = 38$
   $3x + 7y = 22$

3. $3x + 5y = 15$
   $5x + 10y = 20$

4. $5x + 6y = 18$
   $3x + 4y = 11$

5. $5x + 5y = 3$
   $7x + 8y = 5$

6. $7x - 3y = 14$
   $9x - 4y = 15$

7. $17x - 8y = 163$
   $23x - 16y = 29$

8. $17x - 8y = 163$
   $23x - 16y = 29$

9. $323x - 467y = 665$
   $433x - 642y = 285$

10. $3859x - 6691y = 5645$
    $5751x - 9976y = 1628$

Applying Your Skills

11. You have a bag of 77 nickels and dimes worth 5 dollars altogether. How many coins of each type do you have?

12. You have a bag of 666 nickels and dimes worth 60 dollars altogether. How many coins of each type do you have?

13. You have a bag of 49 dimes and quarters worth 10 dollars altogether. How many coins of each type do you have?

14. You have a bag of 555 nickels and dimes worth 90 dollars altogether. How many coins of each type do you have?

15. You buy two dozen hamburgers and cheeseburgers for a small party and pay a total of $19.80. If hamburgers cost 75 cents each and cheeseburgers cost 95 cents each (including tax), how many of each did you buy?

16. You buy 100 dozen chicken and beef sandwiches for a big class reception and pay a total of $1604.60. If chicken sandwiches cost $1.25 cents each and beef sandwiches cost $1.45 cents each (including tax), how many of each did you buy?

17. A package store sold 696 bottles of wine and liquor during a single day, with total receipts of $5298. If the price of wine was $5.75 per bottle and the price of liquor was $11.75 per bottle, how many bottles of each were sold that day?

18. A automobile dealer sold 47 cars one month for a total of $718,500. Some were small cars that sold for $13,500 each, and some were large cars that sold for $18,750 each. How many of each were sold that month?

19. Suppose both a 6% alcohol solution and a 12% alcohol solution are available. How many gallons of each must be mixed in order to get 30 gallons of a 10% alcohol solution?

20. Suppose both a 4% alcohol solution and a 14% alcohol solution are available. How many gallons of each must be mixed in order to get 50 gallons of a 7% alcohol solution?

For Exercises 21 and 22, you need to know that 24-karat gold is pure gold, that 12-karat gold is an alloy that is half gold, that 17-karat gold is an alloy that consists of 17 parts gold and 7 parts of another metal, and so on.

21. Suppose that both pure gold and 14-karat gold are available. How many grams of each must be mixed to get 120 grams of 17-karat gold?

22. Suppose that both 8-karat gold and 16-karat gold are available. How many grams of each must be mixed to get 120 grams of 13-karat gold?

For Exercises 23 and 24, you need to know that brass is an alloy of copper and zinc. For instance, 75% brass is 75% copper and 25% zinc; 90% brass is 90% copper and 10% zinc; and so on.

23. Suppose that both 75% brass and 90% brass are available. How many pounds of each must be mixed to get 150 pounds of 87% brass?

24. Suppose that both 65% brass and 95% brass are available. How many pounds of each must be mixed to get 90 pounds of 77% brass?

6.3 LINEAR SYSTEMS OF EQUATIONS

In Section 6.2 we talked about solving two linear equations in two unknowns. Now we talk about solving three or more linear equations in the same number of unknowns. For instance, three linear equations in the three unknowns x, y, z take the form

$$ax + by + cz = p$$
$$dx + ey + fz = q \qquad (1)$$
$$gz + hy + kz = r,$$

where the coefficients on the left-hand sides and the constants on the right-hand sides are given numbers. A **solution** of this system is simply an ordered triple (x, y, z) of numbers that satisfy all three of the equations.

In this section we discuss methods of solving the system in (1) that are analogous to the methods of Section 6.2 for pairs of linear equations:

- the method of elimination,
- the use of determinants, and
- the use of matrices.

Graphical and numerical tabulation methods are not as useful for three or more equations as they are for two equations in two unknowns. One reason is that the graph of a single linear equation in three variables is a *plane* in space (rather than a line in the plane). Therefore, a solution of the system in (1) is a common point of intersection of three planes in space, and this situation is too complicated to represent effectively with a simple graphing calculator.

The Method of Elimination

The three equations in (1) constitute three *pairs* of equations—the first and second equations, the first and third equations, and the second and third equations. If we eliminate x in two of these pairs, we are left with two equations in the two unknowns y and z that can be solved using any one of the methods discussed in Section 6.2. We can then substitute the values of y and z that are found into one of the original equations to solve finally for the value of x.

EXAMPLE 1 Solving a Linear System Using Elimination

Use the method of elimination to solve the system

$$\begin{aligned} x + 2y + z &= 4 \\ 3x + 8y + 7z &= 20 \\ 2x + 7y + 9z &= 23. \end{aligned} \qquad (2)$$

SOLUTION

First, we observe that we can eliminate x by subtracting 3 times the first equation from the second equation:

$$(3x + 8y + 7z) - 3 \cdot (x + 2y + z) = 20 - 3 \cdot (4).$$

The result is the equation

$$2y + 4z = 8. \qquad (3)$$

Next, we eliminate x again by subtracting 2 times the first equation from the third equation:

$$(2x + 7y + 9z) - 2 \cdot (x + 2y + z) = 23 - 2 \cdot (4).$$

This gives

$$3y + 7z = 15. \qquad (4)$$

Equations (3) and (4) constitute a linear system

$$2y + 4z = 8$$
$$3y + 7z = 15 \tag{5}$$

in the two unknowns y and z. We might ordinarily use 2×2 determinants to solve this system. However, note that the first equation here can be divided by 2 and then solved for

$$y = 4 - 2z. \tag{6}$$

Substitution in the second equation in (5) then gives

$$3(4 - 2z) + 7z = 15,$$
$$12 - 6z + 7z = 15,$$
$$z = 3.$$

Substitution of $z = 3$ in (6) gives

$$y = 4 - 2(3) = -2.$$

Now that we know the values $y = -2$ and $z = 3$, substitution in the first equation in (2) yields

$$x + 2 \cdot (-2) + (3) = 4,$$
$$x - 4 + 3 = 4,$$
$$x = 5.$$

Consequently, we have found the solution $x = 5, y = -2, z = 3$ of the original system (2).

We just substituted the values of y and z we found into the original first equation to find the value of x. It is therefore a good check to verify that our three values $x = 5, y = -2, z = 3$ satisfy the other two equations in (2):

$$3(5) + 8(-2) + 7(3) = 15 - 16 + 21 = 20$$
$$2(5) + 7(-2) + 9(3) = 10 - 14 + 27 = 23$$

● Thus all is well, and the ordered triple $(5, -2, 3)$ is the solution to this system.

The precise steps in the solution of Example 1 should *not* be memorized. Instead, carry out the basic approach for three equations in three unknowns however you wish, along the following lines:

- First, eliminate one of the unknowns from some pair of the given equations.
- Then, eliminate the same unknown from another pair of the given equations.
- Solve the resulting pair of equations in two unknowns.
- Finally, substitute the values found into one of the original equations to find the value of the unknown that was eliminated.

In short, you start with the idea of eliminating one of the unknowns and then solving the equations that remain. As for the details, just "hammer it out" as you go, any which way you can.

In principle, the method of elimination can be applied to systems of four linear equations in four unknowns, to five linear equations in five unknowns, and so on. But in practice it is usually easier and quicker for systems larger than 3×3 to use the determinant and matrix methods that we discuss next.

The Determinant Method

Although it would be extremely tedious, one could carry out the method of elimination with the general 3×3 linear system in (1) to solve for the values of the three unknowns x, y, z in terms of the coefficients and constants that appear in the three equations. The resulting formulas are best expressed in terms of 3×3 determinants.

The **coefficient determinant** of the system in (1) is defined by

$$\Delta = \begin{vmatrix} a & b & c \\ d & e & f \\ g & h & k \end{vmatrix} = +a \begin{vmatrix} e & f \\ h & k \end{vmatrix} - b \begin{vmatrix} d & f \\ g & k \end{vmatrix} + c \begin{vmatrix} d & e \\ g & h \end{vmatrix}. \tag{7}$$

(We write the Greek capital Δ for D for determinant.) It remains to evaluate the three 2×2 determinants to get the value of Δ. Observe both

- the $+ - +$ pattern of signs on the right-hand side in (7), and that
- each element of the first row of the 3×3 determinant is multiplied by the 2×2 "subdeterminant" that remains when that element's row and column are deleted.

EXAMPLE 2 | Finding the Coefficient Determinant

Find the coefficient determinant of the linear system

$$x + 2y + z = 4$$
$$3x + 8y + 7z = 20 \tag{8}$$
$$2x + 7y + 9z = 23.$$

of Example 1.

SOLUTION

The value of the coefficient determinant of the system is

$$\Delta = \begin{vmatrix} 1 & 2 & 1 \\ 3 & 8 & 7 \\ 2 & 7 & 9 \end{vmatrix} = +(1) \begin{vmatrix} 8 & 7 \\ 7 & 9 \end{vmatrix} - (2) \begin{vmatrix} 3 & 7 \\ 2 & 9 \end{vmatrix} + (1) \begin{vmatrix} 3 & 8 \\ 2 & 7 \end{vmatrix}$$

$$= +(72 - 49) - 2(27 - 14) + (21 - 16)$$
$$= +23 - 2(13) + 5$$
$$= 23 - 26 + 5,$$

so $\Delta = 2$.

274 CHAPTER 6 Polynomial Models and Linear Systems

Here we have expanded the determinant "along its first row," but a 3 × 3 determinant can be expanded along any row or column. We need only remember the checkerboard pattern

$$\begin{vmatrix} + & - & + \\ - & + & - \\ + & - & + \end{vmatrix} \tag{9}$$

of signs that must be attached to the particular row or column elements we use as coefficients of the corresponding three 2 × 2 subdeterminants. For instance, to expand the coefficient determinant of Example 2 along its second column (instead of along its first row), we attach the signs that appear in the second column in (9). This gives

$$\Delta = \begin{vmatrix} 1 & 2 & 1 \\ 3 & 8 & 7 \\ 2 & 7 & 9 \end{vmatrix} = -(2)\begin{vmatrix} 3 & 7 \\ 2 & 9 \end{vmatrix} + (8)\begin{vmatrix} 1 & 1 \\ 2 & 9 \end{vmatrix} - (7)\begin{vmatrix} 1 & 1 \\ 3 & 7 \end{vmatrix}$$

$$= -2(27 - 14) + 8(9 - 2) - 7(7 - 3)$$

$$= -2(13) + 8(7) - 7(4)$$

$$= -26 + 56 - 28 = 2,$$

so we get the same value $\Delta = 2$ of Example 2 by an entirely different calculation. You should check this value by calculating it still another way—by expanding along the first column or along the third row, for instance.

The **determinant formulas** for solution of the 3 × 3 linear system

$$ax + by + cz = p$$
$$dx + ey + fz = q$$
$$gx + hy + kz = r \tag{10}$$

are

$$x = \frac{\begin{vmatrix} p & b & c \\ q & e & f \\ r & h & k \end{vmatrix}}{\Delta}, \quad y = \frac{\begin{vmatrix} a & p & c \\ d & q & f \\ g & r & k \end{vmatrix}}{\Delta}, \quad z = \frac{\begin{vmatrix} a & b & p \\ d & e & q \\ g & h & r \end{vmatrix}}{\Delta}. \tag{11}$$

The way to remember these formulas is the same as with the 2 × 2 determinant formulas in Section 6.2. We observe that, in solving for any one of the unknowns, *that* unknown's column of coefficients are replaced (in its numerator determinant) by the right-hand-side constants in the system (10). Thus

- the numerator determinant for x is obtained from the coefficient determinant upon replacing the first-column coefficients a, d, g by p, q, r;
- the numerator determinant for y is obtained from the coefficient determinant upon replacing the second-column coefficients b, e, h by p, q, r;
- the numerator determinant for z is obtained from the coefficient determinant upon replacing the third-column coefficients c, f, k by p, q, r.

Consequently, there is no need to remember the particular letters that appear in (11). Instead, we think solely of replacing columns of left-hand side coefficients with the column of right-hand-side constants.

Recall that the coefficient determinant Δ must be nonzero in order for us to be able to divide by it in (11). If $\Delta = 0$, then it turns out that the system has *either* infinitely many different solutions *or* no solution at all. In the applications we consider, we need not be concerned with these exceptional cases.

Determinant calculations by hand are particularly error-prone, so it's good practice to calculate each one two different ways (by expanding along different rows and columns). This can be tedious, so the use of a calculator is an attractive alternative. In this case, the accurate entry of each determinant is crucial because the entry of the nine elements of a 3×3 determinant is itself somewhat error-prone. We will illustrate the use of the calculator to evaluate the determinants in the next example.

EXAMPLE 3 Solving a Linear System Using Determinants

Solve the system

$$x + 2y + z = 4$$
$$3x + 8y + 7z = 20 \qquad (12)$$
$$2x + 7y + 9z = 23$$

using determinants.

SOLUTION

Unnecessary labor—and additional possibility of keystroke errors—can be avoided by an efficient approach to entry of the four determinants that are needed for the solution of a 3×3 system. Figure 6.3.1 shows the entry of the 3×3 array or matrix **[D]** of coefficients in the 3×3 system (12). As in Section 6.2, the available calculator names for matrices are selected from the **MATRIX NAMES** menu, and are automatically enclosed with square brackets. (The TI-84 **MATRIX** command is **MATRX** on the older TI-83 calculator.)

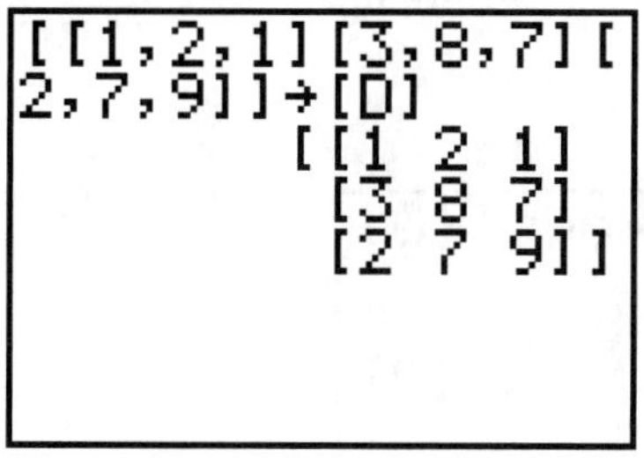

FIGURE 6.3.1 The coefficient matrix of the 3×3 system in (12).

Instead of entering separately the three numerator matrices in (11), we first make three separate copies **[A]**, **[B]**, and **[C]** of the coefficient matrix (Fig. 6.3.2). Then we need only edit each of these matrices, "adjusting" the appropriate column of each to get the numerator matrices required for the x, y, and z determinant formulas. For instance, Fig. 6.3.3 shows the original first column of **[A]** replaced with the column of right-hand-side constants in (12).

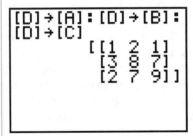

FIGURE 6.3.2 Making three copies of the coefficient matrix.

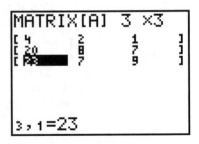

FIGURE 6.3.3 The numerator matrix **[A]** for x.

Figures 6.3.4 and 6.3.5 show the numerator matrices for y and z. Editing the three matrices is much quicker than entering them from scratch.

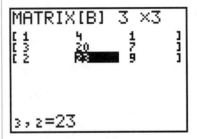

FIGURE 6.3.4 The numerator matrix **[B]** for y.

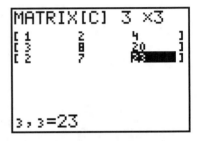

FIGURE 6.3.5 The numerator matrix **[C]** for z.

Once they're entered, we divide out the determinants in (11) to calculate the desired values of the unknowns $x, y,$ and z. Since the numerator matrices **[A]**, **[B]**, and **[C]** correspond to the unknowns $x, y,$ and z and the denominator matrix is **[D]**, the determinant formulas in (7) take the calculator forms

$$x = \frac{\text{det}(\mathbf{[A]})}{\text{det}(\mathbf{[D]})}, \quad y = \frac{\text{det}(\mathbf{[B]})}{\text{det}(\mathbf{[D]})}, \quad z = \frac{\text{det}(\mathbf{[C]})}{\text{det}(\mathbf{[D]})}$$

that we see in Fig. 6.3.6.

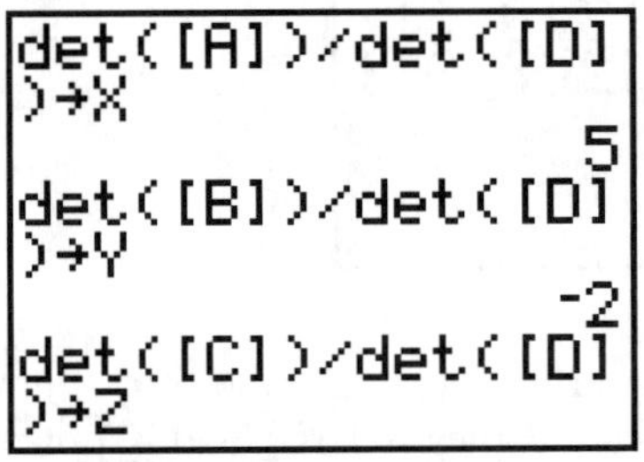

FIGURE 6.3.6 Finally, calculating the unknowns $x, y,$ and z.

Thus, the solution to the linear system is the triple $x = 5, y = -2, z = 3$, the same solution we obtained when we solved the system by elimination.

The key to "making math easy" often lies in making the notation work for us (rather than the other way around). Surely it's easy to remember that the numerator matrices **[A]**, **[B]**, and **[C]** correspond to the unknowns x, y, and z and that the denominator matrix is **[D]**.

As we saw in Section 6.2, the solution of a system of three linear equations by elimination involves some ingenuity; the solution using determinants works easily (as long as the coefficient determinant is not zero). Just substitute in the determinant formula (11) or (12), and grind it out (either by hand or by calculator).

Using Matrices

Figures 6.3.1–6.3.6 illustrate the calculator entry and manipulation of 3×3 arrays or *matrices*. In general, matrices are used to solve linear systems of three or more equations in the same way as with two equations in two unknowns (Section 6.2). The solutions of the 3×3 system

$$ax + by + cz = p$$
$$dx + ey + fz = q \qquad (13)$$
$$gx + hy + kz = r$$

are determined by its 3×3 **coefficient matrix**

$$\mathbf{A} = \begin{bmatrix} a & b & c \\ d & e & f \\ g & h & k \end{bmatrix} \qquad (14)$$

and its 3×1 matrix

$$\mathbf{B} = \begin{bmatrix} p \\ q \\ r \end{bmatrix} \qquad (15)$$

of right-hand side constants. Observe that the three *rows*

$$[a \quad b \quad c],$$
$$[d \quad e \quad f],$$
$$[g \quad h \quad k]$$

of the coefficient matrix **A** contain (respectively) the coefficients of the unknowns in the three separate equations in (13). The three *columns*

$$\begin{bmatrix} a \\ d \\ g \end{bmatrix} \begin{bmatrix} b \\ e \\ h \end{bmatrix} \begin{bmatrix} c \\ f \\ k \end{bmatrix}$$

of **A** are (respectively) the coefficients of x, the coefficients of y, and the coefficients of z in the equations.

In terms of its coefficient and constant matrices, the linear system in (16) can be abbreviated by writing

$$\begin{bmatrix} a & b & c \\ d & e & f \\ g & h & k \end{bmatrix} \begin{bmatrix} x \\ y \\ z \end{bmatrix} = \begin{bmatrix} p \\ q \\ r \end{bmatrix},$$

or simply

$$\mathbf{A}\begin{bmatrix} x \\ y \\ z \end{bmatrix} = \mathbf{B}, \tag{16}$$

where

$$\begin{bmatrix} x \\ y \\ z \end{bmatrix}$$

is a 3×1 *unknown matrix* consisting of the three unknown numbers x, y, and z. Just as the solution of the single equation

$$Ax = B \tag{17}$$

(where A, B, and x represent ordinary numbers) can be written in the form

$$x = \frac{B}{A} = A^{-1}B, \tag{18}$$

we can symbolically write

$$\begin{bmatrix} x \\ y \\ z \end{bmatrix} = \mathbf{A}^{-1}\mathbf{B} \tag{19}$$

to represent the solution matrix for the 3×3 linear system in (13).

After the coefficient matrix $\mathbf{A}$ has been entered, the calculator can find the 3×3 "inverse matrix" $\mathbf{A}^{-1}$. Then the product of the matrices $\mathbf{A}^{-1}$ and $\mathbf{B}$ can be calculated, so that the result is the 3×1 **solution matrix** $\begin{bmatrix} x \\ y \\ z \end{bmatrix}$ consisting of the three unknown values x, y, and z.

EXAMPLE 4   Solving a Linear System Using Matrices

Use matrices to solve the linear system

$$x + 2y + z = 4$$
$$3x + 8y + 7z = 20$$
$$2x + 7y + 9z = 23.$$

SOLUTION

This system has coefficient and constant matrices given by

$$\mathbf{A} = \begin{bmatrix} 1 & 2 & 1 \\ 3 & 8 & 7 \\ 2 & 7 & 9 \end{bmatrix} \quad \text{and} \quad \mathbf{B} = \begin{bmatrix} 4 \\ 20 \\ 23 \end{bmatrix},$$

which can be entered from the **MATRIX EDIT** menu.

Once the matrices **[A]** and **[B]** have been entered, we need only enter **[A]**$^{-1}$***[B]** [as in Equation (19)] using the **MATRIX NAMES** menu to calculate the solution (Fig. 6.3.7).

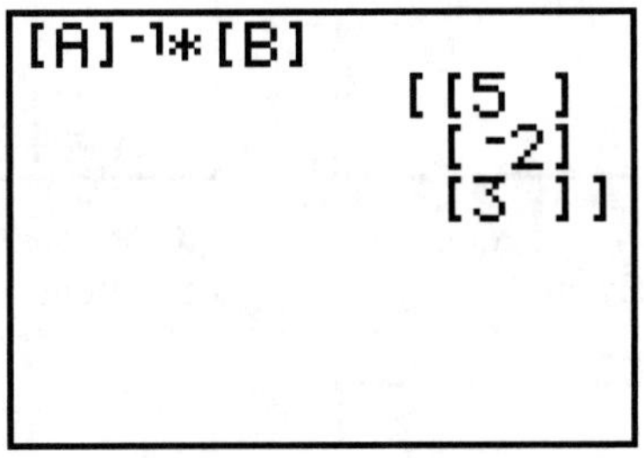

```
[A]-1*[B]
           [[5 ]
            [-2]
            [3 ]]
```

FIGURE 6.3.7 Calculating the values $x = 5$, $y = -2$, $z = 3$ of the three unknowns in Example 4.

The result

$$\begin{bmatrix} x \\ y \\ z \end{bmatrix} = \begin{bmatrix} 5 \\ -2 \\ 3 \end{bmatrix}$$

● tells us that the solution of our system is given by $x = 5$, $y = -2$, and $z = 3$.

To solve any other linear system of three equations in three unknowns, we need only re-enter the elements of the coefficient and constant matrices **[A]** and **[B]** for our new system, then re-calculate the solution matrix **[A]**$^{-1}$***[B]**. In fact, we can solve in exactly the same way a system of four equations in four unknowns or a system of five or more equations in the same number of unknowns.

EXAMPLE 5   Solving a Larger Linear System

Solve the linear system

$$3w - 2x + 7y + 5z = 505$$
$$2w + 4x - y + 6z = 435$$
$$5w + x + 7y - 3z = 286$$
$$4w - 6x - 8y + 9z = 445$$

of four equations in the four unknowns w, x, y, and z.

SOLUTION

Here we need only enter the coefficient matrix **[A]** and the constant matrix **[B]**, then calculate **[A]**$^{-1}$***[B]** as indicated in Figs. 6.3.8–6.3.10.

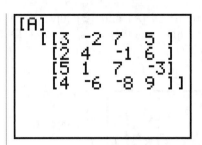

FIGURE 6.3.8 The 4 × 4 coefficient matrix **A** in Example 5.

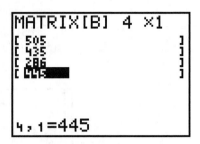

FIGURE 6.3.9 The 4 × 1 matrix **B** of contents in Example 5.

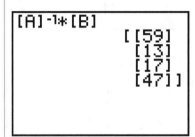

FIGURE 6.3.10 Finding the solution values in Example 5.

● Thus we find that $w = 59, x = 13, y = 17$, and $z = 47$.

Applications

In Section 6.2 we discussed applications that lead to two linear equations in two unknowns. If you're armed with a powerful calculator, applications with three unknowns are essentially the same. The only difference is that you must continue writing relations among the three unknowns until you have three equations to solve.

EXAMPLE 6 Solving a Value Problem

You are walking down the street minding your own business when you spot a small but heavy leather bag lying on the sidewalk. It turns out to contain U.S. Mint American Eagle gold coins of the following types:

- one-half-ounce gold coins that sell for $285 each,
- one-quarter-ounce gold coins that sell for $150 each, and
- one-tenth-ounce gold coins that sell for $70 each.

A bank receipt found in the bag certifies that it contains 258 such coins with a total weight of 67 ounces and a total value of exactly $40,145. How many coins of each type are there?

SOLUTION

If x is the number of half-ounce coins, y is the number of quarter-ounce coins, and z is the number of tenth-ounce coins in the bag, then the fact that there are 258 coins in all means that

$$x + y + z = 258.$$

If we multiply number of coins of each type by its weight in ounces and add up these weights, we get the equation

$$\frac{1}{2}x + \frac{1}{4}y + \frac{1}{10}z = 67$$

because we know that the coins weigh 67 ounces altogether. Finally, we add up the dollar values of the coins of each type and get the equation

$$285x + 150y + 70z = 40145.$$

This system of 3 equations in three unknowns has coefficient and constant matrices

$$\mathbf{A} = \begin{bmatrix} 1 & 1 & 1 \\ \frac{1}{2} & \frac{1}{4} & \frac{1}{10} \\ 285 & 150 & 70 \end{bmatrix} \quad \text{and} \quad \mathbf{B} = \begin{bmatrix} 258 \\ 67 \\ 40145 \end{bmatrix}.$$

Once we have entered these two matrices by way of the **MATRIX EDIT** menu, the usual calculation (Fig. 6.3.11) reveals that our bounty consists of 67 half-ounce gold coins, 96 quarter-ounce gold coins, and 95 tenth-ounce gold coins.

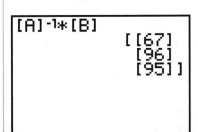

FIGURE 6.3.11 The numbers of gold bullion coins in Example 6.

As an example of a mixing problem involving three equations in three unknowns, let's consider a paint store that prepares whatever color of paint the customer orders by mixing appropriate amounts of pure red paint, pure green paint, and pure blue paint.

EXAMPLE 7 | Solving a Mixture Problem

A commercial customer orders 81 gallons of paint containing equal amounts of red paint, green paint, and blue paint—and hence could be prepared by mixing 27 gallons of each. However, the store wishes to prepare this order by mixing three types of paint that are already available in large quantity:

- a *reddish* paint that is a mixture of 50% red, 25% green, and 25% blue paint;
- a *greenish* paint that is 12.5% red, 75% green, and 12.5% blue paint; and
- a *bluish* paint that is 20% red, 20% green, and 60% blue paint.

How many gallons of each are needed to prepare the customer's order?

SOLUTION

If x gallons of reddish paint, y gallons of greenish paint, and z gallons of bluish paint are mixed, then the resulting mixture contains

- $0.5x$ gallons of pure red paint, because reddish paint is 50% red; plus
- $0.125y$ gallons of pure red paint, because greenish paint is 12.5% red; plus
- $0.2z$ gallons of pure red paint, because bluish paint is 20% red.

These three amounts must add up to the 27 gallons of pure red paint needed for the customer's desired mixture, so it follows that x, y, and z must satisfy the equation

$$0.5x + 0.125y + 0.2z = 27.$$

Similarly, the amounts $0.25x$ gallons of pure green paint in the reddish paint, $0.75y$ gallons of pure green paint in the greenish paint, and the $0.2z$ gallons of pure green paint in the bluish paint must add up to the 27 gallons of pure green paint that is needed for the desired mixture. Thus, x, y, and z must satisfy the second equation

$$0.25x + 0.75y + 0.2z = 27.$$

Finally, the amounts $0.25x$ gallons of pure blue paint in the reddish paint, $0.125y$ gallons of pure blue paint in the greenish paint, and the $0.6z$ gallons of pure blue paint in the bluish paint must add up to the 27 gallons of pure blue paint that is needed for the desired mixture. Thus, x, y, and z must satisfy the third equation

$$0.25x + 0.125y + 0.6z = 27.$$

These three equations comprise a 3×3 system with coefficient and constant matrices

$$\mathbf{A} = \begin{bmatrix} 0.5 & 0.125 & 0.2 \\ 0.25 & 0.75 & 0.2 \\ 0.75 & 0.125 & 0.6 \end{bmatrix} \quad \text{and} \quad \mathbf{B} = \begin{bmatrix} 27 \\ 27 \\ 27 \end{bmatrix}.$$

Figure 6.3.12 shows the result when we enter these matrices in the calculator and calculate **[A]**$^{-1}$***[B]**.

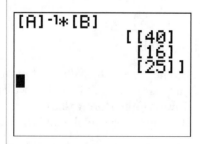

FIGURE 6.3.12 The solution matrix for Example 7.

Thus, the calculator solution gives the amounts $x = 40$ gallons of reddish paint, $y = 16$ gallons of greenish paint, and $z = 25$ gallons of bluish paint that must be mixed to prepare the customer's 81 gallons of equally mixed red/green/blue paint. ●

6.3 Exercises | Building Your Skills

Each of the determinants in Exercises 1–4 is most easily evaluated by expansion along a carefully chosen row or column. Find the value of each determinant.

1. $\begin{vmatrix} 0 & 3 & -2 \\ -3 & 11 & 17 \\ 0 & 5 & 7 \end{vmatrix}$

2. $\begin{vmatrix} 5 & 4 & 10 \\ 4 & 5 & 20 \\ 0 & 0 & 5 \end{vmatrix}$

3. $\begin{vmatrix} 5 & 77 & 4 \\ 0 & 3 & 0 \\ 4 & 99 & 5 \end{vmatrix}$

4. $\begin{vmatrix} 21 & 31 & 5 \\ 11 & 10 & 0 \\ 12 & 11 & 0 \end{vmatrix}$

Each of the linear systems in Exercises 5–8 is especially easy to solve by elimination because a single operation combining a carefully selected pair of equations suffices to reduce the given 3×3 system to a 2×2 system. Solve each system in this manner.

5. $x + 4y - z = 5$
   $x + 3y + z = 8$
   $2y + z = 4$

6. $2x - 3y + z = 7$
   $3y + 2z = 11$
   $2x + 2y + z = 12$

7. $7x + 3y + z = 39$
   $5x + 3y + 2z = 36$
   $2x + 3z = 23$

8. $7x + 5y + 3z = 18$
   $4x + 2z = 14$
   $2x + 5y + 6z = 17$

Use either determinants or matrix inverses (with your calculator) to solve the systems in Exercises 9–13.

9. $3x + 4y + 7z = 52$
   $6x - 8y + 5z = 89$
   $2x + 5y + 4z = 23$

10. $5x + 5y - 9z = 15$
    $7x + 6y - 3z = 65$
    $4x - 8y + 5z = 25$

11. $17x + 42y - 36z = 213$
    $13x + 45y - 34z = 226$
    $12x + 47y - 35z = 197$

12. $231x + 157y - 241z = 420$
    $323x + 181y - 375z = 412$
    $542x + 161y - 759z = 419$

13. $81w - 97x + 67y - 19z = 82$
    $12w - 15x - 79y - 82z = 86$
    $42w - 54x + 14y - 29z = 91$
    $13w + 27x + 63y + 25z = 78$

Applying Your Skills

14. Suppose you find a smaller bag of U.S. American Eagle half-ounce, quarter-ounce, and tenth-ounce gold coins valued as in Example 6, with half-ounce gold coins

worth $285 each, quarter-ounce gold coins worth $150 each, and tenth-ounce gold coins worth $70 each. If this bag contains a total of 58 coins with a total weight of 17 ounces and a total value of $10,065, how many gold coins of each type are there?

15. Now you really strike it rich! You find a bag containing one-ounce U.S. American Eagle gold coins valued at $550 each, together with half-ounce and quarter-ounce coins valued as in Exercise 16. If this bag contains a total of 365 coins with a total weight of exactly 11 pounds and a total value of $100,130, how many gold coins of each type are there?

16. A commercial customer orders 51 gallons of paint that contains equal amounts of red paint, green paint, and blue paint, and hence could be prepared by mixing 17 gallons of each pure color of paint. However, the store wishes to prepare this order by mixing three types of already mixed paint that are available in large quantity—the reddish paint and the bluish paint of Example 7, and a greenish paint that is 1/6 red, 2/3 green, and 1/6 blue paint. How many gallons of each are needed?

17. Now the paint store receives a really big order—for 244 gallons of paint that is 1/2 red paint, 1/4 green paint, and 1/4 blue paint. The store has three already-mixed types of paint available in large quantity—the greenish paint and the bluish paint of Example 7, plus a reddish paint that is 2/3 red paint, 1/6 green paint, and 1/6 blue paint. How many gallons of each must be mixed in order to fill this order?

18. A tour busload of 45 people attended two Florida theme parks on successive days. On Day 1 the entrance fee was $15 per adult, $8 per child, and $12 per senior citizen, and the total charge was $558. On Day 2 the entrance fee was $20 per adult, $12 per child, and $17 per senior citizen, and the total charge was $771. How many adults, children, and senior citizens were on this tour bus?

19. For some crazy reason, the lunches bought at the first theme park were totaled separately for the adults, children, and seniors. The adults ordered 34 hot dogs, 15 French fries, and 24 soft drinks for a total bill of $70.85. The children ordered 20 hot dogs, 14 French fries, and 15 soft drinks for a total bill of $46.65. The senior citizens ordered 11 hot dogs, 10 French fries, and 12 soft drinks for a total bill of $30.00. What were the prices of a hot dog, an order of French fries, and a soft drink?

20. Carter has a weekend business selling Beanie Babies at flea markets in Missouri. She keeps very close track of the changing prices of these popular collectibles. In April a customer spilled coffee on a page of her records, and information on Happy the hippo, Hippity the bunny, and Hoot the owl was lost. The following table gives the numbers of each Beanie Baby sold and the total receipts from three different markets. How much did Carter charge in April for each of these Beanie Babies?

	HAPPY	HIPPITY	HOOT	RECEIPTS
Jefferson City	5	8	6	$488
Columbia	4	9	10	$653
Cape Girardeau	8	6	8	$562

21. An office supply company with four locations sells TI-82, TI-83, TI-85, and TI-86 calculators. During the month of August, the stores reported the following sales:

	TI-82	TI-83	TI-85	TI-86	RECEIPTS
Store 1	4	94	10	32	$14130
Store 2	8	80	3	29	$12100
Store 3	6	50	2	4	$5970
Store 4	5	63	7	21	$9655

At what price did the company sell each model calculator?

22. A fast food restaurant sells four types of sandwiches—hamburgers, cheeseburgers, roast beef, and chicken—and has four cash registers. At the end of each day, each cash register tallies the number of each type of sandwich sold and the total sandwich receipts for the day. The four cash register operators work at different speeds, and one day's totals were as follows:

	HAMBURGERS	CHEESEBURGERS	ROAST BEEF	CHICKEN	RECEIPTS
Register 1	37	44	17	23	$232.99
Register 2	28	35	13	17	$178.97
Register 3	32	39	19	21	$215.99
Register 4	47	51	25	29	$294.38

What was the price of each of the four types of sandwiches?

23. The fast food restaurant of Exercise 22 adds a ham sandwich to its menu, and due to increased business it also adds a fifth cash register and reduces prices. After this expansion, one day's totals were as follows:

	HAMBURGERS	CHEESEBURGERS	ROAST BEEF	CHICKEN	HAM	TOTAL
Register 1	41	49	22	26	19	$292.79
Register 2	34	39	18	20	16	$236.73
Register 3	36	43	23	24	18	$270.70
Register 4	49	52	26	31	24	$340.19
Register 5	52	55	24	28	25	$341.64

What were the new prices of the five types of sandwiches?

6.4 POLYNOMIAL DATA MODELING

The following table shows the total world population (in billions) at 5-year intervals. Actual world populations are shown for the years 1960–1995. The figures listed for the years 2000–2025 are future populations predicted in the late 1990s by the United Nations on the basis of detailed demographic analysis of population trends on a country-by-country basis throughout the world.

YEAR	WORLD POPULATION (BILLIONS)	PERCENTAGE GROWTH
1960	3.039	
1965	3.345	1.94%
1970	3.707	2.08%
1975	4.086	1.97%
1980	4.454	1.74%
1985	4.851	1.72%
1990	5.279	1.71%
1995	5.688	1.50%
2000	6.083	1.35%
2005	6.468	1.23%
2010	6.849	1.15%
2015	7.227	1.08%
2020	7.585	0.97%
2025	7.923	0.88%

Each entry in the third column of this table gives the average annual percentage growth rate during the preceding 5-year period. For instance, to find the average growth rate r from 1960 to 1965, we solve the equation

$$3.345 = 3.039(1 + r)^5$$

for

$$r = \left(\frac{3.345}{3.039}\right)^{1/5} - 1 \approx 0.01937 \approx 1.94\%.$$

We see that the world population grew at an annual rate of about 2% during the 1960s, but the rate of growth has slowed since then, and it is expected to slow even more during the early decades of the twenty-first century.

Thus the growth of the world population at the present time in history is *not* natural or exponential in character. In this section we explore the possibility of fitting world population data with polynomial models that can be used to predict future populations.

Linear and Quadratic Models

First, recall from Section 2.2 that two data points determine a linear function whose graph is a straight line. Thus, given any two year–population pairs (t_1, P_1) and (t_2, P_2), there exists a linear function

$$P(t) = P(t_1) + a(t - t_1)$$

such that $P(t_1) = P_1$ and $P(t_2) = P_2$. If we simplify this point-slope form, we get the slope-intercept form

$$P(t) = b + at. \tag{1}$$

SECTION 6.4 Polynomial Data Modeling

As in the following example, we need only solve two linear equations in two unknowns to find the values of the coefficients a and b in (1).

EXAMPLE 1 | Finding a Linear Population Model

Find a linear function to fit the 1980 and 1990 world population values, and use the model to predict the year 2000 world population.

SOLUTION

If we substitute first the data point (1980, 4.454) and then the data point (1990, 5.279) into (1), we get the two equations

$$\begin{aligned} 4.454 &= b + 1980a \\ 5.279 &= b + 1990a. \end{aligned} \qquad (2)$$

We note that the b-terms are eliminated if we subtract the first equation from the second one. The result is $10a = 5.279 - 4.454 = 0.825$, so $a = 0.0825$. Substitution of this value in the first equation in (2) then gives

$$b = 4.454 - 1980(0.0825) = -158.896.$$

With these coefficients, our linear population model (1) is

$$P(t) = -158.896 + 0.0825t. \qquad (3)$$

You should always check your solution of a system of equations. The way we found b appears to guarantee that the first equation in (2) is satisfied, and

$$-158.896 + 0.0825(1990) = 5.279,$$

so the second one is satisfied also. (The first time we ourselves worked Example 1, a numerical error in recording the value of a was revealed when we performed this check.)

Notice that we have not "reset the clock" as we often do. So if we substitute $t = 2000$ in (3), we get

$$P(2000) = -158.896 + 0.0825(2000) = 6.104 \text{ billion}$$

as our "linear extrapolation" for the world population in the year 2000. This is a bit more than the U.N. prediction of 6.083 billion. (This prediction dates from late in 1999, but you can find the actual year 2000 world population by consulting the web site **www.census.gov**, and thus see whose prediction is better, ours or the U.N.'s.)

Just as two data points determine a linear function whose graph is a straight line, three given data points determine a quadratic function with parabolic graph (as in Chapter 5). Thus, given any three year–population pairs $(t_1, P_1), (t_2, P_2)$, and (t_3, P_3), there exists a quadratic function

$$P(t) = c + bt + at^2 \qquad (4)$$

such that $P(t_1) = P_1, P(t_2) = P_2$, and $P(t_3) = P_3$. As in the following example, we need to solve three linear equations in three unknowns to find the values of the coefficients a, b, and c in (4).

EXAMPLE 2 Finding a Quadratic Population Model

Find a quadratic function to fit the 1970, 1980, and 1990 world population values, and use the model to predict the year 2000 world population.

SOLUTION

If we substitute the three data points (1970, 3.707), (1980, 4.454), and (1990, 5.279) into (4), we get the three equations

$$3.707 = c + 1970b + 1970^2 a$$
$$4.454 = c + 1980b + 1980^2 a \qquad (5)$$
$$5.279 = c + 1990b + 1990^2 a.$$

This is a linear system of equations with the 3×1 unknown and constant matrices

$$\mathbf{X} = \begin{bmatrix} c \\ b \\ a \end{bmatrix} \quad \text{and} \quad \mathbf{B} = \begin{bmatrix} 3.707 \\ 4.454 \\ 5.279 \end{bmatrix}$$

and with the 3×3 coefficient matrix

$$\mathbf{A} = \begin{bmatrix} 1 & 1970 & 1970^2 \\ 1 & 1980 & 1980^2 \\ 1 & 1990 & 1990^2 \end{bmatrix}.$$

After we store the coefficient and constant matrices in the calculator entry as **A** and **B**, Fig. 6.4.1 then shows the solution for the unknown matrix **X** (which we save as **[G]** because our calculator allows for saving only the 10 matrices **[A]** through **[J]**).

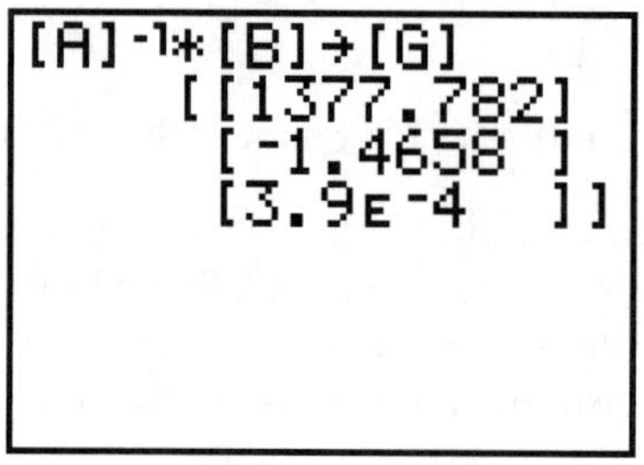

```
[A]-1*[B]→[G]
     [[1377.782]
      [-1.4658  ]
      [3.9E-4   ]]
```

FIGURE 6.4.1 Calculating the coefficients in (5).

We see that the components of $\mathbf{X} = [G]$ are $c = 1377.782$, $b = -1.4658$, and $a = 0.00039$. Substituting these coefficient values in (4), we get the quadratic population model

$$P(t) = 1377.782 - 1.4658t + 0.00039 t^2. \qquad (6)$$

To facilitate checking our solution of the equations in (5), we first save the coefficient values $c = 1377.782$, $b = -1.4658$, and $a = 0.00039$ as the calculator values **A**, **B**, and **C**. Then we can quickly check the equations individually as indicated in Fig. 6.4.2. Alternatively, we can check the system "all at once" by matrix multiplication as indicated in Fig. 6.4.3 (where the formal matrix product **AX** is entered in calculator notation as **[A]*[G]**).

```
A+B*T+C*T²          1970
                   3.707
1980→T
                    1980
A+B*T+C*T²
                   4.454
1990→T
```

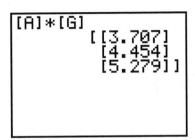

FIGURE 6.4.2 Checking our solution of the equations in (5).

FIGURE 6.4.3 Matrix check of our solution of system (5).

If we substitute $t = 2000$ in (6), the result is

$$P(2000) = 1377.782 - 1.4658(2000) + 0.00039(2000)^2 = 6.182 \text{ billion.}$$

If we regard the U.N. prediction of 6.083 billion as the "correct" figure for the year 2000, then we see that our quadratic prediction of 6.182 billion (Example 2) is a *worse* approximation than our linear prediction (Example 1) of 6.104 billion. This contradicts the common belief that a higher-degree approximation is always better. This is not always true. One reason that Example 2 did not yield a better approximation may be that we used more-ancient data—going back to 1970—than in Example 1, where we went back only to 1980. In Exercises 1 and 2 we ask you to see whether the use of more recent 1980–1985–1990 or 1985–1990–1995 data yields more accurate predictions for the world population in the year 2000.

Higher-Degree Models

Just as three data points determine a quadratic function, four given data points determine a cubic function, and five given data points determine a quartic function. For instance, given any four year–population pairs $(t_1, P_1), (t_2, P_2), (t_3, P_3)$, and (t_4, P_4), there exists a cubic function

$$P(t) = d + ct + bt^2 + at^3 \tag{7}$$

such that $P(t_1) = P_1, P(t_2) = P_2, P(t_3) = P_3$, and $P(t_4) = P_4$. Note that the number of coefficients is always one greater than the degree of the polynomial function. So the method in higher-degree cases is the same as in Examples 1 and 2. However many data points we start with, we set up a polynomial of degree *one less* than the number of given data points. Then substitution of the given data yields a linear system with the right number of equations to solve for the coefficients in our polynomial.

EXAMPLE 3 Finding a Cubic Population Model

Find a cubic function to fit the 1980, 1985, 1990, and 1995 world population values, and use the model to predict the year 2000 world population.

SOLUTION

If we substitute the four data points (1980, 4.454), (1985, 4.851), (1990, 5.279), and (1995, 5.688) into (7), we get the four equations

$$4.454 = d + 1980c + 1980^2 b + 1980^3 a$$
$$4.851 = d + 1985c + 1985^2 b + 1985^3 a$$
$$5.279 = d + 1990c + 1990^2 b + 1990^3 a \qquad (8)$$
$$5.688 = d + 1995c + 1995^2 b + 1995^3 a.$$

This is a 4×1 linear system $\mathbf{AX} = \mathbf{B}$ with

$$\mathbf{A} = \begin{bmatrix} 1 & 1980 & 1980^2 & 1980^3 \\ 1 & 1985 & 1985^2 & 1985^3 \\ 1 & 1990 & 1990^2 & 1990^3 \\ 1 & 1995 & 1995^2 & 1995^3 \end{bmatrix} \quad \text{and} \quad \mathbf{B} = \begin{bmatrix} 4.454 \\ 4.851 \\ 5.279 \\ 5.688 \end{bmatrix}$$

and with the 4×1 unknown matrix $\mathbf{X}$ comprised of the coefficients d, c, b, and a in (7).

Once these matrices are entered into the calculator, the matrix computation shown in Fig. 6.4.4 stores the coefficient values $d = 523703.828$, $c = -790.42222$, $b = 0.39762$, and $a = -0.00006666666667$ (copying all digits shown to get the greatest possible accuracy) as elements of the 4×1 calculator matrix **[G]**.

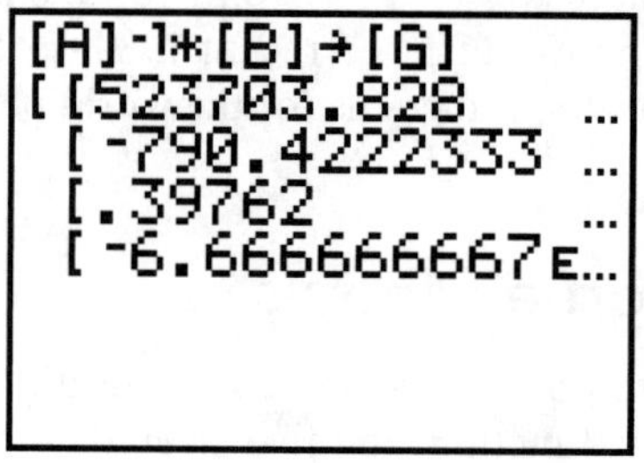

FIGURE 6.4.4 Calculating the coefficients in (7).

Thus our cubic population model is

$$P(t) = 523703.828 - 790.42222t + 0.39762t^2 - 0.0000666666667t^3. \qquad (9)$$

Just as in Fig. 6.4.3, we can (and should) readily check "all at once" that the cubic function in (9) fits the 1980–1985–1990–1995 data with which we started.

It is quite tedious—as well as error-prone—to retype numerical solutions. Figure 6.4.5 shows a much more efficient way to store the numerical values of the coefficients in (9)—by "picking out" the appropriate elements of the solution matrix **[G]**. Figure 6.4.6 then shows the result

$$P(2000) = 523703.828 - 790.42222(2000)$$
$$+ 0.39762(2000)^2 - 0.0000666666667(2000)^3 \approx 6.028 \text{ billion}$$

• people, obtained by substituting $t = 2000$ in (9).

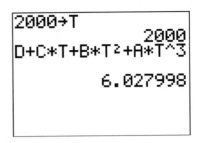

FIGURE 6.4.5 Storing the coefficients in the cubic model (9).

FIGURE 6.4.6 Cubic polynomial prediction of the year 2000 world population.

The following table compares our linear, quadratic, and cubic predictions with the "correct" U.N. prediction for the year 2000. It also includes the prediction of the quartic polynomial in Example 4 given later. Each "error" in the third column of this table is the amount by which the corresponding prediction underestimates (positive error) or overestimates (negative error) the U.N. prediction. We see that the cubic prediction is better than the quadratic prediction but worse than the linear prediction. The table also displays the prediction from a quartic (fourth degree) polynomial; this quartic prediction is even worse. Thus there apparently is no correlation between the degree of the polynomial model and the accuracy of its predictions.

	YEAR 2000 PREDICTION	ERROR
Linear	6.104	−0.021
Quadratic	6.182	−0.099
Cubic	6.028	+0.055
Quartic	5.976	+0.107
United Nations	6.083	

In order to fit a fourth-degree population model of the form

$$P(t) = e + dt + ct^2 + bt^3 + at^4 \tag{10}$$

to the 1975–1980–1985–1990–1995 world population data, we need to solve the linear system

$$\begin{bmatrix} 1 & 1975 & 1975^2 & 1975^3 & 1975^4 \\ 1 & 1980 & 1980^2 & 1980^3 & 1980^4 \\ 1 & 1985 & 1985^2 & 1985^3 & 1985^4 \\ 1 & 1990 & 1990^2 & 1990^3 & 1990^4 \\ 1 & 1995 & 1995^2 & 1995^3 & 1990^4 \end{bmatrix} \begin{bmatrix} e \\ d \\ c \\ b \\ a \end{bmatrix} = \begin{bmatrix} 4.086 \\ 4.454 \\ 4.851 \\ 5.279 \\ 5.688 \end{bmatrix} \tag{11}$$

to find the values of the coefficients a, b, c, d, and e in (10). Some of the entries in this matrix are so large ($1995^4 = 15{,}840{,}599{,}000{,}625$) that your calculator is unable to solve the system. Using mathematical software capable of storing more digits, we find that

$$P(t) = -53{,}568{,}386.224 + 108{,}075.0227667t - 81.76541333333t^2 \qquad (12)$$
$$+0.02749333333333t^3 - 0.00000346666666666t^4.$$

Figure 6.4.7 shows the U.N. world population data points for the years 1960–2025, together with the plots of the linear, quadratic, cubic, and quartic population functions.

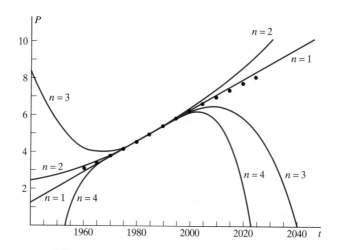

FIGURE 6.4.7 Plot of world population data points and the interpolating polynomials of degrees $n = 1$ through $n = 4$.

It looks as though the more work we do to find a polynomial fitting selected data points, the less we get for our effort. It certainly is true in this figure that—outside the interval from 1975 to 1995—the higher is the degree of the polynomial, the worse it appears to fit the given data points. The issue here is the difference between

- *interpolating* data points *within* the interval of given points being fitted, and
- *extrapolating* data points *outside* this interval.

All four of our polynomials appear to do a good job of interpolation, but, somewhat paradoxically, the higher is the degree, the worse is the apparent accuracy of extrapolation. The highly questionable accuracy of data extrapolation outside the interval of interpolation has significant implications. For instance, consider a news report that when a certain alleged carcinogen was fed to mice in sufficient amounts to kill an elephant, the mice got cancer. It is then argued that moderate amounts of this carcinogen may cause cancer in humans. Or that if 1 part per billion of this carcinogen in the environment kills 1 person, then 1 part per million (1000 times as much) will kill 1000 people. Such arguments are common, but who knows? They may well be cases of extrapolation beyond the range of accuracy. The bottom line is that interpolation is fairly safe—though hardly fail-safe—but extrapolation is risky.

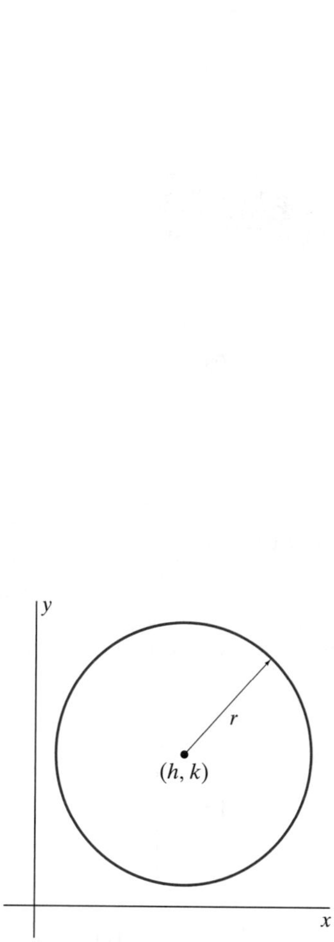

Geometric Applications

FIGURE 6.4.8 A circle with radius r and center (h, k).

In contrast with population prediction, there are situations in which curve fitting is exact. For instance, the fact that "two points determine a line" in the plane means that, when we fit the linear function $y = a + bx$ to a given pair of points, we get precisely the one and only straight line in the plane that passes through these points. Similarly, "three points determine a circle," meaning that there is one and only one circle in the plane that passes through three given points. In order to find this particular circle (Fig. 6.4.8), we need to recall that the equation of a circle with center (h, k) and radius r is

$$(x - h)^2 + (y - k)^2 = r^2. \tag{13}$$

EXAMPLE 4 Finding the Equation of a Circle

Find the equation of the circle that is determined by the points $P(-1, 5)$, $Q(5, -3)$, and $R(6, 4)$.

SOLUTION

Substitution of the xy-coordinates of each of the three points P, Q, and R into (10) gives the three equations

$$\begin{aligned}(-1 - h)^2 + (5 - k)^2 &= r^2 \\ (5 - h)^2 + (-3 - k)^2 &= r^2 \\ (6 - h)^2 + (4 - k)^2 &= r^2.\end{aligned} \tag{14}$$

Expansion and collection of coefficients in these three equations gives

$$\begin{aligned}h^2 + k^2 + 2h - 10k + 26 &= r^2 \\ h^2 + k^2 - 10h + 6k + 34 &= r^2 \\ h^2 + k^2 - 12h - 8k + 52 &= r^2.\end{aligned} \tag{15}$$

These equations may at first look a bit formidable because of the three square-terms in each. However, subtraction of any two of them eliminates all these square-terms! Thus subtraction of the second and third equations in (12) from the first one yields the linear system

$$\begin{aligned}12h - 16k - 8 &= 0 \\ 14h - 2k - 26 &= 0\end{aligned}$$

of just two equations in the two unknowns h and k. Just two equations should seem easy by now; we get $h = 2$ and $k = 1$. Substitution of these two values into the first equation in (14) now gives $r^2 = (-3)^2 + (4)^2 = 25$, so $r = 5$. Thus our circle has center $(2, 1)$ and radius 5 (Fig. 6.4.9).

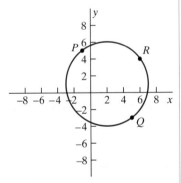

FIGURE 6.4.9 The circle through the three points of Example 4.

Three points in the plane also determine a **central conic** with equation of the form

$$ax^2 + bxy + cy^2 = 1. \tag{16}$$

A typical central conic is an ellipse (or "flattened circle") that has been rotated about the origin.

EXAMPLE 5 — Finding a Central Conic

Find the central conic that passes through the same three points $P(-1, 5)$, $Q(5, -3)$, and $R(6, 4)$ of Example 4.

SOLUTION

Substitution of the xy-coordinates of each of the three points P, Q, and R into (16) gives the linear system of three equations

$$a - 5b + 25c = 1$$
$$25a - 15b + 9c = 1$$
$$36a + 24b + 16c = 1 \qquad (17)$$

in the three unknowns a, b, and c. In Exercise 3 we ask you to solve this system by the method of elimination to obtain fractions (rather than decimal numbers) for the coefficients. You should obtain the values

$$a = \frac{277}{14212}, \quad b = -\frac{172}{14212}, \quad \text{and } c = \frac{523}{14212}.$$

If we substitute these coefficient values in (16) and multiply the result by 14212, we get the desired equation

$$277x^2 - 172xy + 523y^2 = 14212 \qquad (18)$$

of our central conic. The computer plot in Fig. 6.4.10 verifies that this rotated ellipse does indeed pass through all three points P, Q, and R.

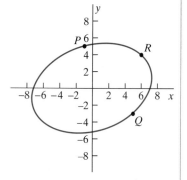

FIGURE 6.4.10 The central conic through the three points of Example 5.

Least-Squares Polynomials

In Section 5.3 we discussed the use of a graphing calculator's "quadratic regression" facility to find the quadratic model $P(t) = P_0 + bt + at^2$ that best fits a given table of tP-data. Here we describe similarly the use of the "cubic regression" and "quartic regression" facilities provided by the **CubicReg** and **QuarticReg** functions listed in the **STAT CALC** menu of Fig. 6.4.11.

FIGURE 6.4.11 Graphing calculator regression functions.

Given a list **L1** of x-data and a corresponding list **L2** of y-data, the command

CubicReg L1, L2, Y3

finds the *cubic* polynomial $y = ax^3 + bx^2 + cx + d$ that best fits the given data—meaning that it minimizes the sum of the squares of the errors or discrepancies between actual and predicted values. This best-fitting cubic polynomial is stored as the **Y=** menu function **Y3**. Similarly, the command

QuarticReg L1, L2, Y4

finds the *quartic* polynomial $y = ax^4 + bx^3 + cx^2 + dx + e$ that best fits the given data, and saves this polynomial as the **Y=** menu function **Y4**.

For greatest accuracy of fit, the calculator should be set (using the **Mode** menu) to display nine significant digits because only the displayed decimal places are stored in the $a, b, c, \ldots$ coefficient values in the **Y=** menu functions.

The given data lists **L1** and **L2** must contain at least four data points to use **CubicReg** and at least five data points to use **QuarticReg**. In the case of exactly four data points **CubicReg** calculates the cubic polynomial that *interpolates* these points (and hence has average error 0), while in the case of exactly five data points **QuarticReg** calculates the quartic polynomial that interpolates these points. With more than four (or five) given data points, **CubicReg** (or **QuarticReg**) finds the polynomial of the indicated degree that *best fits* these points, and hence has least possible average error.

EXAMPLE 6

Finding Polynomial Models for World Population

Find the quadratic, cubic, and quartic polynomials that best fit the 1960–1990 world population data listed in the table at the beginning of this section.

SOLUTION

For variety we choose $t = 0$ in 1950, so the years 1960, 1965, ..., 1985, 1990 correspond to the values $t = 10, 15, \ldots, 35, 40$. We store these t-values in the list **L1** and the corresponding world population values in the list **L2**. Then Figs. 6.4.12–6.4.14 show the results of the successive commands

QuadReg L1, L2, Y2
CubicReg L1, L2, Y3
QuarticReg L1, L2, Y4

that save the optimal quadratic polynomial as **Y2**, the optimal cubic polynomial as **Y3**, and the optimal quartic polynomial as **Y4** (with nine-place accuracy selected as suggested previously).

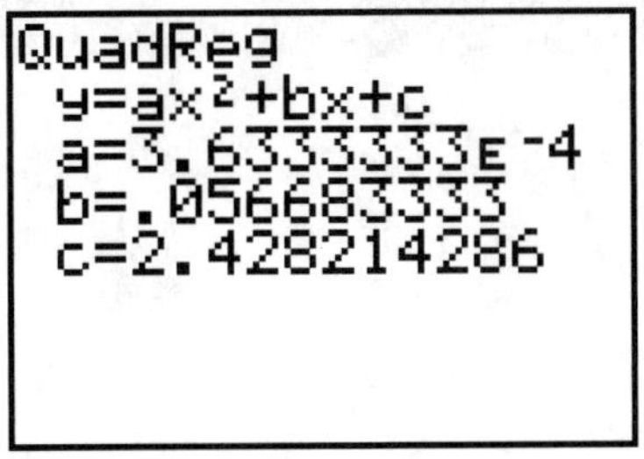

```
QuadReg
y=ax²+bx+c
a=3.6333333E-4
b=.056683333
c=2.428214286
```

```
CubicReg
y=ax³+bx²+cx+d
a=-2.888889E-6
b=5.8000000E-4
c=.051772222
d=2.460714286
```

FIGURE 6.4.12 The best-fitting quadratic polynomial Y_2.

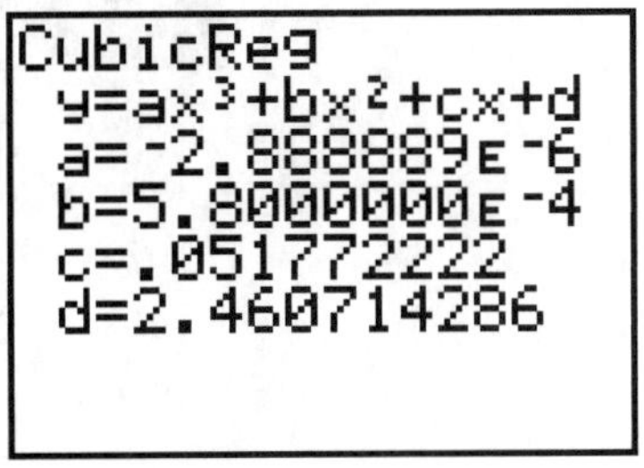

FIGURE 6.4.13 The best-fitting cubic polynomial Y_3.

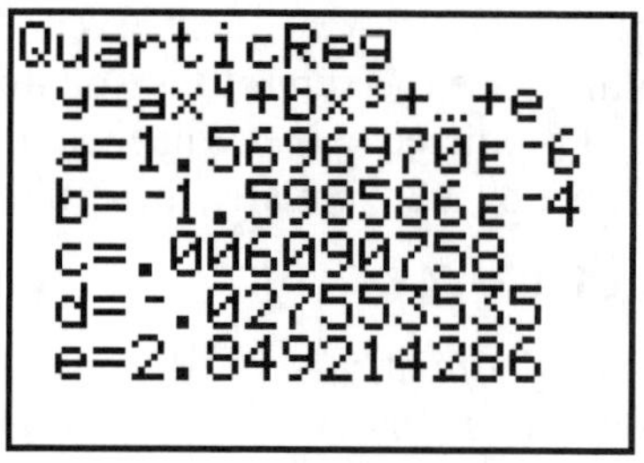

CHAPTER 6 Polynomial Models and Linear Systems

```
QuarticReg
 y=ax^4+bx^3+...+e
 a=1.5696970E-6
 b=-1.598586E-4
 c=.006090758
 d=-.027553535
 e=2.849214286
```

FIGURE 6.4.14 The best-fitting quartic polynomial Y_4.

Thus we find the optimal quadratic polynomial

$$P_2(t) = 0.00036333333\, t^2 + 0.056683333\, t + 2.428214286, \qquad (19)$$

the optimal cubic polynomial

$$P_3(t) = -0.000002888889\, t^3 + 0.00058\, t^2 + 0.051772222\, t + 2.460714286, \qquad (20)$$

and the optimal quartic polynomial

$$P_4(t) = 0.000001569697\, t^4 - 0.0001598586\, t^3 + 0.006090758\, t^2 \\ -0.027553535\, t + 2.849214286. \qquad (21)$$

Figure 6.4.15 shows the resulting table of values of **Y2** and **Y3**—that is, $P_2(t)$ and $P_3(t)$. Upon scrolling this calculator screen to the right we get the table of values of **Y3** and **Y4**—that is, $P_3(t)$ and $P_4(t)$—shown in Fig. 6.4.16.

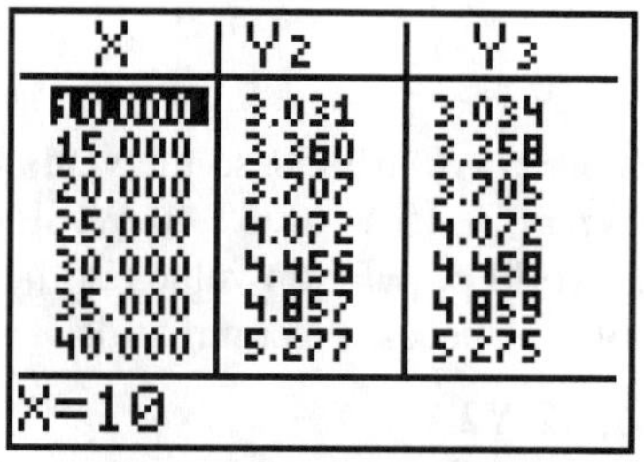

X	Y2	Y3
10.000	3.031	3.034
15.000	3.360	3.358
20.000	3.707	3.705
25.000	4.072	4.072
30.000	4.456	4.458
35.000	4.857	4.859
40.000	5.277	5.275

X=10

FIGURE 6.4.15 Quadratic (**Y2**) and cubic (**Y3**) polynomial values.

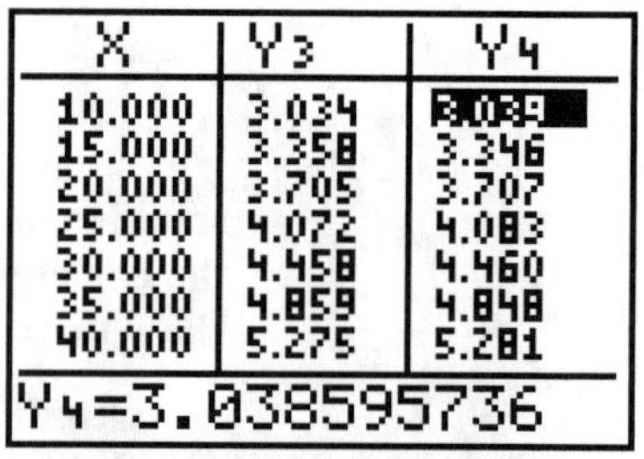

X	Y3	Y4
10.000	3.034	3.039
15.000	3.358	3.346
20.000	3.705	3.707
25.000	4.072	4.083
30.000	4.458	4.460
35.000	4.859	4.848
40.000	5.275	5.281

Y4=3.038595736

FIGURE 6.4.16 Cubic (**Y3**) and quartic (**Y4**) polynomial values.

The following table assembles these results:

YEAR	t	ACTUAL POPULATION	QUADRATIC	CUBIC	QUARTIC
1960	10	3.039	3.031	3.034	3.039
1965	15	3.345	3.360	3.358	3.346
1970	20	3.707	3.707	3.705	3.707
1975	25	4.086	4.072	4.072	4.083
1980	30	4.454	4.456	4.458	4.460
1985	35	4.851	4.857	4.859	4.848
1990	40	5.279	5.277	5.275	5.281

Comparing the quadratic, cubic, and quartic predictions with the actual populations for the years 1960–1990, we calculate the sums of squares of errors

$$SSE_2 = (0.008)^2 + (-0.015)^2 + (0.000)^2 + (0.014)^2 + (-0.002)^2 \\ + (-0.006)^2 + (0.002)^2 = 0.000529,$$

$$SSE_3 = (0.005)^2 + (-0.013)^2 + (0.002)^2 + (0.014)^2 + (-0.004)^2 \\ + (-0.008)^2 + (0.004)^2 = 0.000490,$$

and

$$SSE_4 = (0.000)^2 + (-0.001)^2 + (0.000)^2 + (0.003)^2 + (-0.006)^2 \\ + (-0.003)^2 + (-0.002)^2 = 0.000059,$$

in the quadratic, cubic, and quartic fits (respectively). The resulting average errors are

$$\text{quadratic average error} = \sqrt{\frac{0.000529}{7}} = 0.009,$$

$$\text{cubic average error} = \sqrt{\frac{0.000490}{7}} = 0.008,$$

and

$$\text{quartic average error} = \sqrt{\frac{0.000059}{7}} = 0.003.$$

We appear to have excellent fits. The predicted year 2000 populations are

$$P_2(50) = 6.171, \quad P_3(50) = 6.138, \quad \text{and} \quad P_4(50) = 6.529,$$

as compared with the United Nation's predicted year 2000 population of 6.083 billion. There is a lesson in the fact that the quartic polynomial gives the worst extrapolation to 2000, even though it gives the best fit (that is, the least average error) for the years 1960–1990. (The lesson is that, when extrapolating outside the region of best fit, "you just never can tell.")

Finally, we show in Fig. 6.4.17 the graphs of our quadratic, cubic, and quartic best-fitting polynomials together with the U.N. data points for 1960–2025. Compare these graphs of best-fitting polynomials with the graphs of interpolating polynomials shown earlier in Fig. 6.4.7. Evidently the best-fitting polynomials "extrapolate

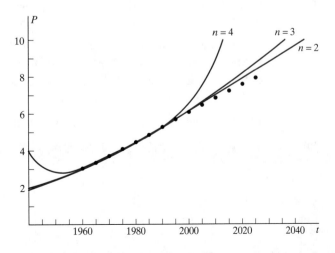

FIGURE 6.4.17 Plot of world population data points and the optimal polynomials of degrees $n = 2$ through $n = 4$.

much better" than the interpolating polynomials. This is reasonable to expect, given that best-fitting polynomials are determined by "more than enough" points, while interpolating polynomials are determined by "just enough" points.

6.4 Exercises

Building Your Skills

1. Use the world population data from 1980, 1985, and 1990 to determine a quadratic model for world population growth. Then use your model to predict the world population in the year 2000. Is this prediction closer to the U.N. prediction of 6.083 billion than the one we found in Example 2 by using the data from 1970, 1980, and 1990?

2. Use the world population data from 1985, 1990, and 1995 to determine a quadratic model for world population growth. Then use your model to predict the world population in the year 2000. Is this prediction closer to the U.N. prediction of 6.083 billion than the one we found in Example 2 by using the data from 1970, 1980, and 1990?

3. Use the method of matrix inverses to solve the linear system indicated in Example 5. Verify that the coefficients you obtain are the same as those shown in the example.

4. Use the method of elimination to solve the linear system indicated in Example 6, using fractions rather than decimal numbers. Verify that the coefficients you obtain are the same as those shown in the example.

5. Find the center and the radius of the circle in the xy-plane that passes through the points $P(4, 2)$, $Q(5, 1)$, $R(-2, -6)$.

6. Find the center and the radius of the circle in the xy-plane that passes through the points $P(-7, 15)$, $Q(0, 22)$, $R(17, 15)$.

7. Find the center and the radius of the circle in the xy-plane that passes through the points $P(3, 5)$, $Q(10, -2)$, $R(-20, -18)$.

8. Find the center and the radius of the circle in the xy-plane that passes through the points $P(43, 41)$, $Q(23, 51)$, $R(-37, -29)$.

Applying Your Skills

Most of the following exercises are based on the U.S. census data in the following table, listed in millions for the census years 1900–1990, by national region described roughly as follows:

- Northeast—from Maine to Pennsylvania and New Jersey
- Midwest—from Ohio to North Dakota to Kansas
- South—from Delaware to Kentucky to Oklahoma and Texas to Florida
- West—from Montana to New Mexico to Hawaii and Alaska

See **http://www.census.gov/population/censusdata/table-16.pdf** for further details.

	1900	1910	1920	1930	1940	1950	1960	1970	1980	1990
Northeast	21.047	25.869	29.662	34.427	35.977	39.478	44.678	49.061	49.137	50.809
Midwest	26.333	29.889	34.020	38.594	40.143	44.461	51.619	56.590	58.867	59.669
South	24.524	29.389	33.126	37.858	41.666	47.197	54.973	62.813	75.367	85.446
West	4.309	7.082	9.214	12.324	14.379	20.190	28.053	34.838	43.171	52.786

In Exercises 9–12 fit (as in Example 1) a linear function to the 1980 and 1990 population values for the indicated region.

9. The Northeast
10. The Midwest
11. The South
12. The West

In Exercises 13–16 fit (as in Example 2) a quadratic function to the 1970, 1980, and 1990 population values for the indicated region.

13. The Northeast
14. The Midwest
15. The South
16. The West

In Exercises 17–20 fit (as in Example 3) a cubic polynomial to the 1960, 1970, 1980, and 1990 population data for the indicated region.

17. The Northeast
18. The Midwest
19. The South
20. The West

In Exercises 21–24 find the cubic polynomial that best fits the 1910, 1930, 1950, 1970, and 1990 population data for the indicated region, and calculate the average error in this optimal cubic.

21. The Northeast
22. The Midwest
23. The South
24. The West

In Exercises 25–28 find the quartic polynomial that best fits the 1900–1990 (all decades) population data for the indicated region, and calculate the average error in this optimal quartic.

25. The Northeast
26. The Midwest
27. The South
28. The West

A sphere in space with **center** (h, k, l) and **radius** r (Fig. 6.4.18) has equation
$$(x - h)^2 + (y - k)^2 + (z - l)^2 = r^2.$$

Four given points in space suffice to determine the values of h, k, l, and r. In Exercises 29 and 30 find the center and radius of the sphere that passes through the four given points $P, Q, R,$ and S. *Hint*: Substitute each given triple of coordinates into the foregoing sphere equation to obtain four equations that h, k, l, and r must satisfy. To solve these equations, first subtract the first one from each of the other three. How many unknowns are left in the three equations that result?

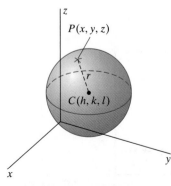

FIGURE 6.4.18 The sphere with center (h, k, l) and radius r.

29. $P(4, 6, 15), Q(13, 5, 7), R(5, 14, 6), S(5, 5, -9)$

30. $P(11, 17, 17), Q(29, 1, 15), R(13, -1, 33), S(-19, -13, 1)$

Chapter 6 | Review

In this chapter, you learned about polynomial functions and models and linear systems of equations. After completing the chapter, you should be able to

- determine whether a function described graphically or symbolically represents a polynomial function;
- find the output value of a polynomial function for a given input value;
- find the input value(s) of a polynomial function for a given output value.
- solve an equation or inequality involving a polynomial function;
- solve a system of two linear equations in two unknowns by the method of elimination;
- solve a system of two or more linear equations in the same number of unknowns by determinant or matrix inverse methods;
- use linear system methods to find a quadratic or cubic interpolating polynomial that fits given data points; and
- use regression methods to find a best-fitting cubic or quartic model for given data.

Review | Exercises

In Exercises 1–6, match each graph both with the most appropriate function model in A–F and with the type of function it represents in I–III. (Choices from I–III may be used more than once.)

A. $f(x) = ax + b$
B. $f(x) = ax^2 + bx + c$
C. $f(x) = ax^3 + bx^2 + cx + d$
D. $f(x) = ax^4 + bx^3 + cx^2 + dx + e$
E. $f(x) = ae^{bx}$
F. $f(x) = a\ln(bx)$

I. polynomial
II. exponential
III. logarithmic

1.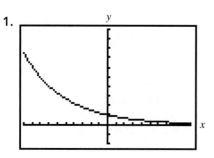

2.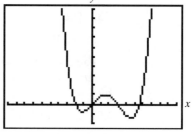

3.

4.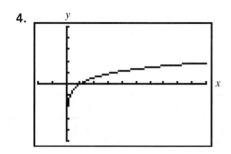

5.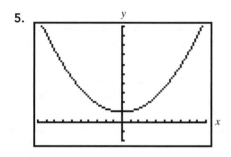

6.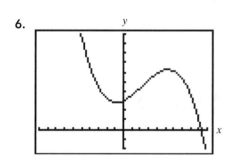

The exact solutions to Exercises 7 and 8 are all integers or simple rational numbers. Approximate the solutions by graphing, then verify the apparent exact solution by substitution the appropriate value in the given equation.

7. $15x^3 - 49x^2 - 104x + 240 = 0$

8. $576x^4 + 120x^3 - 6650x^2 - 2375x + 6250 = 0$

9. During fall registration, the student government association sold hot dogs, chips and soft drinks to raise money for a scholarship fund. The table following gives the numbers of each item sold and the total receipts for each day.

	HOT DOGS	CHIPS	SOFT DRINKS	RECEIPTS
Wednesday	54	32	80	$177.00
Thursday	64	50	95	$216.00
Friday	45	22	50	$128.50

If a student purchased two hot dogs, one bag of chips, and a soft drink, how much was she charged?

10. The following data give percentage of all music sold (recorded music and music video) that was rap music in the given years.

Year (t)	1993	1994	1995	1996
Percentage rap music R	9.2	7.9	6.7	8.6

Source: World Almanac and Book of Facts 1998.

a. Use linear system methods to determine the interpolating cubic polynomial for these data.

b. Use your model to predict the percentage of all music sold that was rap music in 1999. Do you believe that this is a good model for predicting rap music sales? Why or why not?

11. The following data indicate the total U.S. lead emissions E in short tons for the indicated years.

Year x	1986	1987	1988	1989	1990	1991	1992	1993	1994	1995
Emissions E (short tons)	7296	6857	6513	6034	5666	5280	4862	4945	5028	4986

Source: World Almanac and Book of Facts 1998.

a. Use the data to determine the best-fitting cubic polynomial model $E_3(x)$, where x is given in years after 1986.

b. Use the data to determine a best-fitting quartic polynomial model $E_4(x)$, where x is given in years after 1986.

c. Both of these models fit the data very well. Which model would you use if you wanted to argue for stronger pollution control laws? Give an extrapolation point using each model to support your answer.

12. (Chapter Opener Revisited). In the discussion that introduced this chapter, we looked at a plot that displayed the percentage of the U.S. population born abroad from 1900 to 1990. The plot was based on the data given in the following table:

Year	1900	1910	1920	1930	1940	1950	1960	1970	1980	1990
Percentage Born Abroad	13.6	14.7	13.2	11.6	8.8	6.9	5.4	4.8	6.2	7.9

Source: World Almanac and Book of Facts 1998.

a. Find the cubic polynomial $P(t)$ that best fits these data, where t represents the number of years after 1900.
b. According to the model, in what year from 1900 to 1950 was the percentage of individuals born abroad largest?
c. According to the model, what was the largest percentage of individuals born abroad from 1900 to 1950?
d. According to the model, in what year from 1950 to 1990 was the percentage of individuals born abroad smallest?
e. According to the model, what was the smallest percentage of individuals born abroad from 1950 to 1990?
f. According to the model, during what years of the twentieth century was less than 10% of the U.S. population born abroad?
g. Use the model to approximate the percentage of the U.S. population that was born abroad in 2000. How does your value compare with the actual 2000 percentage of 10.4%?

INVESTIGATION Cost Curves and Minimum Average Cost

When a company manufactures a product, it incurs two basic types of cost. The first type of cost is fixed cost. This is the cost that is commonly referred to as "overhead" (or "fixed cost"), and might include such things as rent or mortgage payment on the property, insurance, telephone, or business license. This cost remains the same no matter how many items are produced. The second kind of cost is called "variable cost" because it depends on the number of items produced. Variable cost could include the cost of materials, workers' salaries, and electricity.

When we speak of the total cost function, we mean the sum of the fixed cost function and the variable cost function. That is,

total cost = total fixed cost + total variable cost,

or, in the notation used in economics, TC = TFC + TVC. (We use the word "total" in these cases to distinguish between these costs and others we will discuss later, and refer to these cost functions as "cost curves.")

Total cost curves are often modeled by cubic polynomials, such as the one whose graph is shown in Fig. 6.R.1.

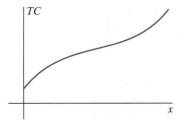

FIGURE 6.R.1 A cubic total cost curve.

Let's think about the features of this kind of curve that make it useful in representing cost:

- First, the y-intercept of this curve is positive. The y-intercept corresponds to the cost for zero output, that is, the fixed cost.
- Second, the function represented by the graph is increasing; in general, as output increases, and so does cost.
- Finally, the shape of the curve is significant—it first curves downward, then curves upward. For

small values of the output x, the curve becomes less steep as increases, while for large values of x, the curve becomes steeper as increases. At small production levels, it is relatively easy to increase production efficiently, making costs increase less rapidly. At a certain point, however, in order to increase output, the company requires more labor and more materials, and costs begin to increase more rapidly.

Brian has decided to turn his hobby of making birdhouses into a small home business, BJD Designer Birdhouses. He has determined his short-run daily cost schedules as indicated in the accompanying table.

OUTPUT	FIXED COST	VARIABLE COST
0	7	0
1	7	11.75
2	7	22.25
3	7	31.75
4	7	40.00
5	7	47.80
6	7	55.30
7	7	63.05
8	7	71.15
9	7	79.75
10	7	89.25

1. Find the total fixed cost function TFC(x) in terms of the output level x for BJD Designer Birdhouses.
2. Find a cubic polynomial function TVC(x) to model the total variable cost as a function of the output x.
3. Find a cubic polynomial function TC(x) to model the total cost as a function of the output x.
   What is the relationship between your answers to (1), (2), and (3)?
   There are two other cost curves that are important in economics. The first is the average variable cost (AVC). The average variable cost is the variable cost per item of all the items being produced, and it is calculated by dividing the total variable cost by the output. That is,

$$\text{AVC}(x) = \frac{\text{TVC}(x)}{x}.$$

For example, $\text{AVC}(8) = \frac{\text{TVC}(8)}{8} = \frac{71}{8} \approx \$8.88$.

4. Use the total variable cost function you found in (2) to find a function for the average variable cost AVC(x). (You might be interested to know that the average variable cost function is an example of a *rational* function, which is the quotient of two polynomial functions.)
   The most important cost curve to an economist is the marginal cost (MC), which is the increase in total cost that results from producing one more item (one more birdhouse, in this case). The easiest way to determine marginal cost is to find the change in cost required to produce each additional unit.
5. Complete the following table to determine a marginal cost schedule for the birdhouse company.

OUTPUT	VARIABLE COST	MARGINAL COST
0	0	11.75
1	11.75	10.50
2	22.25	
3	31.75	
4	40.00	
5	47.80	
6	55.30	
7	63.05	
8	71.15	
9	79.75	
10	89.25	—

6. Find a quadratic function model MC(x) that represents the marginal cost as a function of the output x.
7. It is a law of economics that the marginal cost curve rises to intersect the average total cost curve at the lowest point of the average cost curve. Graph the average total cost curve and the marginal cost curve and find their point of intersection. If Brian wants to minimize his average cost per birdhouse, how many birdhouses should he produce per day?

CHAPTER 7 | BOUNDED GROWTH MODELS

7.1 LIMITED POPULATIONS

7.2 FITTING LOGISTIC MODELS TO DATA

7.3 DISCRETE MODELS AND CHAOS

How do you know when a certain disease reaches epidemic proportions? For the layperson, a disease appears to be an epidemic when it is either already widespread or spreading at an alarming rate. Since the 1980s, you have probably heard references to the "AIDS epidemic" and comments about deaths from AIDS increasing "exponentially." But is this indeed the case?

The chart in Fig. 7.0.1 shows the cumulative deaths from HIV (the virus that causes AIDS) for the years 1987–1997. We can see from this chart that from 1987 to about 1994, deaths from HIV were increasing at an increasing rate. This portion of the graph displays the upward-curving shape typical of an exponential function. But after about 1994, while the total number of deaths was still increasing, it was increasing at a decreasing rate, and may appear to be "leveling off." This portion of the graph curves downward, and looks more logarithmic than exponential.

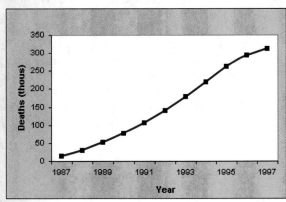

FIGURE 7.0.1 Cumulative HIV deaths.
Source: National Center for Health Statistics.

Because these data represents cumulative deaths, the function modeling the data can never be decreasing. Thus, the graph cannot have the peaks and valleys usually associated with polynomial functions. These data provide a good example of a bounded growth function, the type of function and model we will study in this chapter.

7.1 LIMITED POPULATIONS

A pristine forest habitat was initially stocked with 10 rabbits. Over the next 3 years the rabbit population in this forest was monitored carefully, with the results shown in the following table:

t MONTHS	P RABBITS	t MONTHS	P RABBITS
0	10		
3	21	21	98
6	40	24	99
9	62	27	100
12	80	30	100
15	91	33	100
18	96	36	100

We see from the first two columns that the rabbits immediately began to reproduce rapidly, as is typical of rabbits. After 6 months the initial population of 10 rabbits had quadrupled to 40 rabbits. In the second 6-month period the rabbit population doubled, from 40 rabbits to 80 rabbits. But then the rate of growth of the population slowed noticeably—only 16 more rabbits were added to the population during the first half of the second year. And we see from the last two columns of the table that the population leveled off at 100 rabbits during the third year.

A possible explanation for this phenomenon is that the forest habitat may contain only enough food to sustain a maximum population of 100 rabbits. At any rate, our rabbits provide an example of a population that is **bounded** or **limited**. Instead of increasing indefinitely as time goes on, it levels off at a finite **limiting population**.

Another example of a limited population is a population of fruit flies in a closed container that can support only so many flies. Similarly, a lake may provide enough food only for a limited population of fish. We will see that many human populations also are limited—for instance, by the area of a country or by the food products it can provide.

The spreadsheet chart in Fig. 7.1.1 shows a plot of the rabbit population data in the foregoing table. It is apparent at a glance that this graph does not resemble any seen in previous chapters. Polynomial, exponential, and logarithmic graphs ultimately continue either to rise or to fall indefinitely; they do not level off.

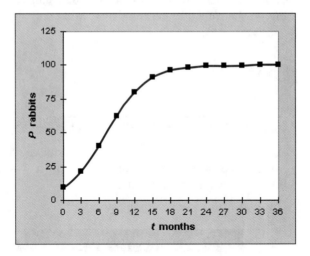

FIGURE 7.1.1 Plot of rabbit population data.

Thus we need some new type of mathematical function that can serve to model a limited population. The "logistic functions" we now define are useful for this purpose.

> **DEFINITION: Logistic Function**
> A **logistic function** is a function of the form
> $$f(x) = \frac{c}{1 + ae^{-bx}}, \tag{1}$$
> where a, b, and c are positive constants and $e \approx 2.71828$ denotes as usual the famous natural base number of Chapter 4. Figure 7.1.2 shows this general logistic function defined with a graphing calculator.

FIGURE 7.1.2 The logistic function $f(x) = \dfrac{c}{1 + ae^{-bx}}$ and its limiting value $y = c$.

EXAMPLE 1 — Verifying the Limiting Value

Use a graph and a table of values to verify that the logistic function
$$f(x) = \frac{12}{1 + 3e^{-x}} \tag{2}$$
has limiting value 12.

SOLUTION

Upon entering the parameters $a = 3$, $b = 1$, and $c = 12$ (as in Fig. 7.1.3), we are prepared to plot or tabulate this function. Figure 7.1.4 shows the resulting graph-table (**G-T** calculator mode) split screen. The logistic function in (2) is plotted on the left in the window $0 < x < 8, 0 < y < 15$, together with the line $y = 12$ (which is plotted also because we defined **Y2 = C** as in Fig. 7.1.2).

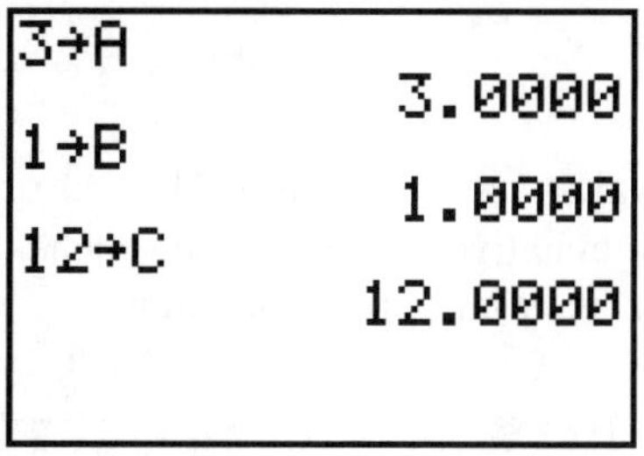

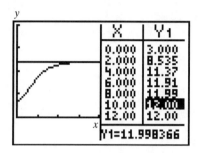

FIGURE 7.1.3 Parameters for the logistic function $f(x) = \dfrac{12}{1 + 3e^{-x}}$.

FIGURE 7.1.4 Calculator graph and table of values of the logistic function $f(x) = \dfrac{12}{1 + 3e^{-x}}$.

We see that the logistic function levels off at the height $c = 12$ in the graph. This is confirmed in the table on the right in Fig. 7.1.4, where we observe that $f(x)$ has reached the approximate value $11.9984 \approx c$ when $x = 10$.

The smooth curve interpolating the data points in Fig. 7.1.1 is the graph of the logistic function

$$P(t) = \frac{100}{1 + 9e^{-0.3t}} \qquad (3)$$

that we obtain with the parameter values $a = 9$, $b = 0.3$, and $c = 100$. Upon

- entering these new parameter values in the manner of Fig. 7.1.3,
- defining the new viewing window $0 < x < 36, 0 < y < 150$, and
- specifying Δ**Tbl** $= 6$ in the table setup menu,

we get the split screen shown in Fig. 7.1.5. [Of course, the calculator is using **X** in place of the independent variable t in (3) and **Y** in place of the dependent variable P].

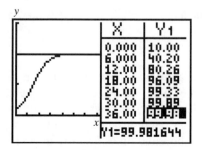

FIGURE 7.1.5 Graph and table for the logistic population model $P(t) = \dfrac{100}{1 + 9e^{-0.3t}}$ of Example 1.

On the left we see the graph apparently leveling off at the height $c = 100$, and on the right we see that $P(t)$ has actually reached the approximate value $99.9816 \approx c$ when $t = 36$ (months).

For each of these two logistic functions in (2) and (3), with the two parameter values $c = 12$ and $c = 100$, we thus see that the value of the function is approaching the fixed number c as x gets larger and larger. Similarly, the value $f(x) = c/(1 + ae^{-bx})$ of *any* logistic function always approaches the fixed value of the numerator constant c as x gets larger and larger. The reason is that the denominator

$$1 + ae^{-bx} = 1 + \frac{a}{e^{bx}}$$

approaches the value 1 as x increases. This, in turn, is so because the exponential $e^{-bx} = 1/e^{bx}$ with negative exponent approaches the value 0 as x gets larger and larger. Thus if x is very large in value, then

$$f(x) = \frac{c}{1 + ae^{-bx}} = \frac{c}{1 + \text{very small}} \approx \frac{c}{1} = c.$$

Hence if the number x is sufficiently large, then the values of $f(x)$ and c are indistinguishable as decimal numbers with a fixed number of decimal places.

This means geometrically that, as illustrated in Figs. 7.1.4 and 7.1.5, the curve $y = f(x)$ approaches the horizontal line $y = c$ as it moves further and further to the right (in the direction of increasing values of x). When x is sufficiently large the logistic curve and the line are visually indistinguishable. As illustrated in the following example, the value of the parameter b in (1) affects **how rapidly** the value $f(x)$ approaches the fixed number c.

The functions

$$f(x) = \frac{1}{1 + 10e^{-100x}} \quad \text{and} \quad g(x) = \frac{1}{1 + 10e^{-0.01x}} \tag{4}$$

are both logistic functions with $a = 10$ and $c = 1$. But $b = 100$ for $f(x)$ and $b = 0.01 = \frac{1}{100}$ for $g(x)$. The graphs of these two functions are shown in Figs. 7.1.6 and 7.1.7.

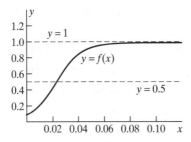

FIGURE 7.1.6 Graph of $f(x) = \dfrac{1}{1 + 10e^{-100x}}$.

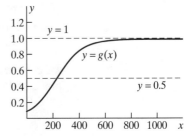

FIGURE 7.1.7 Graph of $g(x) = \dfrac{1}{1 + 10e^{-0.01x}}$.

These two graphs may look the same at first glance. Both $f(x)$ and $g(x)$ approach the same value 1 as x increases. However, look at the different scales on the x-axes. Evidently $f(x) \approx 1$ when $x \geq 0.1$, while $g(x) \approx 1$ when $x \geq 1000$. Because 1000 equals ten thousand times 0.1, we might say that $f(x)$ (with $b = 100$) approaches 1 ten thousand times faster than $g(x)$ (with $b = 0.01$) approaches 1.

EXAMPLE 2 Finding Input Values

The line $y = 0.5$ is also shown in Figs. 7.1.6 and 7.1.7. Find the points where the logistic curves $y = f(x)$ and $y = g(x)$ cross this horizontal line. That is, find x_1 and x_2 such that $f(x_1) = 0.5$ and $g(x_2) = 0.5$.

SOLUTION

The desired values can be found either graphically or symbolically. We illustrate the two approaches by finding x_1 graphically and x_2 symbolically.

Graphic Approach Figure 7.1.8 shows the graphs $y = 1/(1 + 10e^{-100x})$ and $y = 0.5$ in the viewing window $0 \leq x \leq 0.1$, $0 \leq y \leq 1.25$. We have used our calculator's intersection-finding facility to locate the intersection of the two graphs. We see that they intersect when $x_1 \approx 0.023$.

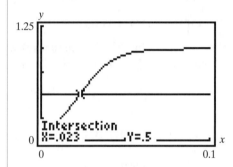

FIGURE 7.1.8 The intersection of the curve $y = \dfrac{1}{1 + 10e^{-100x}}$ and the line $y = 0.5$.

Symbolic Approach We use the natural logarithms of Chapter 4 to find x_2 by solving the equation $g(x) = 0.5$, that is,

$$\frac{1}{1 + 10e^{-0.01x}} = 0.5.$$

If we multiply both sides of the equation by $1 + 10e^{-0.01x}$, we get

$$1 = 0.5(1 + 10e^{-0.01x}).$$

Then dividing both sides of the equation by 0.5 (or, equivalently, multiplying both sides by 2), we get

$$2 = 1 + 10e^{-0.01x}.$$

Upon subtracting 1 and then dividing by 10, we get

$$0.1 = e^{-0.01x} \quad \text{or}$$
$$e^{-0.01x} = 0.1$$

Now we take natural logarithms of both sides. Because $\ln(e^{-0.01x}) = -0.01x$ (remembering that the natural exponential and logarithm functions undo each other), this gives

$$-0.01x = \ln(0.1),$$

so finally (after multiplying by -100) our calculator gives the desired value

$$x_2 = -100 \ln(0.01) \approx 230.26.$$

In summary, we see that

- the logistic function $f(x)$ with the larger value $b = 100$ starts with $f(0) = \frac{1}{11}$ and reaches the value $\frac{1}{2}$ when $x \approx 0.023$, while
- the logistic function $g(x)$ with the smaller b value 0.01 starts with $g(0) = \frac{1}{11}$ and doesn't reach the value $\frac{1}{2}$ until $x \approx 230$.

It would be good practice for you to reverse our procedure; that is, to find x_1 symbolically and x_2 graphically. Of course you should get the same numerical results.

Example 2 illustrates the important principle that

The larger is the parameter b in the logistic function $f(x) = \dfrac{c}{1 + ae^{-bx}}$, the faster $f(x)$ approaches the limiting value c as x increases.

Logistic Population Models

In Fig. 7.1.1 the markers indicate the actual rabbit population data points given by our initial data, while the smooth curve is the graph of the logistic function

$$P(t) = \frac{100}{1 + 9e^{-0.3t}}.$$

Hence this function is a **logistic model** that closely approximates the actual rabbit population in Example 2.

In the case of the general logistic population model

$$P(t) = \frac{c}{1 + ae^{-bt}} \tag{5}$$

it is customary to write $M = c$ for the limiting population that $P(t)$ approaches, so

$$P(t) = \frac{M}{1 + ae^{-bt}}. \tag{6}$$

If $P(0) = P_0$ denotes the initial population, then substitution of $t = 0$ gives

$$P_0 = \frac{M}{1 + a}$$

(because $e^0 = 1$). If we solve this last equation for

$$a = \frac{M}{P_0} - 1 = \frac{M - P_0}{P_0}$$

and substitute in (6), we get

$$P(t) = \frac{M}{1 + \dfrac{M - P_0}{P_0} e^{-bt}}.$$

Multiplication of numerator and denominator by P_0 finally gives the subsequent formula in (7).

SECTION 7.1 Limited Populations

DEFINITION: Logistic Population Model

The **logistic population model** with *limiting population M*, *initial population* P_0, and *rate constant b* is

$$P(t) = \frac{MP_0}{P_0 + (M - P_0)e^{-bt}}. \tag{7}$$

In the case of a population occupying a specific habitat, the limiting population M is sometimes called the **carrying capacity** of this environment. It is the maximal population that the environment has the capacity to support (or "carry").

We can check our derivation of this model first by substituting $t = 0$ to obtain

$$P(0) = \frac{MP_0}{P_0 + (M - P_0) \cdot 1} = \frac{MP_0}{M} = P_0,$$

so the model does, indeed, correspond to an initial population of P_0. Second, when t is so large that $e^{-bt} \approx 0$, we see that

$$P(t) \approx \frac{MP_0}{P_0 + (M - P_0) \cdot 0} = \frac{MP_0}{P_0} = M.$$

Thus the model does, indeed, correspond to a limiting population of M.

If the initial and limiting populations P_0 and M are given, then a knowledge of the population at some single future time suffices to determine the value of the rate constant b in (7).

EXAMPLE 3 When Can You Fish?

You own a small lake with a carrying capacity of 300 fish. You initially stock it with 50 fish, and plan to permit fishing to begin when the lake contains 200 fish. After 1 year there are 100 fish in the lake. Assuming a logistic fish population, how long must you wait to begin fishing?

SOLUTION

We are given the initial and limiting population values. But we must find the value of the rate constant b in (7) before we can proceed to answer the question. Thus there are two steps to the solution—first finding b, and then finding when $P(t) = 200$.

Step 1 Substitution of $M = 300$ and $P_0 = 50$ in the general logistic population model (7) gives

$$P(t) = \frac{15000}{50 + 250 e^{-bt}}. \tag{8}$$

In order to have a "complete" population model that we can use for predictive purposes, we obviously need also the value of b. But we are given that $P(12) = 100$, so substitution of $t = 12$ in (8) yields the equation

$$\frac{15000}{50 + 250e^{-12b}} = 100 \tag{9}$$

the rate constant b must satisfy. Figure 7.1.9 shows a graphing calculator set up to solve this equation graphically, using **X** as the independent variable b and **Y** as the dependent variable.

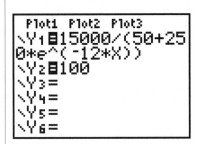

FIGURE 7.1.9 A graphing calculator set up to solve Equation (9).

Some experimentation is required to determine an appropriate viewing window. We are talking about populations in the hundreds, so the y-scale should be measured in hundreds. But it turns out that the desired value of $b = x$ is quite small (in particular, smaller than 1). Figure 7.1.10 shows the resulting plot in the window $0 \leq x \leq 1$, $0 \leq y \leq 400$. The calculator's intersection-finding facility has been applied, and we see that $b = 0.0764$ (rounded off to four decimal places).

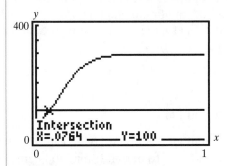

FIGURE 7.1.10 Solving the equation $\frac{15000}{50 + 250e^{-12b}} = 100$.

Substitution of this value of b in (8) yields the logistic model

$$P(t) = \frac{15000}{50 + 250e^{-0.0764t}} \tag{10}$$

for our fish population.

Step 2 In order to find when $P(t) = 200$ we need to solve the equation

$$\frac{15000}{50 + 250e^{-0.0764t}} = 200. \tag{11}$$

We could do this graphically (much as in Step 1), but let's do this step symbolically. Cross-multiplication gives

$$15000 = 200(50 + 250e^{-0.0764t}) = 10000 + 50000e^{-0.0764t},$$

that is,

$$3 = 2 + 10e^{-0.0764t} \quad \text{(upon division by 5000),} \quad \text{or}$$
$$10e^{-0.0764t} = 1.$$

Hence

$$e^{-0.0764t} = \tfrac{1}{10}$$
$$-0.0764t = \ln(\tfrac{1}{10})$$
$$t = -\frac{\ln(0.1)}{0.0764} \approx 30.1385$$

● (in months). Consequently, we must wait just over $2\tfrac{1}{2}$ years (after initial stocking) before beginning to fish our lake.

For practice, you might rework Example 3, solving Equation (9) symbolically in Step 1 and solving Equation (11) graphically in Step 2.

There's nothing to prevent us from stocking the lake of Example 3 with an initial fish population greater than its carrying capacity of $M = 300$ fish. If we substitute $M = 300$ and $b = 0.0764$ in (7), we get the logistic population function

$$P(t) = \frac{300 P_0}{P_0 + (300 - P_0)e^{-0.0764t}} \tag{12}$$

that models the fish population of the lake if it is stocked initially with P_0 fish. Figure 7.1.11 shows the resulting population graphs with a variety of initial populations ranging from 25 to 600 fish when $t = 0$. We see that, whatever is the initial population P_0—either less than or greater than the carrying capacity $M = 300$—the fish population levels off at the lake's carrying capacity of 300 fish. If there are too many fish initially, then the fish population $P(t)$ decreases in order to approach M, while it increases in order to approach M is there are too few fish initially. In short, nature compensates for any initial imbalance.

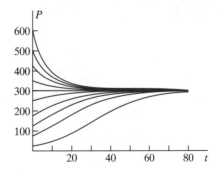

FIGURE 7.1.11 The logistic population $P(t) = \dfrac{300 P_0}{P_0 + (300 - P_0)e^{-0.0764t}}$ with different initial population values P_0.

This behavior is characteristic of all logistic populations. Whatever is the initial population $P_0 > 0$, the resulting population $P(t)$ approaches the same limiting population M as t increases.

EXAMPLE 4 Finding an Input Value

Suppose the lake of Example 3 is stocked initially with 500 fish. Under natural conditions — that is, with no fishing taking place — how long will it be until the lake contains only 350 fish?

SOLUTION

With $M = 300$, $P_0 = 500$, and $b = 0.0764$ in Equation (7), we find that the fish population after t months is given by

$$P(t) = \frac{150000}{500 - 200e^{-0.0764t}}. \tag{13}$$

Substitution of $P = 350$ then yields the equation

$$\frac{150000}{500 - 200e^{-0.0764t}} = 350 \tag{14}$$

that we need to solve for t. Figure 7.1.12 shows the graph—in the window $0 \leq x \leq 36, 0 \leq y \leq 600$—of the logistic function on the left-hand side and the horizontal line defined by the right-hand side of this equation.

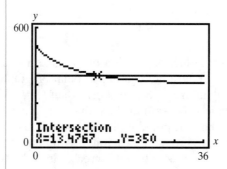

FIGURE 7.1.12 Solving the equation $\frac{150000}{500 - 200e^{-0.0764t}} = 350$.

We have used our calculator's intersection-finding facility, and see that the four-decimal-place solution of Equation (14) is $t = 13.4767$. Thus it's about $13\frac{1}{2}$ months until only 350 fish are left in the lake.

Summary

This section contains just two basic formulas—the formula

$$f(x) = \frac{c}{1 + ae^{-bx}} \tag{1}$$

for a purely mathematical *logistic function* with limiting value c, and the formula

$$P(t) = \frac{MP_0}{P_0 + (M - P_0)e^{-bt}} \tag{7}$$

for a *logistic population model* with limiting value M, initial population P_0, and rate constant b. It is neither necessary nor desirable that you remember the relationship between the parameters $a, b,$ and c in (1) and those in (7). It is important that you know how to

- fit the logistic population model (7) to given population information, as in Step 1 of Example 3, and how to
- find when a given value of a logistic function or population is attained, as in Examples 2 and 4 or in Step 2 of Example 3.

7.1 Exercises | Building Your Skills

The following exercises can be solved either graphically or symbolically. For each of the logistic functions given in Exercises 1–4, find x such that f(x) is 75% of the limiting value c.

1. $f(x) = \dfrac{200}{1 + 2e^{-0.1x}}$

2. $f(x) = \dfrac{300}{1 + 4e^{-0.05x}}$

3. $f(x) = \dfrac{4}{1 + 0.5e^{-2x}}$

4. $f(x) = \dfrac{10000}{1 + 7e^{-17x}}$

In each of Exercises 5–10 find a formula of the form in Equation (7) for the logistic population model that has the given limiting population M and the given initial population P_0 and satisfies the additional population condition that is given.

5. $M = 250, P_0 = 50,$ and $P(10) = 150$

6. $M = 1000, P_0 = 200,$ and $P(25) = 400$

7. $M = 7, P_0 = 1,$ and $P(1) = 5$

8. $M = 17, P_0 = 3,$ and $P(100) = 5$

9. $M = 100, P_0 = 300,$ and $P(10) = 200$

10. $M = 2000, P_0 = 5000,$ and $P(50) = 4000$

11. Determine when the population of Exercise 7 is half of its carrying capacity.

12. Determine when the population of Exercise 8 is half of its carrying capacity.

13. Determine when the population of Exercise 9 is 1.5 times its carrying capacity.

14. Determine when the population of Exercise 10 is 1.5 times its carrying capacity.

Applying Your Skills

15. A lake has a carrying capacity of 500 fish. On July 1, 2000, the lake is stocked with 50 fish. Ten months later there are 100 fish in the lake. Assuming a logistic fish population, in what month of what year will there be 350 fish in the lake?

16. A large game preserve has a carrying capacity of 300 foxes. On July 1, 2000, the preserve is stocked with 20 foxes. A year later there are 50 foxes in the preserve. Assuming a logistic fox population, in what month of what year will there be 150 foxes in the game preserve?

Exercises 17 and 18 deal with the spread of an incurable contagious disease in a city with given constant population M. Let P(t) be the population of infected individuals at time t. Assume that P(t) is a logistic population with limiting population M (meaning that everyone in the city eventually gets it).

17. The city has a constant population of 240 thousand people. On July 1, 2000, the infected population numbers 25 thousand, and on January 1, 2001 it numbers 40 thousand. In what month of what year will half of the city's population be infected?

18. The city has a constant population of 600 thousand people. On July 1, 2000, a quarter of the people are infected, and 5 months later one third are infected. In what month of what year will three-quarters of the city's population be infected?

Exercises 19 and 20 deal with the spread of a rumor in a city with a constant population M. The rumor spreads just like an incurable infectious disease. The population P(t) of those who have heard it at time t is logistic, and eventually everyone in the city hears it.

19. Two people start a rumor in a city with a population of 20 thousand people. After 4 weeks 1 thousand people have heard this rumor. How many weeks (since its inception) will it take for half the people in the city to have heard this rumor?

20. On February 1, 2000, one tenth of the people in a city of 120 thousand people have heard a certain rumor, and 2 months later half of the city's people have heard it. In what month of what year will 90% of the city's population have heard this rumor?

Exercises 21 and 22 deal with advertising. The spread of knowledge of a new product is similar to the spread of a rumor. Assume the number P(t) of people who've heard of the product at time t is a logistic population, and that if the advertising campaign continues indefinitely, then eventually everyone hears of it.

21. An advertising campaign starts when only 5% of the people in the United States (population 270 million) have heard of an all-new detergent. After 4 weeks of advertising 15% of the people know about this new detergent. The soap company plans to advertise it until 35% of the people have heard of it. How many weeks will this take?

22. Now it's a new presidential candidate that's being advertised. Initially, only 1% of the people in the United States recognize his name. After 6 weeks of political advertising, 10% of the people have heard of him. How many *more* weeks will it take for 60% of the people to have heard of this candidate?

7.2 FITTING LOGISTIC MODELS TO DATA

The following table shows the population of Albuquerque, New Mexico, as recorded in the census counts of 1960, 1970, 1980, and 1990.

t	YEAR	POPULATION P (THOUSANDS)	TEN-YEAR CHANGE	AVERAGE ANNUAL CHANGE (THOUSANDS/YEAR)
0	1960	201		
10	1970	245	41	4.1
20	1980	333	88	8.8
30	1990	385	52	5.2

We see that the population of Albuquerque increases more rapidly during the decade of the 1970s than in either the preceding decade of the 1960s or in the succeeding decade of the 1980s. Obviously the growth is not linear, and it also seems pointless to fit this population with a natural (exponential) growth model, which would predict growth at an ever-increasing rate. Similarly, we would not expect a quadratic model to fit Albuquerque's population well. (Can you explain why not?)

Figure 7.2.1 shows the Albuquerque tP-population data entered in a graphing calculator for analysis. Figure 7.2.2 shows an **xyline Stat Plot** of these data. Do you see some resemblance with the logistic function graphs of Section 7.1? There's at least a hint of a "leveling off" that suggests that we try a logistic population model of the form

$$P(t) = \frac{c}{1 + ae^{-bt}}. \tag{1}$$

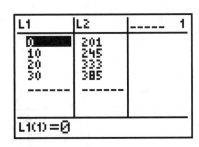

FIGURE 7.2.1 STAT EDIT calculator screen showing the Albuquerque population data in table form.

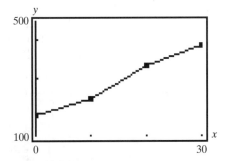

FIGURE 7.2.2 xyline Stat Plot of the Albuquerque population data.

EXAMPLE 1   Finding a Logistic Population Model

Find a logistic model for the Albuquerque population data.

SOLUTION

As we have seen, the calculator has many different types of regression. Fortunately, we see a **Logistic** selection among the choices on the **Stat Calc** menu. Having stored the *t*-data as list **L1** and the *P*-data for Albuquerque as list **L2** as shown in Fig. 7.2.2, the command

Logistic L1, L2, Y1

produces the result shown in Fig. 7.2.3. (Of course the calculator uses x instead of t for the independent variable and y instead of P for dependent variable.)

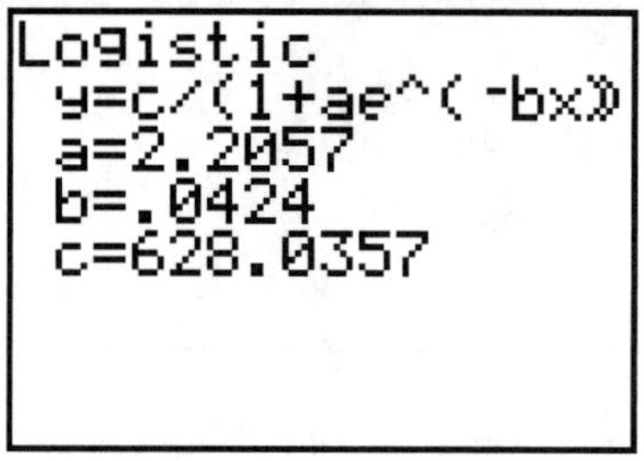

```
Logistic
y=c/(1+ae^(-bx))
a=2.2057
b=.0424
c=628.0357
```

FIGURE 7.2.3 Optimal logistic model resulting from the TI-83 **STAT CALC Logistic L₁, L₂, Y₁** command.

Thus we obtain the logistic model

$$P(t) = \frac{628.0357}{1 + 2.2057 e^{-0.0424 t}}. \qquad (2)$$

The logistic model in (2) predicts that the population of Albuquerque will not continue to grow indefinitely but will eventually level off with a limiting population of about 628 thousand people (because $c = 628.0357$ and P is measured in thousands). The **Logistic** command above also stores the optimal logistic function as **Y1**, so we can immediately generate the graph shown in Fig. 7.2.4, with the window $0 \leq t \leq 140, 0 \leq P \leq 700$ corresponding to the period from 1960 to 2100. We see that the population of Albuquerque continues to increase noticeably until about the middle of the 21st century, but then levels off conspicuously during the second half of the century. The table of population values pictured in Fig. 7.2.5 shows that the population in the year 2050 (when $t = 90$) is just under 600 thousand, and that by the end of the twenty-first century the population has reached 624 thousand (just 4 thousand short of the limiting population).

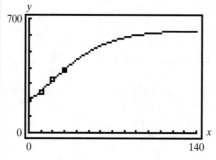

FIGURE 7.2.4 Graph of the optimal logistic model in (2).

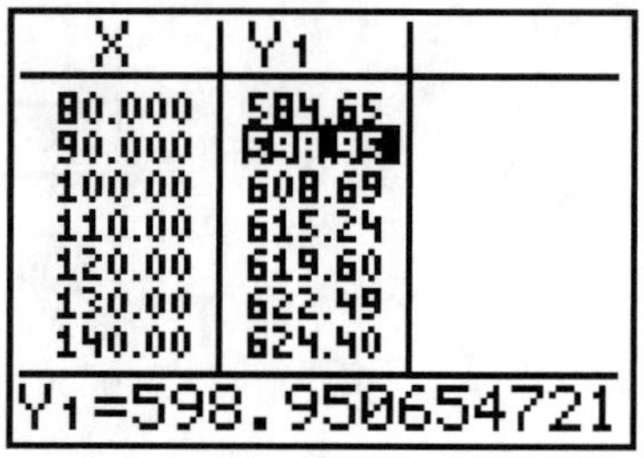

FIGURE 7.2.5 Table of logistic population values.

Would you bet real money that the population of Albuquerque will never reach 630 thousand? This would certainly be unwise! Real confidence in any such mathematical prediction is unwarranted, because the population function in (2) is merely a least-squares model based on our limited population 1960–1990 data.

Almost anything could actually happen in the next century—ranging from a vast increase in the population due to immigration to a decimation of the population by war or disease.

EXAMPLE 2 Finding the Average Error

Find the average error in the 1960–1990 population figures predicted by the logistic model in (2).

SOLUTION

The following table gives the actual and predicted Albuquerque populations (in thousands) for the years 1960–1990 (that is, for $t = 0$ to $t = 30$), along with the corresponding errors.

t	ACTUAL POPULATION	LOGISTIC $P(t)$	ERROR E
0	201	196	5
10	245	257	-12
20	333	323	10
30	385	388	-3

The sum of the squares of the errors shown in the table is

$$\text{SSE} = (5)^2 + (-12)^2 + (10)^2 + (-3)^2 = 278.$$

Since there are $n = 4$ data points, the average error in the optimal model is therefore given by

$$\text{average error} = \sqrt{\frac{\text{SSE}}{n}} = \sqrt{\frac{278}{4}} = \sqrt{69.5} \approx 8.337.$$

Thus the average error in the predicted population is almost 8.4 thousand people. Can you see that this is about 2% for the range of populations involved?

EXAMPLE 3 Finding an Input Value

When should the city planners of Albuquerque anticipate a population of a half million?

SOLUTION

For a graphical solution we need only define **Y2 = 500** since **Y1** was already defined as the logistic function in (2). This gives the plot shown in Fig. 7.2.6.

As indicated there, the calculator's intersecting-finding facility gives $t = 50.786$ when $P = 500$ (thousand). Adding 50 years to the starting date of 1960, we get 2010 for the year in which the predicted population hits a half million.

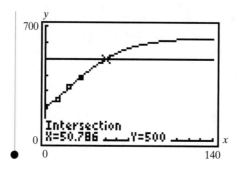

FIGURE 7.2.6 Finding graphically when the predicted population of Albuquerque is 500 thousand.

Political Modeling

We turn next to a political modeling example. Political campaigning is a form of advertising—in this case, advertising the candidate. The same general principle—that only bounded growth in approval is possible—applies to advertising any product, be it a new, improved detergent or a new line of jeans.

In politics, it is commonly observed that when an attractive candidate starts campaigning with only limited initial voter approval, the percentage $P(t)$ of voters favoring this candidate at first increases steadily with time t as his or her name recognition spreads. However, this percentage certainly can never exceed 100%. Consequently, neither a linear nor an exponential nor any kind of polynomial model for $P(t)$ is plausible. Of the different types of mathematical models we've encountered, only a (bounded growth) logistic model for the candidate's voter approval percentage $P(t)$ makes sense.

EXAMPLE 4 | Finding a Logistic Political Model

The following table summarizes voter approval of a gubernatorial candidate who begins the campaign as the favorite of only 10% of the voters in the state.

Months Campaigning	0	1	2	3	4
Voter Approval	10%	13%	17%	22%	27%

After the first month of campaigning, polling indicates that 13% of the voters favor this candidate, and after 4 months 27% of them favor him. It looks like voter approval is taking off, but eventually it must level off. Assuming a logistic model for the percentage $P(t)$ of voters favoring this candidate after t months, determine whether it is possible for this candidate to win the election. If so, how long must the campaign last in order for him to gain majority voter approval?

SOLUTION

To find the optimal logistic model, we first enter (as in Fig. 7.2.7) the polling data shown in the table, using percentage points rather than decimal percentages to record voter approval. Then the command **Logistic L1, L2, Y1** produces the results shown in Fig. 7.2.8.

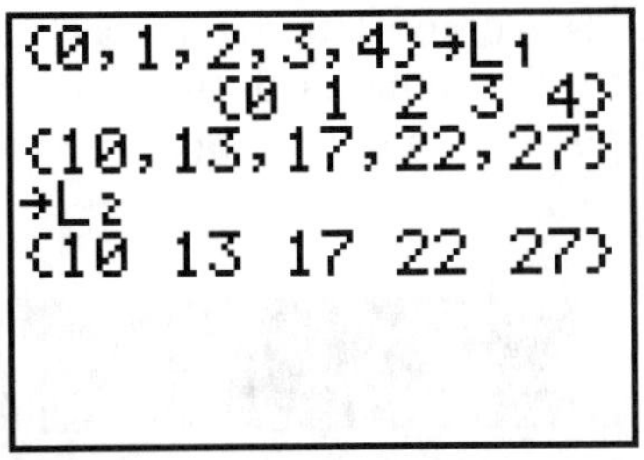

FIGURE 7.2.7 Entering the polling data of Example 4.

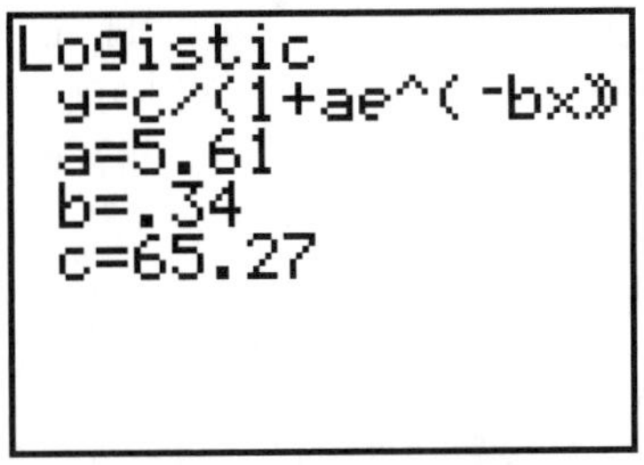

FIGURE 7.2.8 Logistic model for voter approval percentage $P(t)$.

Thus our optimal logistic model is

$$P(t) = \frac{65.27}{1 + 5.61 e^{-0.34t}}. \tag{3}$$

It follows that, if the campaign lasts long enough, our candidate can win election in a landslide with 65% of the vote.

Since the logistic function in (3) is automatically entered as **Y1**, we can immediately tabulate values to follow the progress of the campaign. The table in Fig. 7.2.9—where we have scrolled down far enough to see values leveling off—indicates that a 65% landslide could take 3 years, and we can certainly hope that no state political campaign will last that long.

However, the practical question is how long the campaign must last in order for the candidate to garner over 50% of the vote. To answer this question, we need only scroll back up in our table to see when **Y1** = $P(t)$ first exceeds 50 (percent). Figure 7.2.10 shows that this occurs after only 9 months.

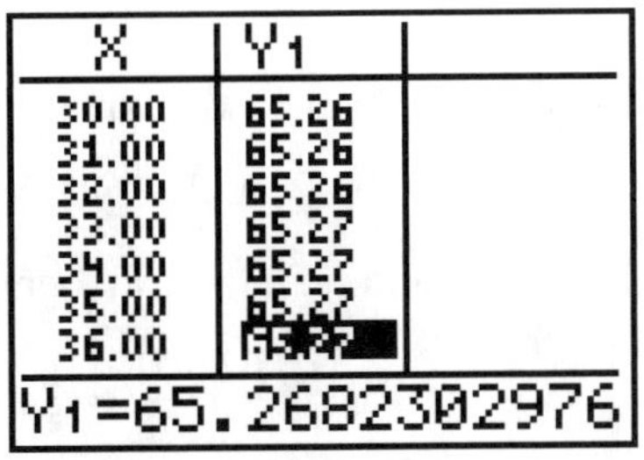

FIGURE 7.2.9 In a 3-year campaign our candidate wins with over 65% of the vote.

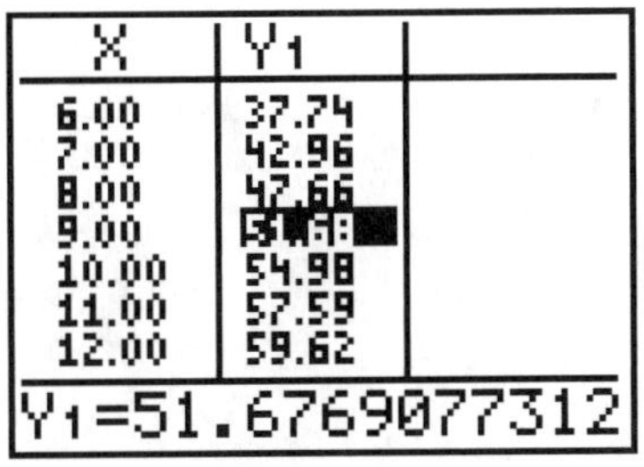

FIGURE 7.2.10 But it only takes 9 months of campaigning for him to gain a majority of the vote.

U.S. Population Projections

Logistic functions were first used around 1840 to model human population growth by the Belgian mathematician and demographer P. F. Verhulst. He sought to use U.S. census population data for the first half of the nineteenth century to predict the future growth of the United States.

322 CHAPTER 7 Bounded Growth Models

Here we will use census data for the last half of the nineteenth century to attempt to project U.S. population growth during the twentieth century. The following table lists population figures for the 1850–1890 period:

YEAR	U.S. POPULATION (MILLIONS)
1850	23.192
1860	31.443
1870	38.558
1880	50.189
1890	62.980

We will try several different fits to the population data shown here and compare the "goodness of fit" in terms of average errors.

EXAMPLE 5   Finding a Continuous Growth Population Model

Find a continuous growth model to fit the U.S. population data for the years 1850 and 1890, and find the average error for this model.

SOLUTION

If we take $t = 0$ in 1850, then $P_0 = P_0 e^0 = P(0) = 23.192$, so

$$P(t) = 23.192 \, e^{rt}.$$

Substitution of $P(40) = 62.980$ gives the equation

$$23.192 \, e^{40r} = 62.980.$$

When we divide by 23.192 and take natural logarithms we get

$$e^{40r} = \frac{62.980}{23.192} \approx 2.7156$$

$$\ln(e^{40r}) = 40r \approx \ln(2.7156) \approx 0.9990$$

$$r \approx \frac{0.9990}{40} \approx 0.0250.$$

Consequently a continuous growth model for U.S. population in the last half of the nineteenth century is

$$P(t) = 23.192 \, e^{0.0250 t}. \tag{4}$$

In each year the population in (4) is multiplied by the factor $e^{0.0250} = 1.0253$. Thus this model predicts continuous growth of 2.50% and an equivalent annual growth of 2.53% in the U.S. population. The following table compares the actual and continuous growth figures for the 1850–1890 period:

YEAR	t	ACTUAL U.S. POPULATION P	CONTINUOUS GROWTH PREDICTION P(T)	ERROR E	E²
1850	0	23.192	23.192	0.000	0.000
1860	10	31.443	29.779	1.664	2.769
1870	20	38.558	38.237	0.321	0.103
1880	30	50.189	49.097	1.092	1.191
1890	40	62.980	63.042	−0.062	0.004

The slight discrepancy in the actual and predicted 1890 populations results from calculator round-off error in the computations leading to (4). At any rate, the sum of the $n = 5$ squared errors in the final column is SSE = 4.067, so the average error is given by

$$\text{average error} = \sqrt{\frac{\text{SSE}}{n}} = \sqrt{\frac{4.067}{5}} \approx 0.902 \text{ million.}$$

EXAMPLE 6   Finding a Natural Growth Population Model

Find the best-fitting natural growth model for the 1850–1890 U.S. population data, and find the average error for this model.

SOLUTION

With the actual t, P data stored in the calculator in **L1** and **L2**, the command **ExpReg L1, L2, Y1** then produces the results shown in Fig. 7.2.11.

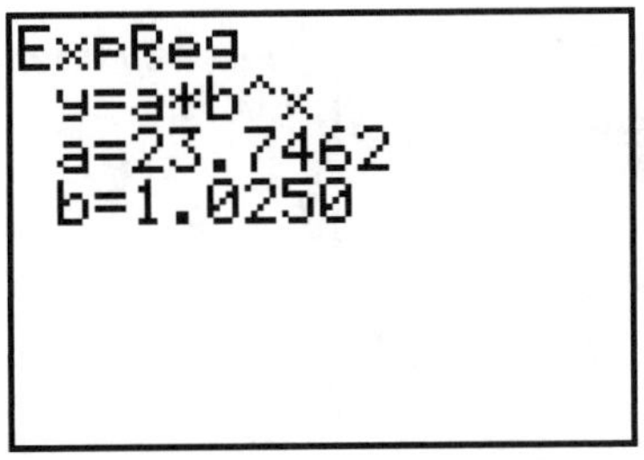

```
ExpReg
y=a*b^x
a=23.7462
b=1.0250
```

FIGURE 7.2.11 The optimal exponential growth model.

Thus our least-squares natural growth model is

$$P(t) = 23.7462 \times 1.0250^t. \tag{5}$$

As compared with the continuous growth model in (4), the optimal model in (5) has the slightly larger initial population $P_0 = 23.7462$ and the slightly smaller annual growth rate of 2.50%. The following table compares the actual and the optimal natural growth figures for the 1850–1890 period:

324 CHAPTER 7 Bounded Growth Models

YEAR	t	ACTUAL U.S. POPULATION P	OPTIMAL NATURAL GROWTH PREDICTION P(T)	ERROR E	E²
1850	0	23.192	23.746	−0.554	0.307
1860	10	31.443	30.397	1.046	1.094
1870	20	38.558	38.911	−0.353	0.125
1880	30	50.189	49.809	0.380	0.144
1890	40	62.980	63.760	−0.780	0.608

The sum of the $n = 5$ squared errors in the final column is SSE $= 2.278$, so the average error is given by

$$\text{average error} = \sqrt{\frac{\text{SSE}}{n}} = \sqrt{\frac{2.278}{5}} \approx 0.675 \text{ million}$$

● (as compared with the previous average error of 0.902 million).

EXAMPLE 7 Finding a Logistic Population Model

Find the best-fitting logistic model for the 1850–1890 U.S. population data.

SOLUTION

With the actual data still stored in **L1** and **L2**, the command **Logistic L1,L2,Y1** produces the result shown in Fig. 7.2.12.

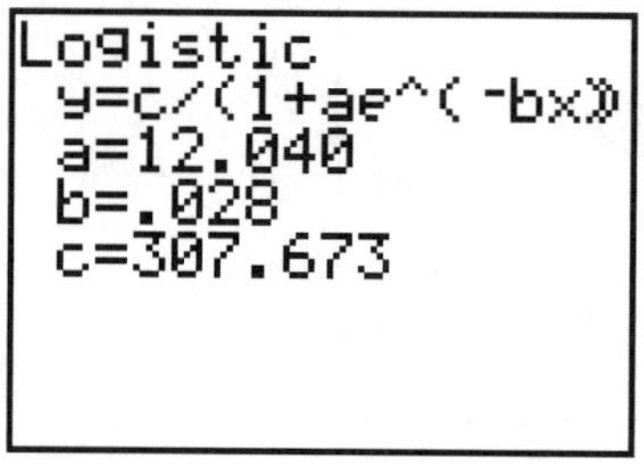

```
Logistic
y=c/(1+ae^(-bx))
a=12.040
b=.028
c=307.673
```

FIGURE 7.2.12 The optimal logistic population model.

Thus the optimal logistic model for U.S. population growth during the 1850–1890 period is

$$P(t) = \frac{307.673}{1 + 12.040\, e^{-0.028t}}. \tag{6}$$

The following table compares the actual and the optimal logistic figures for the 1850–1890 period:

YEAR	t	ACTUAL U.S. POPULATION P	OPTIMAL LOGISTIC GROWTH PREDICTION P(T)	ERROR E	E^2
1850	0	23.192	23.595	−0.403	0.162
1860	10	31.443	30.464	0.979	0.959
1870	20	38.558	39.058	−0.500	0.250
1880	30	50.189	49.642	0.547	0.290
1890	40	62.980	62.428	10.552	0.304

The sum of the $n = 5$ squared errors in the final column is SSE = 1.974, so the average error is given by

$$\text{average error} = \sqrt{\frac{\text{SSE}}{n}} = \sqrt{\frac{1.974}{5}} \approx 0.628 \text{ million}.$$

This is only a small improvement over the optimal exponential average error of 0.675 million, but there is nevertheless a vast difference between the implications of our exponential and logistic models for the U.S. population. The optimal exponential model in (5) predicts a never-ending increase in the United States population, whereas the optimal logistic model in (6) predicts that this population will eventually level off at a bit under 307 million people.

The following table compares the actual population figures with the optimal exponential and logistic population model figures for the United States during the twentieth century:

YEAR	ACTUAL U.S. POPULATION	NATURAL GROWTH MODEL	LOGISTIC MODEL
1900	76.212	81.618	77.518
1910	92.228	104.478	94.845
1920	106.022	133.741	114.125
1930	123.203	171.200	134.840
1940	132.165	219.150	156.280
1950	151.326	280.531	177.625
1960	179.323	359.103	198.071
1970	203.302	459.682	216.944
1980	226.542	588.432	233.780
1990	248.710	753.243	248.346

Undoubtedly, it is something of a fortuitous coincidence that after the passage of a century, the logistic population model predicts the 1990 U.S. population right on the nose at 248+ million people. Indeed, you can verify that the average error in the 10 logistic population figures in this table is about 14 million (nowhere near as good as the final figure suggests). But the exponential model prediction for 1990 is a U.S. population of over three-quarters of a billion, way off the chart. The spreadsheet chart in Fig. 7.2.13 compares these actual, natural growth model, and logistic model figures visually.

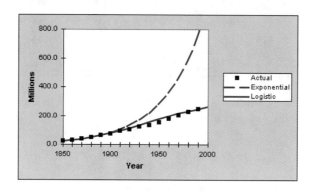

FIGURE 7.2.13 Comparing the actual, exponential, and logistic populations.

EXAMPLE 8 Finding an Input Value

When does the logistic model predict that the U.S. population will hit 300 million?

SOLUTION

The logistic model in (6) is still stored as **Y1**, so we plot **Y1** and **Y2 = 300** in the window $0 \leq x, y \leq 400$. Then we use the **intersect** command as shown in Fig. 7.2.14.

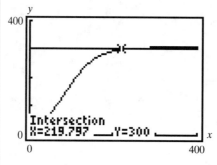

FIGURE 7.2.14 Finding when the predicted U.S. population is 300 million.

Thus the predicted U.S. population hits 300 million 219.797 years after 1850, that is—adding 219 to 1850—during the year 2069. Remember that all this comes from the five original data points marked by the small squares in the lower left of Fig. 7.2.13.

7.2 Exercises | Building Your Skills

In each of Exercises 1–6, use the logistic regression facility of your calculator to find the logistic function of the form $f(x) = c/(1 + ae^{-bx})$ that best fits the given data (with a, b, and c rounded off accurate to three decimal places), and find the average error in the fit.

1.

x	3	4	6
y	4	5	6

2.

x	20	40	60
y	18	30	40

3.

x	1	2	3	4
y	9	12	14	15

4.

x	15	30	50	80
y	19	50	124	201

5.

x	1	5	6	7	13
y	8	20	23	26	35

6.

x	5	20	40	60	70
y	43	107	294	543	637

Each of the tables in Exercises 7–12 gives the 1960–1990 census data for a U.S. city.

    a. Use the logistic regression facility of your calculator to find the logistic model of the form $P(t) = c/(1 + ae^{-bt})$ that best fits the given data (with a, b, and c rounded off accurate to three decimal places).
    b. Predict the population of the city in the year 2025.
    c. Find the city's limiting population, and determine when its population will be 90% of its limiting population

7. Oklahoma City, OK

Year	1960	1970	1980	1990
Population (thousands)	324	368	404	445

8. Tucson, AZ

Year	1960	1970	1980	1990
Population (thousands)	213	263	331	405

9. Anchorage, AK

Year	1960	1970	1980	1990
Population (thousands)	44.2	48.1	174	226

10. Lexington, KY

Year	1960	1970	1980	1990
Population (thousands)	62.8	108	204	225

11. Aurora, CO

Year	1960	1970	1980	1990
Population (thousands)	48.5	75.0	159	222

12. Tulsa, OK

Year	1960	1970	1980	1990
Population (thousands)	262	330	361	367

Applying Your Skills

13. The following table summarizes voter approval, after a given number of weeks of campaigning, of a mayoral candidate who begins the campaign as the favorite of only 12% of the voters in the city. Assuming a logistic model for the percentage $P(t)$ of voters favoring this candidate after t weeks, determine whether it is possible for this candidate to win the election. If so, how long must the campaign last in order for him to gain majority voter approval?

Weeks of Campaigning	0	1	2	3	4
Voter Approval	12%	15.5%	19.5%	24%	29%

14. The following table summarizes voter approval, after a given number of months of campaigning, of a presidential candidate who begins the campaign as the favorite of only 5% of the voters in the country. Assuming a logistic model for the percentage $P(t)$ of voters favoring this candidate after t months, determine whether it is possible for this candidate to win the election. If so, how long must the campaign last in order for him to gain majority voter approval?

Months of Campaigning	0	3	6	9	12
Voter Approval	5%	8.5%	13.7%	20.7%	28.8%

In Exercises 15 and 16 a lake is initially stocked with fish, and thereafter the fish population grows logistically. The given table lists the number of fish in the lake as recorded by a careful count during the first year. Fit the data logistically in order to predict the lake's limiting population of fish. How long will it take for the number of fish to reach 80% of the limiting population?

15.

Months	0	3	6	9	12
Fish	30	39	49	60	70

16.

Months	0	3	6	9	12
Fish	50	73	103	137	173

In Exercises 17 and 18, a contagious disease begins spreading among the people in a city. The given table lists the percentage of diseased people as monitored during the first month. Fit the data logistically in order to predict the limiting percentage of people who eventually will suffer this disease. How long will it take for the percentage of diseased people to reach 95% of the limiting percentage?

17.

Weeks	0	1	2	3	4
Diseased People	4%	7%	12%	18%	24%

18.

Weeks	0	1	2	3	4
Diseased People	2.0%	2.5%	3.2%	4.0	4.9%

In Exercises 19 and 20:
  a. First fit a logistic population model to the given U.S. population data.
  b. What limiting population does this model predict?
  c. In what year does it predict the U.S. population will hit 300 million?

19.

Year	1900	1920	1940	1960	1980
Population (millions)	76.2	106.0	132.2	179.3	226.5

20.

Year	1950	1960	1970	1980	1990
Population (millions)	151.3	179.3	203.3	226.5	248.7

21. The following table gives the 1950–1990 population data for China. Fit a logistic model to these data, and use the model to determine China's limiting population.

Year	1950	1960	1970	1980	1990
Population (millions)	563	651	820	985	1139

In Exercises 22 and 23:

a. First fit a logistic population model to the given world population data.
b. What limiting population does this model predict?
c. In what year does it predict the world population will hit 9 billion?

22.

Year	1950	1960	1970	1980	1990
Population (billions)	2.556	3.039	3.707	4.454	5.277

23.

Year	1975	1980	1985	1990	1995
Population (billions)	4.086	4.454	4.851	5.277	5.682

7.3 DISCRETE MODELS AND CHAOS

In Section 7.1 we introduced the logistic model

$$P(t) = \frac{MP_0}{P_0 + (M - P_0)e^{-bt}}. \qquad (1)$$

This model describes a bounded population P that begins with $P(0) = P_0$ and varies smoothly and continuously as a function of time t. Ultimately it approaches the limiting population M.

Let's consider now a rodent population in a fixed environment that has a short annual breeding season, so that births occur only once per year. Suppose also that deaths of the rodents occur only during the breeding season (perhaps due to exhaustion). Such a rodent population would *not* vary smoothly and continuously like the model in (1). Instead, it would vary by a sequence of annual "discrete jumps"

- from P_0 to the population P_1 after 1 year; then
- from P_1 to the population P_2 after 2 years; then
- from P_2 to the population P_3 after 3 years; and so on.

Thus the growth of the rodent population in succeeding years could be described by listing a sequence of populations

$$P_0, P_1, P_2, P_3, \ldots, P_n, P_{n+1}, \ldots, \qquad (2)$$

where P_n denotes the number of rodents after n years.

Note that each of the populations P_n appearing in (2) would be an *integer*—the precise number of rodents after n years. This is so because the typical happy rodent couple has exactly one or two (or perhaps five) baby rodents each year, but never the 2.7 children of the mythical average rural human family. The resulting discrete population values contrast with the fractional values that the logistic model in (1) typically predicts.

In the case of a discrete population it is useful to focus on the **change**

$$\Delta P_n = P_{n+1} - P_n \qquad (3)$$

in the population from year n to year $n+1$. Thus ΔP_n is the net number of births minus deaths that occur in the rodent population during the nth year's breeding season.

Because a larger number of births occur in a larger population, it is reasonable to expect the change ΔP_n to be proportional to the population P_n itself. On the other hand, if the population is to approach the finite limiting population M, then the number of births must decrease as the population nears M. This would happen if the change ΔP_n were also proportional to the difference $M - P_n$, which is small when P_n is close to M. The assumption that ΔP_n is proportional to both P_n and $M - P_n$ means that

$$\Delta P_n = kP_n(M - P_n), \tag{4}$$

where the proportionality constant k depends on the characteristics of the specific population and its supporting environment.

If we write Equation (3) in the form

$$P_{n+1} = P_n + \Delta P_n \tag{5}$$

and then substitute (4), we get the *discrete population model*

$$P_{n+1} = P_n + kP_n(M - P_n). \tag{6}$$

The constant k in (6) is sometimes called the *growth rate parameter* for the population. Note that, once the values of the constants k and M are known, this discrete model gives each succeeding population P_{n+1} not in terms of the time variable, but rather in terms of the preceding population P_n.

The initial population P_0 must be given, so that (6) with $n = 0$ then yields

$$P_1 = P_0 + kP_0(M - P_0).$$

Now that we know P_1, we see that Equation (6) with $n = 1$ gives

$$P_2 = P_1 + kP_1(M - P_1).$$

Substituting $n = 2$ in (6), we next calculate

$$P_3 = P_2 + kP_2(M - P_2).$$

Obviously, we can continue in this way to calculate successive population values, one at a time. For instance, once we have calculated $P_0, P_1, P_2, P_3, \ldots, P_{10}$, we can substitute $n = 10$ in (6) and write

$$P_{11} = P_{10} + kP_{10}(M - P_{10}).$$

The sequence of successive population values we obtain depends upon the initial population P_0 we start with, and so when its value is known we specify the initial population as a part of the discrete model, as in (7) in the next example.

EXAMPLE 1 Calculating Successive Population Values

If we start with $P_0 = 5$ rodents in a forest environment that will support $M = 100$ rodents, then $k = 0.005$ in (6) gives the discrete population model

$$P_0 = 5, \quad P_{n+1} = P_n + 0.005 P_n(100 - P_n). \tag{7}$$

Use this model to find the rodent population after 1 year, 2 years, and 3 years.

SOLUTION

Starting with $P_0 = 5$, we can calculate the successive population values one at a time:

$$P_1 = 5 + 0.005(5)(100 - 5) \approx 7.375.$$
$$P_2 = 7.375 + 0.005(7.375)(100 - 7.375) \approx 10.791.$$
$$P_3 = 10.791 + 0.005(10.791)(100 - 10.791) \approx 15.604.$$

Even though we're discussing a discrete population, our model still predicts nonintegral decimal population values. Rounding to the nearest whole number, we conclude that our population numbers 7 rodents after 1 year, 11 rodents after 2 years, and 16 rodents after 3 years. ●

However, it is much faster and more efficient to set up an iteration using our calculator. Figure 7.3.1 shows the initial setup, with the values $k = 0.005, M = 100$, and $P_0 = 5$. The first line shown in Fig. 7.3.2 implements the formula in (7). When this line is entered, the value of P_1 is calculated and stored as the variable **P**. If the **ENTER** key is pressed again, then this command is re-executed, so that the value of P_2 is calculated and now stored as the variable **P**. Each successive press of the **ENTER** key calculates another population value. Figure 7.3.2 shows the first six resulting population values P_1, P_2, P_3, P_4, P_5, and P_6.

```
0.005→K
              .005
100→M
           100.000
5→P
             5.000
```

```
P+K*P*(M-P)→P
             7.375
            10.791
            15.604
            22.188
            30.821
            41.481
```

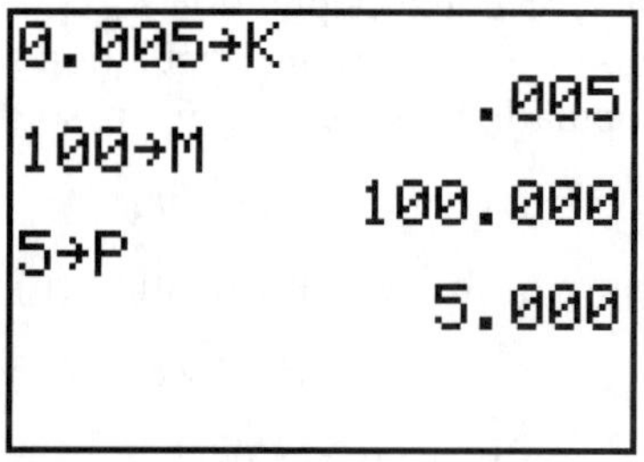

FIGURE 7.3.1 Initial setup for Example 1.

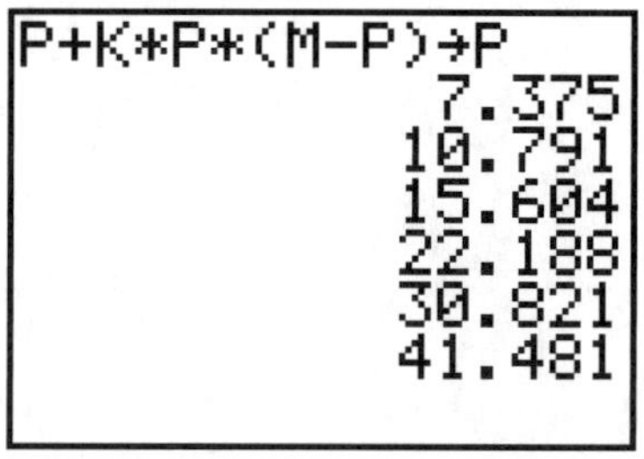

FIGURE 7.3.2 The first six population values, P_1, P_2, P_3, P_4, P_5, and P_6.

We see that the first three values calculated match the three values we found "by hand," and, rounding off, we find that the rodent population numbers 22 rodents after 4 years, 31 rodents after 5 years, and 41 rodents after 6 years.

We count key presses carefully, one at a time, and continue to re-execute our population update command. Figure 7.3.3 shows the populations $P_{14}, P_{15}, P_{16}, P_{17}, P_{18}, P_{19}$, and P_{20} that are visible after 20 executions.

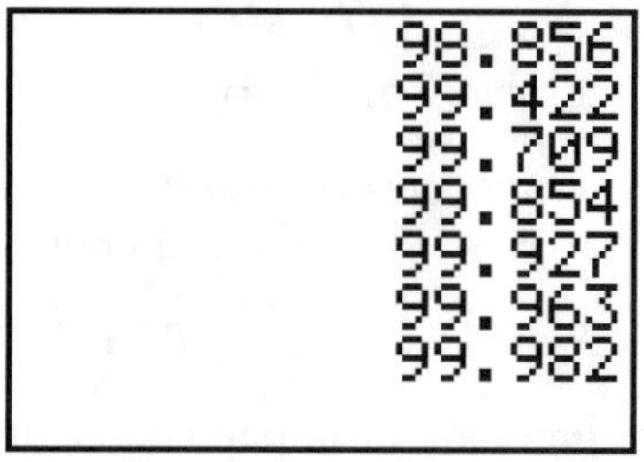

```
            98.856
            99.422
            99.709
            99.854
            99.927
            99.963
            99.982
```

FIGURE 7.3.3 The last seven population values, $P_{14}, P_{15}, P_{16}, P_{17}, P_{18}, P_{19}$, and P_{20}.

We see that after 20 years the populations predicted by the discrete model in (7) are, indeed, leveling off at the limiting population of 100 rodents.

The model in (6) actually is "discrete" in that it provides for the calculation of a discrete sequence of predicted populations, not that these predicted population values are necessarily integers. Sometimes the discrete model and the logistic model in (1) are distinguished by calling

$$P(t) = \frac{MP_0}{P_0 + (M - P_0)e^{-bt}}$$

a *continuous* logistic model and

$$P_{n+1} = P_n + kP_n(M - P_n)$$

a *discrete* logistic model. The following example indicates that continuous and discrete logistic models can predict comparable results.

EXAMPLE 2 Comparing Discrete and Continuous Models

Compare the populations predicted by the discrete model

$$P_0 = 5, \quad P_{n+1} = P_n + 0.005P_n(100 - P_n), \tag{7}$$

of Example 1 and by the continuous model

$$P(t) = \frac{500}{5 + 95e^{-0.45t}}. \tag{8}$$

SOLUTION

The following table shows the values (rounded to integers) for the discrete model and the continuous model for values of $n = t$ from 0 through 20:

TIME $n = t$	DISCRETE MODEL (7)	CONTINUOUS MODEL (8)	TIME $n = t$	DISCRETE MODEL (7)	CONTINUOUS MODEL (8)
0	5	5			
1	7	8	11	92	88
2	11	11	12	96	92
3	16	17	13	98	95
4	22	24	14	99	97
5	31	33	15	99	98
6	41	44	16	100	99
7	54	55	17	100	99
8	66	66	18	100	99
9	77	75	19	100	100
10	86	83	20	100	100

Careful examination of the data in this table shows that 16 of the 21 pairs of population predictions differ by only 1 or 2 rodents, with the remaining 5 pairs of predictions differing by 3 or 4 rodents. The spreadsheet chart shown in Fig. 7.3.4 shows that the discrete model data points exhibit the same "sigmoid plot" shape that is associated with a logistic population curve. In particular, the population first increases at an increasing rate, then increases at a decreasing rate as it levels off to approach the limiting population.

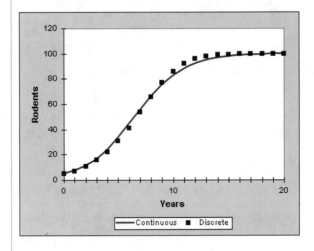

FIGURE 7.3.4 Comparing the discrete and continuous model population predictions in Example 2.

You may be interested to know that the continuous model here was obtained by substituting $M = 100$ and $P_0 = 5$ in (1), then adjusting the value of b to obtain the greatest correspondence (as measured by the sum of squares of differences) between the populations predicted by the two models.

Not every discrete population model increases steadily toward a limiting value. If $P_0 = 15$, $M = 120$, and $k = 0.015$, then Equation (6) gives the discrete population model

$$P_0 = 15, \qquad P_{n+1} = P_n + 0.015 P_n (120 - P_n). \tag{9}$$

A calculator calculation as in Example 1 yields the data shown in the following table:

YEAR n	DISCRETE MODEL (9)	YEAR n	DISCRETE MODEL (9)
0	15		
1	39	11	122
2	86	12	119
3	130	13	121
4	111	14	119
5	126	15	121
6	115	16	119
7	124	17	120
8	117	18	120
9	123	19	120
10	118	20	120

Here we see a different phenomenon than a population that simply increases toward and gradually levels off as it approaches its limiting population. In this case, the population quickly shoots up to a population of 130 rodents after 3 years—10 rodents more than the limiting population of 120 rodents. It then oscillates back and forth over and under the limiting population before leveling off and approaching it. This oscillation is charted in Fig. 7.3.5.

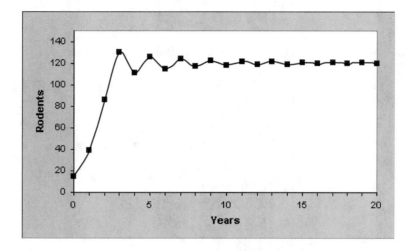

FIGURE 7.3.5 The oscillating discrete population of Example 2.

Period Doubling and Chaos

The following example shows that a discrete bounded population may *never* level off!

EXAMPLE 3 An Oscillating Population Model

Suppose our rodents have the same limiting population $M = 100$ as in Example 1, the initial population $P_0 = 25$, and the larger growth rate parameter $k = 0.021$. Carry out 50 successive iterations to determine the behavior of this model.

SOLUTION

With $P_0 = 25$ and $k = 0.021$, the function modeling this population is

$$P_0 = 25, \quad P_{n+1} = P_n + 0.0021 P_n (100 - P_n). \tag{10}$$

Figure 7.3.6 shows the calculator setup for performing the iteration, and Fig. 7.3.7 displays the population values P_{44}, P_{45}, P_{46}, P_{47}, P_{48}, P_{49}, and P_{50}.

```
0.021→K
              .021
100→M
          100.000
25→P
           25.000
P+K*P*(M-P)→P
```

```
112.86
 82.37
112.86
 82.37
112.86
 82.37
112.86
```

FIGURE 7.3.6 Initial setup for Example 4.

FIGURE 7.3.7 The last seven population values P_{44}, P_{45}, P_{46}, P_{47}, P_{48}, P_{49}, and P_{50} in Example 4.

We see that these predicted population values do not level off and approach the limiting population value $M = 100$. Instead, we have

$$P_{45} = P_{47} = P_{49} = \cdots = 82.37 \approx 82$$

and

$$P_{46} = P_{48} = P_{50} = \cdots = 112.86 \approx 113.$$

In Fig. 7.3.8 the population values for $n = 0$ to $n = 20$ are plotted.

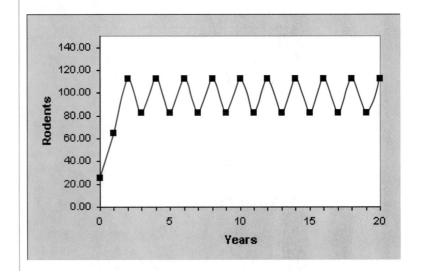

FIGURE 7.3.8 The oscillating discrete population of Example 3.

Thus we see, both numerically and graphically, that the discrete model in (10) ultimately predicts a population of 82 rodents in each odd-numbered year, but a population of 113 rodents in each even-numbered year!

We describe the situation in Example 3 by saying that the population ultimately oscillates between 82 rodents and 113 rodents with a "period" of 2 years. The *period* of a population oscillation is simply the number of generations—two in this case—that it takes for each population value to be repeated.

Now that we know about oscillating populations, it is better to think of the "limiting population" M as the maximum *stable* population that the environment can support. It simplifies further discussion to rewrite the logistic iteration (6) as follows. First we write

$$P_{n+1} = P_n + k P_n (M - P_n)$$
$$= (1 + kM) P_n - k P_n^2$$
$$P_{n+1} = r P_n - k P_n^2,$$

where $r = 1 + kM$. If we substitute

$$P_n = \frac{r}{k} x_n \quad \text{and} \quad P_{n+1} = \frac{r}{k} x_{n+1}$$

in the last equation, we get

$$\frac{r}{k} x_{n+1} = r \left(\frac{r}{k} x_n \right) - k \left(\frac{r}{k} x_n \right)^2 = \frac{r^2}{k} (x_n - x_n^2).$$

SECTION 7.3 Discrete Models and Chaos

Finally, division by the common factor r/k yields the simple-looking iteration

$$x_{n+1} = rx_n(1 - x_n). \tag{11}$$

From here on, we focus our attention on the final iterative formula in (11). It is, in essence, an idealized version of our original discrete population model. Beginning with a given initial value x_0 and a *growth parameter* r, we can use (11) to generate as usual a sequence of values

$$x_0, x_1, x_2, x_3, \ldots, x_n, x_{n+1}, \ldots. \tag{12}$$

The nth value x_n is commonly interpreted as a *fractional population*—that is, as a fraction of the maximum possible stable population. For instance, if $x_n = 0.5$, then the nth population is half of the maximum stable population M. In effect, we're measuring populations in terms of a unit that corresponds to the maximum stable population M that the environment can support.

We therefore call (11) a **discrete fractional population model with growth parameter** r. Because we substituted $r = 1 + kM > 1$ along the way in deriving (11), only values of r greater than 1 are pertinent to discrete fractional population models.

EXAMPLE 4 Verifying the Limiting Value

Starting with the initial value $x_0 = 0.5$, verify that if $r = 1.5$, then x_n approaches the limiting value 1/3.

SOLUTION

We set up our calculator as indicated in Fig. 7.3.9 to carry out the iteration $x_{n+1} = rx_n(1 - x_n)$ and display three decimal places.

FIGURE 7.3.9 Calculator setup for the iteration $x_{n+1} = rx_n(1 - x_n)$.

Figures 7.3.10 and 7.3.11 show the first 13 iterates, and we can see that (correct to three decimal places) we have reached the limiting value of 1/3.

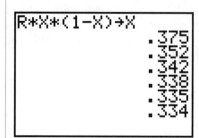

FIGURE 7.3.10 The first six iterates in Example 4.

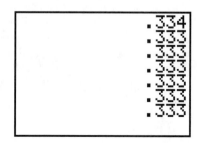

FIGURE 7.3.11 The next seven iterates.

When using our continuous logistic model, we found that the limiting value did not depend on the initial value. Might the same situation occur with our discrete model? The next example investigates this question.

EXAMPLE 5 Finding the Limiting Value

Using the initial values $x_0 = 0.1$ and $x_0 = 0.99$, determine the limiting value for the discrete fractional population model in Example 4.

SOLUTION

Setting up our calculator as we did in Example 4, with 0.1 stored as our initial x, Fig. 7.3.12 shows iterates x_7 through x_{13}. We see that we are again reaching the limiting value 1/3.

Now using 0.99 as our initial x, Fig. 7.3.13 shows iterates x_{17} through x_{23}. Once again, we seem to be approaching the same limiting value of 1/3.

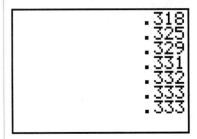

FIGURE 7.3.12 Iterates x_7 through x_{13} with $x_0 = 0.1$.

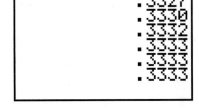

FIGURE 7.3.13 Iterates x_{17} through x_{23} with $x_0 = 0.99$.

Thus, when using the iteration $x_{n+1} = rx_n(1 - x_n)$, you apparently can begin with any initial value x_0 between 0 and 1. In our examples we will use the initial value $x_0 = 0.5$.

What happens if you change the value of r? It turns out that changing the value of r has quite an impact on the iterates. In Exercise 13, you will see that, as long as r is between 1 and 3, x_n eventually approaches a single limiting value. (If r is close to 1 or 3, you have to be very patient and continue iterating long enough to see the limiting value.)

If the value of the growth parameter is greater than 3, then oscillation sets in. If $r = 3.25$ then x_n ultimately oscillates between the two values 0.4953 and 0.8124. This period-2 oscillation is illustrated in Fig. 7.3.14, where we have carried out 1000 iterations, but show only the last dozen.

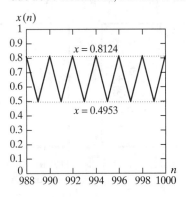

FIGURE 7.3.14 Graph of $x(n) = x_n$ showing the period-2 oscillations obtained with $r = 3.25$.

And with still larger values of r something even stranger occurs. When the growth rate is increased to $r = 3.5$, the fractional population x_n ultimately cycles repeatedly through the *four* distinct values 0.501, 0.875, 0.383, and 0.827 (Fig. 7.3.15).

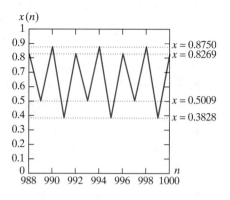

FIGURE 7.3.15 Graph of $x(n) = x_n$ showing the period-4 cycle of iterates obtained with $r = 3.5$.

In Exercise 14, you will verify that if $r = 3.55$, then the fractional population x_n ultimately cycles repeatedly through the *eight* distinct values 0.506, 0.887, 0.355, 0.813, 0.541, 0.882, 0.370, and 0.828.

Obviously things now are changing quite rapidly as the growth parameter r increases. This is the phenomenon of **period doubling**—from a period-2 oscillation to a period-4 oscillation to a period-8 oscillation—for which the innocuous-looking iteration $x_{n+1} = rx_n(1 - x_n)$ has become famous in recent years.

As the growth parameter is increased beyond $r = 3.56$, period doubling (from period 16 to period 32 to period 64 and so on) occurs so rapidly that utter chaos appears to break out somewhere near $r = 3.57$ (Fig. 7.3.16).

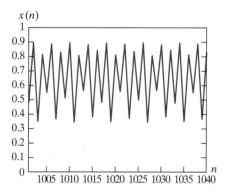

FIGURE 7.3.16 With $r = 3.57$: Chaos!

However many successive iterates with $r = 3.57$ you look at, no periodic oscillations are evident. The earlier periodicity seems to have disappeared, and the population appears to be changing (from one year to the next) in some essentially random fashion, with no apparent pattern. Indeed, the deterministic and predictable population growth that is observed with smaller growth parameters now seems to have degenerated into a nondeterministic and nonpredictable process of seemingly random change from one year to the next.

This is the famous phenomenon of *chaos* that not only mathematics students but also ordinary TV viewers and newspaper readers have heard about. Although chaos was first discovered (around 1970) in connection with seemingly simple population growth problems, it is now known that the world around us is full of complex chaotic phenomena, ranging, for instance, from the weather to the stock market.

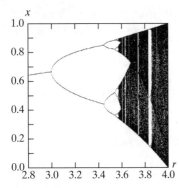

FIGURE 7.3.17 The pitchfork diagram.

Figure 7.3.17 illustrates the phenomenon of "period doubling toward chaos" that occurs with the iteration $x_{n+1} = rx_n(1 - x_n)$ as the growth parameter r is increased. It provides a visual representation of the way in which the behavior of the iteration depends upon the value of the growth parameter. For each value in the range $2.8 \leq r \leq 4.0$ a computer was programmed to first carry out 1000 iterations to achieve "stability." Then the next 250 values of x generated by the iteration were plotted on the vertical axis. That is, the computer screen "pixel" at the point (r, x) was "turned on" (and printed black in the figure).

The descriptively named "pitchfork diagram" that results then shows at a glance whether a given value of r corresponds to an oscillation or to chaos. If the resolution in the picture suffices to make it clear that only finitely many values of x are plotted above a given value of r, then we see that an oscillation through finitely many fractional population values eventually occurs for the specified value of the growth parameter.

For instance, as we scan Fig. 7.3.17 from left to right, we see a single limiting population until $r \approx 3$, then a period-2 oscillation until $r \approx 3.45$, then an oscillation of period 4, then one of period 8, and so forth, rapidly approaching the darkness of chaos. But note the vertical bands of "white space" that first appear in the diagram between $r = 3.6$ and $r = 3.7$ and then again between $r = 3.8$ and $r = 3.9$. These bands represent regions where (periodic) order returns from the preceding chaos, only to again degenerate into chaos as r increases further.

The Mandelbrot Set

The pitchfork diagram of Fig. 7.3.17 is one of two chaos/fractal pictures that are so emblematic of our era that they sometimes appear in elaborately printed and expensive picture books that are displayed on decorative reception room tables. The other is based on the innocuous-looking iteration

$$x_0 = 0, \quad x_{n+1} = x_n^2 + c, \tag{13}$$

where c is a constant. We describe how this iteration leads ultimately to the construction of perhaps the most complicated and fascinating geometric object that has yet appeared in the long history of mathematics. The famous **Mandelbrot set** was discovered in 1980 by the French-born mathematician Benoit Mandelbrot, who was then working at the IBM Watson Research Center in New York.

The first five iterates defined by (13) are

$$\begin{aligned}
x_1 &= c, \\
x_2 &= c^2 + c, \\
x_3 &= (c^2 + c)^2 + c, \\
x_4 &= ((c^2 + c)^2 + c)^2 + c, \\
x_5 &= (((c^2 + c)^2 + c)^2 + c)^2 + c.
\end{aligned}$$

If you were asked to calculate such iterates for a specific value of c, you would want to use a calculator rather than pencil and paper. Figure 7.3.18 shows a calculator setup for this kind of iteration when $c = -2$.

You will see in Exercise 17 that there is no clear pattern to what happens to the iterates as c changes. Sometimes the value of x_n gets larger and larger, sometimes the successive iterates approach a limiting value, sometimes the iterates

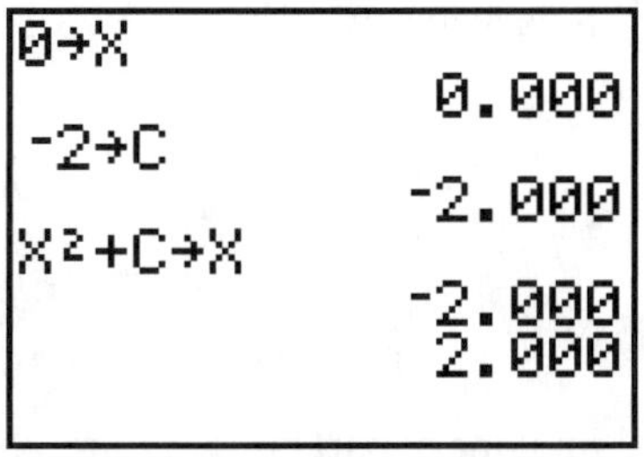

FIGURE 7.3.18 Calculator setup for the iteration $x_0 = 0$, $x_{n+1} = x_n^2 + c$.

```
0→X
              0.000
-2→C
             -2.000
X²+C→X
             -2.000
              2.000
```

oscillate, and sometimes they are exactly the same value. But it turns out that the important question is this: For what values of the constant c in (13) do the resulting iterates get larger and larger, and for what values do they remain "bounded"?

The answer, which you will discover, is that the iterates remain bounded if and only if the constant c is in the range $-2 \leq c \leq 0.25$. This result is interesting, though perhaps not exciting. The real excitement stemmed from Mandelbrot's idea to let c be a *complex* constant, and to replace (13) with the iteration

$$z_0 = 0, \quad z_{n+1} = z_n^2 + c \tag{14}$$

that generates a sequence $z_1, z_2, z_3, \ldots, z_n, z_{n+1}, \ldots$ of complex numbers. Recall that a complex number is a number of the form $z = x + yi$, where $i = \sqrt{-1}$. Complex numbers are added and multiplied in the usual way of algebra, collecting real and imaginary terms separately and using the fact that $i^2 = -1$. For instance,

$$(3 + 7i) + (8 - 2i) = (3 + 8) + (7i - 2i) = 11 + 5i,$$

$$(3 + 7i) \cdot (8 - 2i) = 3(8 - 2i) + (7i)(8 - 2i)$$

$$= (24 - 6i) + (56i - 14i^2)$$

$$= 24 - 6i + 56i + 14 = 38 + 50i.$$

EXAMPLE 6

With $c = 1 + i$, find the first five iterates for $z_0 = 0$, $z_{n+1} = z_n^2 + c$.

SOLUTION

Performing the indicated iteration yields

$$z_1 = (0)^2 + (1 + i) = 1 + i,$$
$$z_2 = (1 + i)^2 + (1 + i) = 1 + 3i,$$
$$z_3 = (1 + 3i)^2 + (1 + i) = -7 + 7i,$$
$$z_4 = (-7 + 7i)^2 + (1 + i) = 1 - 97i,$$
$$z_5 = (1 - 97i)^2 + (1 + i) = -9407 - 193i.$$

In general, if the nth iterate defined by (14) is denoted by

$$z_n = x_n + iy_n$$

and the complex constant c is given by

$$c = a + bi,$$

then the next iterate is given by

$$\begin{aligned} z_{n+1} &= z_n^2 + c \\ &= (x_n + iy_n)^2 + (a + bi) \\ &= (x_n^2 + 2ix_n y_n - y_n^2) + (a + bi) \\ z_{n+1} &= (x_n^2 - y_n^2 + a) + i(2x_n y_n + b). \end{aligned}$$

Thus the real and imaginary parts of z_{n+1} are given by $x_{n+1} = x_n^2 - y_n^2 + a$ and $y_{n+1} = 2x_n y_n + b$.

We can therefore describe the **complex Mandelbrot iteration** by forgetting about all these complex numbers and simply starting with the origin $z_0 = (0, 0)$ and a fixed point $c = (a, b)$ in the xy-plane. We then use the iteration

$$x_{n+1} = x_n^2 - y_n^2 + a, \quad y_{n+1} = 2x_n y_n + b \tag{15}$$

to generate a sequence

$$(x_1, y_1), (x_2, y_2), (x_3, y_3), \ldots, (x_n, y_n), \ldots$$

of points in the plane.

Some advanced graphing calculators can carry out the complex Mandelbrot iteration directly. Figure 7.3.19 shows a TI-86 calculator setup for carrying out the iteration with $c = (0, 0.5)$. After 50 entries (Fig. 7.3.20) we see that the iterates are approaching the point $(-0.1360, 0.3931)$.

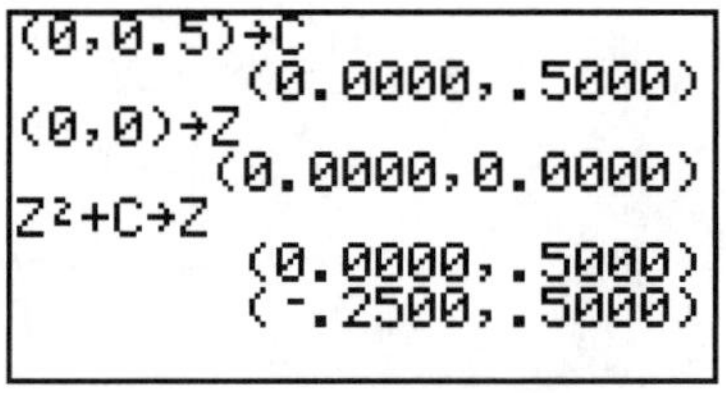

FIGURE 7.3.19 TI-86 calculator setup for the complex Mandelbrot iteration.

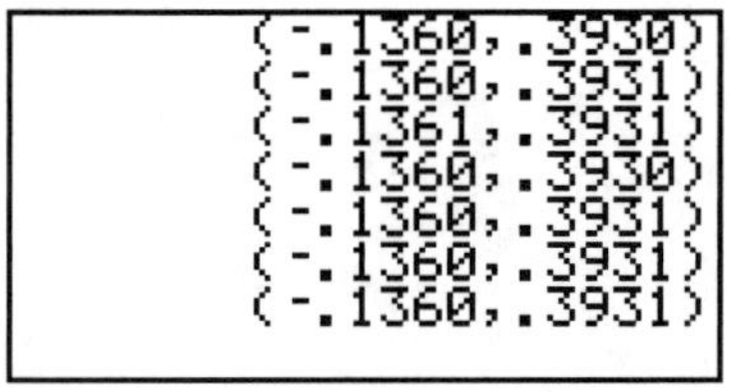

FIGURE 7.3.20 With $c = (0, 0.5)$ the iterates approach the point $(-0.1360, 0.3931)$.

We now ask what happens for a given constant point $c = (a, b)$ in the plane. Does the resulting sequence of iteration points generated using (15) "diverge" further and further from the origin—as appears may be the case in Example 6—or does it remain "bounded" and stay fairly close to the origin (as in Fig. 7.3.20)? It turns out that the sequence of iterates either stays inside the circle $x^2 + y^2 = 4$ of radius 2 or it goes off to "infinity."

The famous Mandelbrot set is the set of all those points $c = (a, b)$ in the xy-plane for which the resulting sequence of iteration points *does* stay within the circle of radius 2. We programmed a computer to divide the xy-square $-1.75 \leq x \leq 0.75, 1.25 \leq y \leq 1.25$ into a 500 by 500 grid of tiny square "pixels." For each of these pixels, its midpoint $c = (a, b)$ was selected and the iteration in (15) carried out for several hundred steps starting each time afresh with $z_0 = (0, 0)$. If the resulting iterates remained inside the circle of radius 2, then the pixel was colored black; otherwise it was colored white.

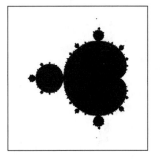

FIGURE 7.3.21 The Mandelbrot set.

Figure 7.3.21 shows the picture of the Mandelbrot set that we obtained in this way. A characteristic feature of the set is that balls protrude from the central heart-shaped figure. From these balls protrude smaller balls, from which protrude still smaller balls, and so on ad infinitum. The Mandelbrot set is the best-known

FIGURE 7.3.22 The Mandelbrot set in the viewing window $-0.65 \leq x \leq -0.40$, $0.45 \leq y \leq 0.70$.

example of a **fractal**—a figure in which similar patterns are replicated at higher and higher levels of magnification. Figure 7.3.22 is a magnification of one of the original balls protruding from the central figure.

Higher resolution is required to do justice to the extraordinary complexity and beauty of the Mandelbrot set. For truly marvelous full-color pictures of the Mandelbrot set with repeated magnifications to show its incredible self-replicating detail at higher and higher resolutions, it is well worth a trip to the library to look at H.-O. Peitgen and P. H. Richter, *The Beauty of Fractals* (Springer-Verlag, 1986). Or you can look at the colored Mandelbrot set at **en.wikipedia.org/wiki/Mandelbrot_set**, where you can point and click to zoom in on more detail. You can also open the picture gallery at **www.mandelbrotset.net** to see why the Mandelbrot set truly is the most beautiful and fabulous object in the entire world of mathematics.

7.3 Exercises | Building Your Skills

In Exercises 1–6, use a discrete logistic model to construct a table of predicted populations as in Example 3. Carry out sufficiently many iterations to see that the rounded-off predicted population approaches the limiting population. Determine whether the predicted populations steadily increase toward the limiting population (as in Examples 1 and 2) or oscillate about it (as in Example 3).

1. $P_0 = 50$, $M = 200$, and $k = 0.005$
2. $P_0 = 100$, $M = 300$, and $k = 0.001$
3. $P_0 = 100$, $M = 500$, and $k = 0.0005$
4. $P_0 = 100$, $M = 200$, and $k = 0.008$
5. $P_0 = 100$, $M = 300$, and $k = 0.006$
6. $P_0 = 200$, $M = 500$, and $k = 0.0037$

In Exercises 7–10, use a discrete logistic model to construct a table of predicted populations. Carry out sufficiently many iterations to see that the rounded-off predicted population oscillates with a period of 2. What are the two values (rounded off accurate to two decimal places) between which the predicted population ultimately oscillates?

7. $P_0 = 50$, $M = 200$, and $k = 0.011$
8. $P_0 = 100$, $M = 300$, and $k = 0.0071$
9. $P_0 = 100$, $M = 500$, and $k = 0.005$
10. $P_0 = 100$, $M = 500$, and $k = 0.0047$

Applying Your Skills

11. Starting with the initial value $x_0 = 0.5$, verify the following:

    a. If $r = 2$, then x_n approaches the limiting value 1/2.
    b. If $r = 2.5$, then x_n approaches the limiting value 3/5.

12. Verify that you still get the same limiting value in each part of Exercise 11 if you use different initial values between 0 and 1, such as $x_0 = 0.1$ and $x_0 = 0.9$.

13. Starting with the initial value $x_0 = 0.5$, try several other values of the growth parameter in the range $1 < r < 3$. How many iterations did it take for your calculator to get to a (three-decimal-place) limiting value?

14. Verify that if $r = 3.55$, then the fractional population x_n ultimately cycles repeatedly through the *eight* distinct values 0.506, 0.887, 0.355, 0.813, 0.541, 0.882, 0.370, and 0.828.

15. Verify that if $r = 3.83$, then the fractional population x_n ultimately cyles repeatedly through the *three* distinct values 0.156, 0.505, and 0.957.

16. Can you find a value of the growth parameter just greater than $r = 3.84$ where a period-6 oscillation occurs?

17. Show that for the Mandelbrot iterations $x_{n+1} = x_n^2 + c$, if $x_0 = 0$ and:

   a. $c = -3$, then the value of x_n gets larger and larger. For instance, $x_6 = 1390967848446$, and thereafter the iterates get *really* big.
   b. $c = -2$ then $x_2 = x_3 = x_4 = \cdots = 2$.
   c. $c = -1$, then successive iterates oscillate between -1 and 0.
   d. $c = -0.5$, then successive iterates approach the number -0.3660.
   e. $c = 0.2$, then successive iterates approach the number 0.2764.
   f. $c = 0.5$, then the value of x_n gets larger and larger.

18. Try enough values of c to corroborate the fact that the Mandelbrot iterates remain bounded if and only if the constant c is in the range $-2 \le c \le 0.25$.

19. Use the equations in (15) directly to verify the results stated in Example 8.

20. If $c = (0,1)$, show that after a couple of iterations the point (x_n, y_n) oscillates between $(0, -1)$ and $(-1, 1)$.

Chapter 7 | Review

In this chapter, you learned about bounded growth function models. After completing the chapter, you should be able to

- determine whether a relation described graphically or symbolically represents a logistic function;
- find the output value of a logistic function for a given input value;
- find the input value of a logistic function for a given output value;
- solve an equation or inequality involving a logistic function;
- find a logistic function that best fits given data;
- use a discrete logistic model to construct a table of function values.

Review | Exercises

In Exercises 1–4, match each graph with the most appropriate function rule in A–D and the type of function it represents in I–IV.

A. $f(x) = ax^3 + bx^2 + cx + d$
B. $f(x) = a\ln(bx)$
C. $f(x) = \dfrac{c}{1 + ae^{-bx}}$
D. $f(x) = ae^{bx}$

I. polynomial
II. exponential
III. logarithmic
IV. logistic

1.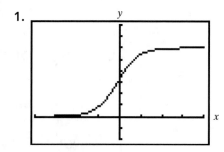

2.

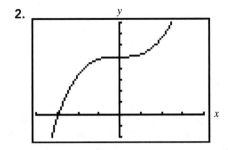

3.

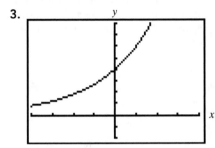

4.

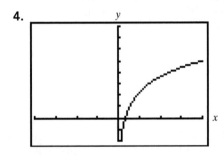

5. Given $f(x) = \dfrac{16}{1 + 1.5e^{-4x}}$, find (a) $f(10)$ and (b) x so that $f(x) = 10$.

6. A lake has a carrying capacity of 800 fish. On July 1, 2000, the lake is stocked with 100 fish. Six months later there are 180 fish in the lake. Assuming a logistic fish population, in what month of what year will there be 600 fish in the lake?

7. Ripley is Sandee House's Ridgeback Hound. When Ripley was a puppy, Sandee measured her weight every few weeks, with the results shown in the following table.

Age (weeks)	8	10	12	14	16	18
Weight (pounds)	14	19	28	46	56	60

a. Find the best-fitting logistic function to model Ripley's weight as a function of her age in weeks.

b. Use your model to predict Ripley's weight at 30 weeks of age.

c. An adult female Ridgeback hound typically weighs about 75 pounds. Based on your model, will Ripley be heavier or lighter than the typical female?

8. Use a discrete logistic model to construct a table of predicted populations for a population with initial value 60, limiting population 180, and growth rate parameter 0.0062. Determine whether the predicted populations steadily increase toward the limiting population or oscillate about it. Give either the limiting value or the values between which it oscillates.

9. (Chapter Opener Revisited) In the discussion that introduced this chapter, we looked at a plot that displayed the cumulative deaths from HIV/AIDS for the years 1987–1997. The plot was based on the data given in the following table:

Year	1987	1988	1989	1990	1991	1995
Cumulative Deaths (thousands)	13.468	30.07	52.152	77.34	106.985	140.461

Year	1993	1994	1995	1996	1997
Cumulative Deaths (thousands)	177.728	219.842	262.557	294.087	310.603

Source: National Center for Health Statistics.

a. Find the logistic function that best fits these data; here t represents the number of years after 1987.

b. What is the limiting number of cumulative deaths from AIDS for this model?

c. Use your model to predict the cumulative deaths from AIDS in 2000.

d. How do your answers in (b) and (c) compare with the June 2000 figure of 438.795 thousand cumulative deaths as reported by www.whitehouse.gov?

e. It appears, based on the 2000 data, that a logistic function is not the best model for cumulative HIV deaths. Find the linear function that best fits these data when the 2000 data point is included.

f. Find the average error in the linear model.

g. Use the model to predict the cumulative AIDS deaths in 1990. How does your answer compare to the actual value reported in the table?

h. Use the model to predict the cumulative AIDS deaths in 2003. How does your answer compare to the actual value of 524.06 thousand cumulative deaths as reported by AIDSHotline.org?

i. Use the model to predict the year in which cumulative AIDS deaths rise to 700 thousand.

INVESTIGATION Sierpinski, Pascal, and Chaos

While many fractals are difficult to construct, there are some that are easy to produce (at least in theory) by repeating simple geometric procedures. In this activity, we will look at fractals based on triangles. It is important to remember that what we generate here are "snapshots" at particular points in the process. It is not possible for us to produce the "real" fractal because it is the result of repeating these procedures indefinitely—a job we could never complete in a finite amount of time.

Taking Away Triangles

We will begin our investigation by drawing an isosceles triangle (one with two equal sides), although you can do this with any shape of triangle you like (Fig. 7.R.1). Then we draw lines connecting the midpoints of the triangle's three sides, creating four smaller triangles that are similar to the large one. We "remove" the middle triangle by coloring it black (Fig. 7.R.2), leaving us with what we call a "stage 1 polygon" consisting of three white triangles.

1. Make an enlarged copy of the step 1 fractal, and continue this process three more times. At each step divide all the white triangles into four similar triangles and color the "middle" triangle of each set of four (Fig. 7.R.3). How many triangles did you color at each step? The remaining white triangles—how many of them are there?—constitute our "stage 2 polygon." The white area remaining at stage 1 is only three-fourth of the original triangle. How much of the original white area is left at stage 2?

If you continued this process indefinitely, you would end up with the fractal known as the Sierpinski Triangle. If you magnify each white triangle in a Sierpinski Triangle, you will see a copy of the whole triangle. (Actually, what you have created is the negative image of how the Sierpinksi Triangle is usually pictured. That is, the "removed" triangles are usually white, and the ones remaining are colored.)

A Triangle of a Different Sort?

Earlier in your study of mathematics, you may have encountered a somewhat different triangle—Pascal's Triangle. While Pascal's Triangle has many interesting applications, we frequently use it to help us to determine the coefficients of the expansion of the binomial $x + y$ when we raise it to whole number powers. Recalling that

- $(x + y)^0 = 1$,
- $(x + y)^1 = x + y$,
- $(x + y)^2 = x^2 + 2xy + y^2$,
- $(x + y)^3 = x^3 + 3x^2y + 3xy^2 + y^3$,

we arrange the coefficients of these terms in a triangular-shaped table, proceeding by descending powers of x and ascending powers of y:

$$\begin{array}{ccccccc} & & & 1 & & & \\ & & 1 & & 1 & & \\ & 1 & & 2 & & 1 & \\ 1 & & 3 & & 3 & & 1 \end{array}$$

We can obtain successive rows by carrying the 1's down the diagonals at the beginning and ending of each row and calculating each of the middle coefficients by adding the numbers directly to the left and right in the row above. Thus, the next row of Pascal's Triangle would be

$$\begin{array}{ccccc} 1 & 1+3 & 3+3 & 3+1 & 1 \end{array}$$

or

$$\begin{array}{ccccc} 1 & 4 & 6 & 4 & 1. \end{array}$$

2. Carefully write out the first 16 rows of Pascal's triangle, using different colors to write odd and even numbers in the table. What pattern seems to be emerging?

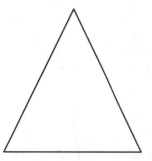

FIGURE 7.R.1

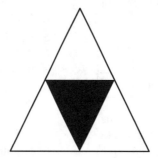

FIGURE 7.R.2

FIGURE 7.R.3

The Chaos Game

Fractals can arise in many surprising ways. Even a random process can be used to produce a highly structured figure. One way to do this is to play what Michael Barnsley, a pioneer in fractal applications, calls the Chaos Game.

We again begin with an isosceles triangle, and label its vertices A, B, and C. We mark a point inside the triangle as our starting point, and "move" it by rolling a die. We will associate each of the numbers (1–6) on the die with one vertex. It doesn't matter which numbers go with which vertex, so for our purposes, we will say 1 and 5 move the point toward A, 2 and 3 move it toward B, and 4 and 6 move it toward C.

Once we roll the die and determine toward which vertex we are to move, we mark the point halfway (on the straight line) between the starting point and the vertex selected. This is our new starting point. We roll again, and mark the point halfway between the second point and the vertex indicated by the die. Continuing on in this fashion, we mark a series of distinct points inside the triangle.

3. Copy the original white triangle we used to make Sierpinski's Triangle. Play five rounds of the Chaos Game, marking a total of six points inside the triangle. Does it appear that a pattern is developing?

As we indicated, creating a fractal requires repeating procedures indefinitely. It would be quite cumbersome to play the Chaos Game by hand long enough to see the fractal develop. The accompanying program, adapted from a program developed by Texas Instruments for their graphing calculators, allows us to speed up the process a great deal.

The program defines the vertices of the triangle as $A = (0, 0)$, $B = (0.5, 1)$ and $C = (1, 0)$. You will be asked by the program to select the coordinates of your initial point. You should choose an x-coordinate between 0 and 1 and a y-coordinate between 0 and 1. (It is entirely possible that the starting point that you select is outside the triangle, but eventually, one of the points will "move" inside and the fractal pattern will begin to develop.)

The program uses a random-number generator rather than a die to determine toward which vertex to move. A whole number between 1 and 6 is chosen at random. As described earlier, if the number is 1 or 5, then the next point is halfway toward the vertex A, (0, 0). If the number is 2 or 3, then the next point is halfway toward the vertex B, (0.5, 1). Finally, if the number is 4 or 6, then the next point is halfway toward the vertex C, (1, 0). The program uses the formula for the midpoint of a line segment to find the new points.

Notice that, just as in the game using the die, (roughly) one-third of the time the point is moved toward each vertex. The advantage of using the calculator is that it can perform the calculations very quickly and plot the points without needing a ruler. It can play many "rounds" of the game in the time it would take us to do just a few.

4. Use the calculator program found at the end of the activity to generate an image. Can you see the fractal pattern clearly when you do 100 iterations? Edit your program to do more iterations by replacing 100 with 500 in the line that says ":For(K,1,100)." Then try 1000 and 2000 iterations. (If you wish to save each of the intermediate fractal pictures, you will need to update the picture number in the last line of the program when you change the number of iterations.) Be patient when you run the program—even your calculator is a bit slow with that many iterations. What pattern do you see developing?

There are many interesting fractal web sites you might wish to explore. Another fractal that can be generated using triangles is called the Koch Snowflake. You might enjoy developing that one too.

Fractal Program

*(Note that $\rightarrow$ indicates the **STO** key on the calculator.)*

FnOff:ClrDraw
PlotsOff:AxesOff
$0 \rightarrow$ Xmin: $1 \rightarrow$ Xmax
$0 \rightarrow$ Ymin: $1 \rightarrow$ Ymax
Disp "CHOOSE 0<X<1"
Input X
Disp "CHOOSE 0<Y<1"
Input Y
For(K,1,100)
randInt(1,6) $\rightarrow$ N
If N = 1 or N = 5
Then
.5X $\rightarrow$ X
.5Y $\rightarrow$ Y
End
If N = 2 or N = 3
Then
.5(.5 + X) $\rightarrow$ X
.5(1 + Y) $\rightarrow$ Y
End
If N = 4 or N = 6
Then
.5(1 + X) $\rightarrow$ X
.5Y $\rightarrow$ Y
End
Pt-On(X,Y)
End
StorePic Pic1

CHAPTER 8
TRIGONOMETRIC MODELS

8.1 PERIODIC PHENOMENA AND TRIGONOMETRIC FUNCTIONS

8.2 TRIGONOMETRIC MODELS AND PERIODIC DATA

Do you know the difference between "weather" and "climate"? Weather generally refers to the state of the atmosphere at a given time, with regard to such characteristics as temperature, humidity, barometric pressure, and cloudiness. Climate refers to a long-term view of weather—that is, climate consists of the composite of the weather conditions of a region, averaged over a number of years. Thus, while scientists expect the weather to change daily, the climate of a certain place remains essentially the same. One of the controversies associated with the issue of "global warming" is whether (or by how much) climates are changing over a period of time.

The chart in Fig. 8.0.1 shows climate data for Topeka, Kansas—specifically, the average maximum temperature over a 24-month period beginning and ending

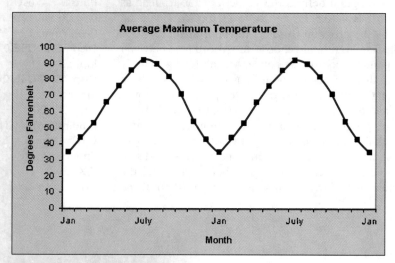

FIGURE 8.0.1 Average maximum temperatures in Topeka, Kansas.
Source: American Automobile Association.

in January. You can see that the graph consists of two annual "cycles" of the same average monthly high temperatures.

As we would expect, the lowest average temperature occurs in the winter (in January) and the highest average temperature occurs in the summer (in July). Because this graph represents temperature averages over many years, the same pattern of values (rounded to the nearest degree) would repeat over and over again as we add more and more months to the data. A function such as this—repeating the same pattern of values over and over again—is called a "periodic function."

Trigonometric functions are typical examples of periodic functions. In this chapter you will study the use of trigonometric functions to model periodic phenomena in the real world.

8.1 PERIODIC PHENOMENA AND TRIGONOMETRIC FUNCTIONS

Many of the phenomena that take place in the world around us fall into one of two broad categories:

- either they involve quantities that increase or decline steadily; or
- they involve quantities that oscillate up and down periodically.

In Chapters 3 and 4 we studied exponential functions that describe processes of (steady) natural growth or decay. In this section we discuss the trigonometric functions that describe oscillatory phenomena and data.

Trigonometric Functions

The following table shows the average temperature in suburban Atlanta for each of the 12 months of the year:

MONTH	AVERAGE TEMPERATURE (°F)
January	41
February	45
March	54
April	62
May	69
June	76
July	79
August	78
September	73
October	62
November	53
December	45

In Fig. 8.1.1 we have plotted these average temperatures month by month for a period of two years. They vary each year up and down from a low of 41°F in January to a high of 79°F in July. This up-and-down repetitive pattern is quite unlike the steadily rising or steadily falling behavior of an exponential function. A new type of function is needed to model temperature oscillations.

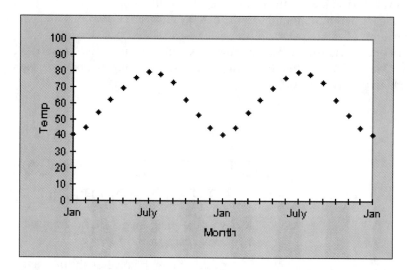

FIGURE 8.1.1 Monthly average temperatures in Atlanta.

Let's take $t = 0$ in January and measure time by the month, so the successive times $t = 1, 2, \ldots, 11$ correspond to the successive months February, March, ..., December. Then $t = 12$ brings us back to January (of the next year), and thereafter the familiar cycle of months repeats in correspondence with the monthly times $t = 12, 13, \ldots, 23$. As time passes,

- the values $t = 0, 12, 24, \ldots$ correspond to January;
- the values $t = 1, 13, 25, \ldots$ correspond to February;
- the values $t = 2, 14, 26, \ldots$ correspond to March; and so forth,

with the values $t = 11, 23, 35, \ldots$ corresponding to December in succeeding years. Whenever we need to be more precise, we'll assume that $t = 0$ means mid-January and $t = 11$ means mid-December.

This monthly average temperature data can be fitted with the *trigonometric function*

$$f(t) = 18.54 \sin(0.52t - 1.57) + 61.33 \tag{1}$$

involving the *sine function* that we introduce shortly. In Section 8.2 we'll describe how one might *find* such a trigonometric "fitting function." In Example 1 we will just explore the way the function given in (1) fits the Atlanta monthly average temperature data.

EXAMPLE 1 | A Trigonometric Model for Temperature Data

Use the average error for the trigonometric function

$$f(t) = 18.54 \sin(0.52t - 1.57) + 61.33$$

to describe how well the function fits the Atlanta monthly average temperature data.

SOLUTION

The following table compares the monthly temperatures predicted by this function with the actual monthly average temperatures. We see that (rounded off to the nearest degree) the discrepancy is no more than a couple of degrees in any month.

t	AVERAGE TEMPERATURE	$f(t)$	ERROR
0	41	42.8	−1.8
1	45	45.2	−0.2
2	54	52.0	2.0
3	62	61.1	0.9
4	69	70.4	−1.4
5	76	77.2	−1.2
6	79	79.9	−0.9
7	78	77.6	0.4
8	73	71.0	2.0
9	62	61.9	0.1
10	53	52.6	0.4
11	45	45.6	−0.6

Note that the temperature predicted by (1) overestimates the average temperature for 6 months and underestimates it for the other 6 months of the year. The sum of the squares of the errors shown in the final column is

$$\text{SSE} = (-1.8)^2 + (-0.2)^2 + (2.0)^2 + \cdots + (0.1)^2 + (0.4)^2 + (-0.6)^2 = 16.99,$$

so the average error in the trigonometric approximation is

$$\text{average error} = \sqrt{\frac{\text{SSE}}{n}} = \sqrt{\frac{16.99}{12}} \approx 1.19.$$

Thus the formula in (1) predicts each month's temperature, on average, to within about a degree, and the trigonometric model is a good fit for the temperature data.

Figure 8.1.2 shows that the smooth curve $y = f(t)$ appears visually to fit the actual data (dots) quite well, confirming the result found using the average error.

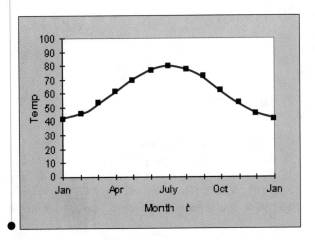

FIGURE 8.1.2 Monthly average temperatures (dots) in Atlanta fitted with the (curve) graph $f(t) = 61.33 + 18.54 \sin(0.52t - 1.57)$.

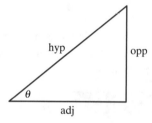

FIGURE 8.1.3 The sides and angle θ of a right triangle.

Equation (1) involves the *sine function*, which, together with its relative the *cosine function*, is needed to describe the oscillations of temperature and other periodic phenomena in nature. The basic trigonometric functions of an angle θ in a right triangle are defined as ratios between pairs of sides of the triangle. As in Fig. 8.1.3—where "adj" stands for adjacent, "opp" for opposite, and "hyp" for hypotenuse—the sine and cosine of the angle θ are defined by

$$\cos\theta = \frac{\text{adj}}{\text{hyp}}, \quad \sin\theta = \frac{\text{opp}}{\text{hyp}}. \tag{2}$$

There are four other trigonometric functions (tangent, cotangent, secant, and cosecant), but here we need only discuss the sine and cosine.

EXAMPLE 2 Finding Sine and Cosine Values

Use Fig. 8.1.4, which shows the common 30°-60°-90° and 45°-45°-90° triangles, to find

a. $\cos 30°$
b. $\sin 30°$
c. $\cos 45°$
d. $\sin 45°$

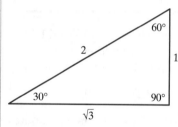 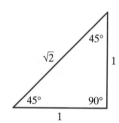

FIGURE 8.1.4 The 30°-60°-90° and 45°-45°-90° triangles.

SOLUTION

From the triangles, we see that

a. $\cos 30° = \dfrac{\sqrt{3}}{2}$,

b. $\sin 30° = \dfrac{1}{2}$,

c. $\cos 45° = \dfrac{1}{\sqrt{2}}$,

d. $\sin 45° = \dfrac{1}{\sqrt{2}}$.

The triangle definitions in (2) apply only to angles less than 90°, but there are circle definitions that apply to angles of any size. Suppose that the initial side of the angle θ is the positive x-axis, so its vertex is located at the origin (Fig. 8.1.5). The angle of rotation (for a positive angle) is then *counterclockwise* from the initial side to the terminal side of the angle. (Negative angles are measured clockwise.) If

$P(x, y)$ is the point at which the terminal side of θ intersects the unit circle (of radius 1), then we define

$$\cos \theta = x, \quad \sin \theta = y. \tag{3}$$

Because x is the adjacent side and y is the opposite side of a right triangle whose hypotenuse is the radius 1 of the unit circle, it follows that the definitions in (3) agree with those in (2) for an *acute* angle less than 90°.

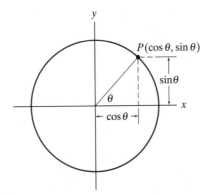

FIGURE 8.1.5 The circle definitions of the sine and cosine.

Radian Measure

In elementary mathematics, angles usually are measured in *degrees*, with 90° in a right angle. However, the trigonometric functions on calculators and computers frequently are based on **radian measure** of angles. Just as miles and meters are different units of length, with 1 mile = 1609.344 meters, degrees and radians are different units of angular measure. The relation between degrees and radians is given by

$$\pi \text{ radians} = 180 \text{ degrees}. \tag{4}$$

Division of both sides by π (and abbreviation of radians to rad, degrees to deg) yields

$$1 \text{ rad} = \frac{180}{\pi} \text{ deg} \approx 57.2958 \text{ deg}, \tag{4'}$$

while division of both sides in (4) by 180 gives

$$1 \text{ deg} = \frac{\pi}{180} \text{ rad} \approx 0.01745 \text{ rad}. \tag{4''}$$

The following table gives degree–radian equivalents and cosines and sines for the common angles of the triangles in Fig. 8.1.4.

DEGREES	RADIANS	COSINE	SINE
0	0	1	0
30	$\pi/6$	$\sqrt{3}/2$	1/2
45	$\pi/4$	$1/\sqrt{2}$	$1/\sqrt{2}$
60	$\pi/3$	1/2	$\sqrt{3}/2$
90	$\pi/2$	0	1

For other angles, you should rely *only* upon memory of the fundamental relation in (4), dividing as necessary as in (4′) and (4″). To recall quickly whether to use $180/\pi$ or $\pi/180$, it helps to remember that a radian is a relatively large angle—almost 60°—while a degree is a quite small angle in terms of radians.

EXAMPLE 3 Degree/Radian Conversions

a. Convert 25° to radians.
b. Convert 0.25 radians to degrees.

Give both an exact answer and a two-decimal-place approximation.

SOLUTION

a. $25° = 25 \times \dfrac{\pi}{180} \text{ rad} = \dfrac{5\pi}{36} \text{ rad} \approx 0.4363 \text{ rad}$

b. $0.25 \text{ rad} = 0.25 \times \dfrac{180}{\pi} \text{ deg} = \dfrac{45}{\pi} \text{ deg} \approx 14.32°$

Figure 8.1.6 shows the right triangles from Example 2, with angles indicated in radian measure. Figure 8.1.7 displays some trig calculations with a graphing calculator set in **radian mode**.

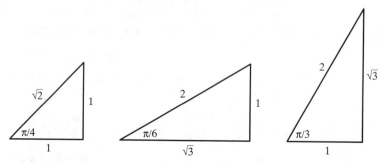

FIGURE 8.1.6 Familiar right triangles.

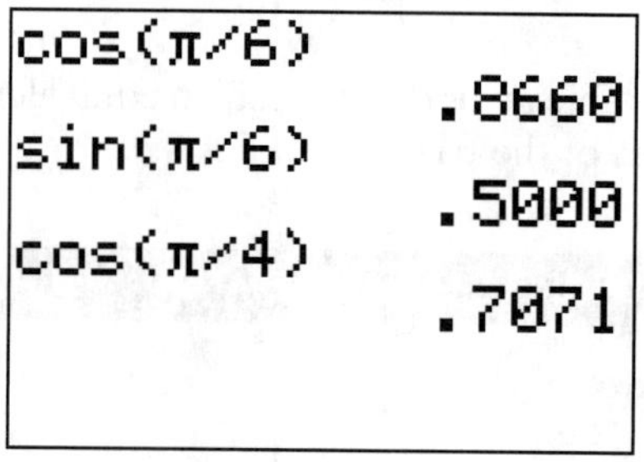

FIGURE 8.1.7 Right-triangle trig calculations.

Because $1/\sqrt{2} \approx 0.7071$ and $\sqrt{3}/2 \approx 0.8660$, we see that these four-place decimal approximations agree with the values obtained from the right triangles of Fig. 8.1.4.

Figure 8.1.8 shows some more trig calculations, illustrating the fact that we can ask for the sine or cosine of any real number (of radians).

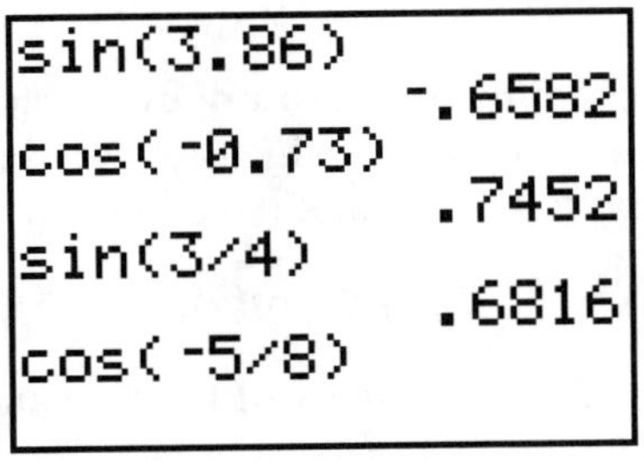

```
sin(3.86)
            -.6582
cos(-0.73)
            .7452
sin(3/4)
            .6816
cos(-5/8)
```

FIGURE 8.1.8 Some more trig calculations.

Observe that the calculator uses the parentheses associated with standard functional notation—$\cos(x)$ and $\sin(x)$. Whenever we use this functional notation with parentheses, it is understood that x is measured in radians. The argument (independent variable) is measured in degrees only when we use the degree symbol explicitly, as in writing $\sin x°$.

Most of us think more naturally of angles in degrees because that's what we first learned in school and what we commonly use in "real life." However, for the purposes of college mathematics you should keep your calculator set in radian mode rather than degree mode. Whenever you want to check quickly the trig mode of your calculator, just remember (from the 30°-60°-90° triangle of Fig. 8.1.4) that the sine of an angle of 30° = $\pi/6$ rad is 1/2:

$$\sin 30° = \sin(\pi/6) = \frac{1}{2}.$$

So we just calculate the sine of 30 to see whether our calculator is in radian mode or in degree mode. If the calculator is in *degree* mode, then it will give 1/2:

sin(30)
     0.5000   (It's in degree mode.)

sin(30)
     −0.9880   (It's in radian mode.)

But if it's in *radian* mode, then it will give 1/2 when we calculate the sine of $\pi/6$:

sin($\pi/6$)
     0.5000   (It's in radian mode.)

sin($\pi/6$)
     0.0091   (It's in degree mode.)

Try all four possibilities out with your calculator—switching trig modes as necessary—to duplicate the results shown. Because it is easy to type in the degree symbol (using the **ANGLE** menu of your calculator) when you need it, it is a good idea to leave your calculator in radian mode.

Trigonometric Graphs and Periodicity

An angle of 2π radians corresponds to one full revolution around the unit circle. Therefore the circle definition in (4) implies that the (radian-based) sine and cosine functions repeat themselves in each new interval of length 2π. In more precise language, these functions have *period* 2π, meaning that

$$\cos(x + 2\pi) = \cos(x), \quad \sin(x + 2\pi) = \sin(x) \tag{5}$$

for every x. Thus addition of 2π to x does *not* change the value of either the sine or the cosine of x. Can you see why it follows that addition of any **even** multiple of π to x leaves both the sine and the cosine unchanged? That is,

$$\cos(x + 2n\pi) = \cos(x), \quad \sin(x + 2n\pi) = \sin(x)$$

for any integer n. This periodicity of the sine and cosine functions is evident in the repetitive character of their graphs (Fig. 8.1.9), where we see each repeatedly varying between the values -1 and $+1$.

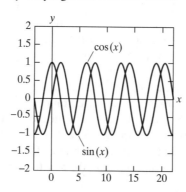

FIGURE 8.1.9 The sine and cosine graphs.

Note also that these graphs are consistent with the facts that

$$\cos(-x) = \cos(x), \quad \sin(-x) = -\sin(x) \tag{6}$$

for all x. [It is therefore said that $\cos(x)$ is an **even** function of x, while $\sin(x)$ is an **odd** function of x—in analogy with the fact that an even power of a negative number is positive, while an odd power of a negative number is negative.] A related useful property of these functions is that

$$\cos(x \pm \pi) = -\cos(x), \quad \sin(x \pm \pi) = -\sin(x). \tag{7}$$

We can summarize (5) and (7) by saying that "Whereas changing x by 2π does not change the value of either the sine and cosine, changing x by π changes (only) the *sign* of each."

EXAMPLE 4 Finding Sine and Cosine Values

Use the properties shown in (5)–(7) to find each of the following:

a. $\cos(2\pi)$

b. $\sin\left(\dfrac{3\pi}{2}\right)$

c. $\sin\left(\dfrac{3\pi}{4}\right)$

d. $\cos\left(\dfrac{7\pi}{6}\right)$

SOLUTION

a. $\cos(2\pi) = \cos(0) = 1$

b. $\sin\left(\dfrac{3\pi}{2}\right) = \sin\left(2\pi - \dfrac{\pi}{2}\right) = \sin\left(-\dfrac{\pi}{2}\right) = -\sin\left(\dfrac{\pi}{2}\right) = -1$

c. $\sin\left(\dfrac{3\pi}{4}\right) = \sin\left(-\dfrac{\pi}{4} + \pi\right) = -\sin\left(-\dfrac{\pi}{4}\right) = \sin\left(\dfrac{\pi}{4}\right) \approx 0.7071$

d. $\cos\left(\dfrac{7\pi}{6}\right) = \cos\left(\dfrac{\pi}{6} + \pi\right) = -\cos\left(\dfrac{\pi}{6}\right) \approx 0.8660$

If you define **Y1 = cos(X)** and **Y2 = sin(X)** on your graphing calculator (in radian mode, as always) then enter

$$\text{Xmin} = -2\pi \qquad \text{Ymin} = -2$$
$$\text{Xmax} = 2\pi \qquad \text{Ymax} = 2$$

to define the viewing window, the resulting graph (Fig. 8.1.10) resembles the artist's sketch in Fig. 8.1.9. Is it clear to you which curve in Fig. 8.1.10 is the graph $y = \cos x$ and which is the graph $y = \sin x$?

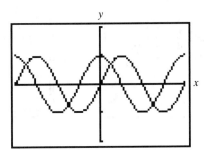

FIGURE 8.1.10 Which is the sine graph, which the cosine graph?

Because

$$\dfrac{\pi}{6} \approx 0.5236 \approx 0.52 \quad \text{and} \quad \dfrac{\pi}{2} \approx 1.5708 \approx 1.57,$$

the Atlanta monthly average temperature model in (1) is equivalent (with two-place accuracy) to

$$f(t) = 61.33 + 18.54 \sin\left(\dfrac{\pi t}{6} - \dfrac{\pi}{2}\right). \tag{8}$$

It is worth verifying that this alleged temperature function is, indeed, *periodic* with a *period* of **12** months—meaning that it predicts the same temperature at time t and at time $t + 12$ twelve months hence. Upon replacing t with $t + 12$ in (8), we get

$$f(t + 12) = 61.33 + 18.54 \sin\left(\dfrac{\pi(t + 12)}{6} - \dfrac{\pi}{2}\right)$$

$$= 61.33 + 18.54 \sin\left(\dfrac{\pi t + 12\pi}{6} - \dfrac{\pi}{2}\right)$$

$$= 61.33 + 18.54 \sin\left(\frac{\pi t}{6} + 2\pi - \frac{\pi}{2}\right)$$

$$= 61.33 + 18.54 \sin\left(\left(\frac{\pi t}{6} - \frac{\pi}{2}\right) + 2\pi\right)$$

$$= 61.33 + 18.54 \sin\left(\frac{\pi t}{6} - \frac{\pi}{2}\right) = f(t).$$

In the last line here we have applied the 2π-periodicity of the sine function—the fact that $\sin(x + 2\pi) = \sin(x)$, using $x = \pi t/6 - \pi/2$.

Thus we have verified that $f(t + 12) = f(t)$, so the predicted average temperature 1 year (12 months) from now is the same as this month's average temperature. To construct your own graph of this function, you must remember that the typical graphing calculator requires use of x as the independent variable and y as the independent variable. Thus the function in (8) is entered as shown in Fig. 8.1.11 [where the original version in (1) is also entered]. Then Fig. 8.1.12 shows the graph of the monthly average temperature function. Do you see that two 12-month periods are shown?

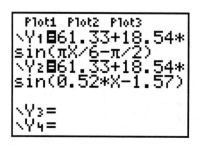

FIGURE 8.1.11 Atlanta's montly average temperature function.

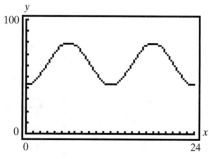

FIGURE 8.1.12 How many 12-month periods of $f(t)$ are shown?

Solving Trigonometric Equations

In previous chapters we frequently used graphing calculator facilities to solve equations numerically. The approach is no different when trigonometric functions are involved.

EXAMPLE 5 Solving a Trigonometric Equation.

Use a graphical technique to solve the equation $x = \cos x$.

SOLUTION

In Figure 8.1.13 we have plotted the graphs $y = x$ and $y = \cos x$. The single point of intersection indicates that the equation $x = \cos x$ has only a single solution.

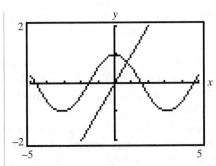

FIGURE 8.1.13 The graphs $y = x$ and $y = \cos x$.

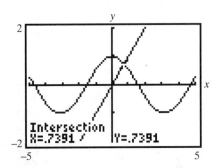

FIGURE 8.1.14 Solving the equation $x = \cos x$ of Example 5.

Indeed, Fig. 8.1.14 shows that we have used our calculator's **CALC intersect** facility to find that the solution of this equation is $x = 0.7391$, accurate to four decimal places.

EXAMPLE 6 Finding All Solutions to an Equation

Find all the solutions of the trigonometric equation $1 - x = 3 \cos x$.

SOLUTION

In Figure 8.1.15 we have plotted the graphs $y = 1 - x$ and $y = 3 \cos x$. Now three points of intersection are visible, so we see that the equation

$$1 - x = 3 \cos x$$

has *three* different solutions. You should use your calculator's **CALC intersect** facility to verify that these three solutions are $x \approx -0.8895$, $x \approx 1.8624$, and $x \approx 3.6380$.

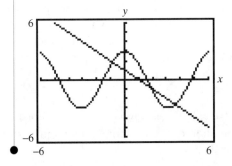

FIGURE 8.1.15 The graphs $y = 1 - x$ and $y = 3 \cos x$ of Example 6.

EXAMPLE 7 When Is the Monthly Average Temperature 70°F?

In which month(s) during the year is the Atlanta monthly average temperature 70°F?

SOLUTION

In order to answer this question, we need to solve the equation

$$61.33 + 18.54 \sin\left(\frac{\pi t}{6} - \frac{\pi}{2}\right) = 70, \tag{9}$$

recalling (from our initial discussion of Atlanta temperatures at the beginning of this section) that $t = 0$ in mid-January and $t = 11$ in mid-December.

In Fig. 8.1.16 we have graphed the left-hand and right-hand sides in Equation (9) for $0 \leq t \leq 12$, and we see two points of intersection. We have used our calculator's intersection-finding facility to find that the first intersection is $t = 3.93 \approx 4$, and you can show similarly that the second intersection is $t = 8.07 \approx 8$. Thus the average temperature is 70°F near the middle of the 4th month after January—that is, in May—and again near the middle of the 8th month after January—that is, in September. More precisely, since $\frac{7}{100}$ of a month is close to 2 days, we conclude that the approximate dates are May 13 and September 17 (why?).

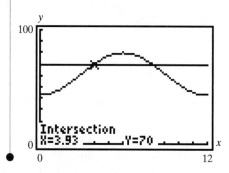

FIGURE 8.1.16 Solving Eq. (9) in Example 7.

Composition—New Functions from Old Ones

In this book you have seen many types of functions, including linear, polynomial, exponential, and logarithmic functions, and now trigonometric functions. Still more varied and complex functions can be "put together" by using these as building-block functions. In addition to adding, subtracting, multiplying, and dividing two given functions, we can also combine functions by letting one function act on the output of the other.

> **DEFINITION: Composition of Functions**
>
> The **composition** of the two functions f and g is the new function $h = f \circ g$ defined by
>
> $$h(x) = f(g(x)). \tag{10}$$
>
> The domain of h consists of all points x for which the right-hand side in (10) is meaningful—that is, x is in the domain of g and $u = g(x)$ is in the domain of f. [The right-hand side in (10) is read "f of g of x."]

Thus the output $u = g(x)$ of the function g is used as the input to the function f (Fig. 8.1.17). We sometimes refer to g as the inner function and to f as the outer function in Equation (10).

SECTION 8.1 Periodic Phenomena and Trigonometric Functions **361**

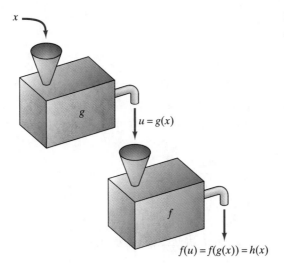

FIGURE 8.1.17 The composition $h = f \circ g$.

EXAMPLE 8 | Composition of Functions

For $f(x) = \sqrt{x}$ and $g(x) = 1 - x^2$, verify that $f(g(x)) \neq g(f(x))$.

SOLUTION

$f(g(x)) = \sqrt{1 - x^2}$, with domain $|x| \leq 1$, whereas
$g(f(x)) = 1 - (\sqrt{x})^2 = 1 - x$, with domain $x \geq 0$.

The $f(g(x))$ notation for compositions is most commonly used in ordinary computations, but the $f \circ g$ notation emphasizes that the composition may be regarded as a new kind of combination of the functions f and g. But Example 8 shows that $f \circ g$ is quite unlike the ordinary product $f \cdot g$ of the two functions f and g because $f \circ g \neq g \circ f$, whereas $f \cdot g = g \cdot f$ [because $f(x) \cdot g(x) = g(x) \cdot f(x)$ whenever $f(x)$ and $g(x)$ are both defined]. So remember that composition is quite different in character from ordinary multiplication of functions.

EXAMPLE 9 | New Functions from Old

If
$$f(x) = x^2 \quad \text{and} \quad g(x) = \cos x,$$
find

a. $f(x) \cdot g(x)$
b. $f(g(x))$
c. $g(f(x))$

SOLUTION

a. $f(x) \cdot g(x) = x^2 \cos x$
b. $f(g(x)) = (\cos x)^2 = \cos^2 x$
c. $g(f(x)) = \cos(x^2) = \cos x^2$

Each of the functions in **(a)**, **(b)**, and **(c)** is defined for all x. But Figs. 8.1.18–8.1.20 illustrate vividly how different these three functions are.

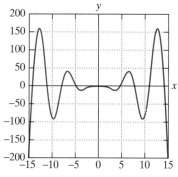

FIGURE 8.1.18 $y = x^2 \cos x$.

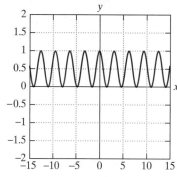

FIGURE 8.1.19 $y = \cos^2 x$.

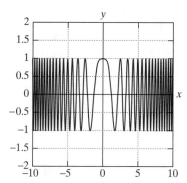

FIGURE 8.1.20 $y = \cos x^2$.

EXAMPLE 10 Creating a Composition of Two Functions

Given the function $h(x) = (x^2 + 4)^{3/2}$, find two functions f and g such that $f(g(x))$.

SOLUTION

It is technically correct—but useless—simply to let $g(x) = x$ and $f(u) = (u^2 + 4)^{3/2}$. We seek a nontrivial answer here. To calculate $(x^2 + 4)^{3/2}$, we must *first* calculate $x^2 + 4$. So we choose $g(x) = x^2 + 4$ as the *inner* function. The *last* step is to raise $u = g(x)$ to the power $\frac{3}{2}$, so we take $f(u) = u^{3/2}$ as the *outer* function. Thus if

$$f(x) = x^{3/2} \quad \text{and} \quad g(x) = x^2 + 4,$$

then $f(g(x)) = f(x^2 + 4) = (x^2 + 4)^{3/2} = h(x)$, as desired.

Example 10 illustrates a useful approach to recognizing a given function $h(x)$ as a composition $f(g(x))$. Instead of just looking at $h(x)$, think of *actually doing it*—that is, calculating $h(x)$ with a calculator. If, starting with a number x, just two calculator functions need to be applied in succession to calculate $h(x)$, then the *first* function applied is the *inner* function g, and the *last* function applied is the *outer* function f.

In this section we have discussed trigonometric functions that oscillate periodically as "time goes by." The following example exhibits a function that combines the steady decrease of a (negative-exponent) exponential function with the oscillation of a trigonometric function.

EXAMPLE 11

Think of the up-and-down vibrations of a car with very poor shock absorbers. These vibrations might be described by the function

$$y(t) = 3 \cdot 2^{-t} \cos(4\pi t) \qquad (11)$$

that gives the car's height y (in inches above or below its normal position) t seconds after it hits a deep pothole. Find:

a. $y(0)$, the car's initial bounce.
b. $y(1)$, $y(2)$, and $y(3)$, and use these values to describe how the height (or "amplitude") of the car's up-and-down oscillations change over time.
c. The period of $y(t)$.

SOLUTION

a. $y(0) = 3 \cdot 2^{-0} \cos(4\pi \cdot 0) = 3 \cdot 1 \cdot 1 = 3$. Thus, when the car hits the pothole, its initial bounce is 3 inches.

b. $y(1) = 3 \cdot 2^{-1} \cos(4\pi \cdot 1) = 3 \cdot \left(\dfrac{1}{2}\right) \cdot 1 = \dfrac{3}{2}$

$y(2) = 3 \cdot 2^{-2} \cos(4\pi \cdot 2) = 3 \cdot \left(\dfrac{1}{4}\right) \cdot 1 = \dfrac{3}{4}$

$y(3) = 3 \cdot 2^{-3} \cos(4\pi \cdot 3) = 3 \cdot \left(\dfrac{1}{8}\right) \cdot 1 = \dfrac{3}{8}$

We see from $y(0)$ and the values shown here that the amplitude of the car's up-and-down oscillations *halves* every second.

c. $4\pi \cdot t = 2\pi$ when $t = \frac{1}{2}$, so the period of the factor $\cos(4\pi t)$ in (11) is $\frac{1}{2}$ second, and hence the car undergoes *two* up-and-down oscillations per second.

Figure 8.1.21 shows the graph of $y(t)$. The curve described in (11) oscillates up and down between the *two* curves $y(t) = \pm 3 \cdot 2^{-t}$. It appears that the car's vibrations die out and are negligible after 7 or 8 seconds.

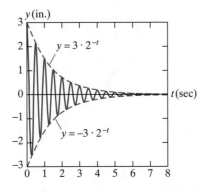

FIGURE 8.1.21 $y = 3 \cdot 2^{-t} \cos(4\pi t)$.

8.1 Exercises | Building Your Skills

Express in radian measure each of the angles given in Exercises 1–4.

1. a. 30° b. 150° c. 210°
2. a. 20° b. 160° c. 340°
3. a. 135° b. 225° c. 315°
4. a. 120° b. 240° c. 300°

Each of the angles in Exercises 5–8 is given in radians. Convert it to degrees.

5. a. $\pi/10$ b. $\pi/9$ c. $\pi/5$
6. a. $2\pi/9$ b. $2\pi/5$ c. $5\pi/6$
7. a. $3\pi/10$ b. $4\pi/9$ c. $7\pi/5$
8. a. $4\pi/15$ b. $11\pi/18$ c. $17\pi/36$

In Exercises 9–12, use trigonometric properties to find the indicated value.

9. a. $\sin\left(\dfrac{5\pi}{6}\right)$ b. $\sin\left(\dfrac{7\pi}{6}\right)$ c. $\sin\left(\dfrac{11\pi}{6}\right)$
10. a. $\sin\left(\dfrac{3\pi}{4}\right)$ b. $\sin\left(\dfrac{5\pi}{4}\right)$ c. $\sin\left(\dfrac{7\pi}{4}\right)$
11. a. $\cos\left(\dfrac{2\pi}{3}\right)$ b. $\cos\left(\dfrac{4\pi}{3}\right)$ c. $\cos\left(\dfrac{5\pi}{3}\right)$
12. a. $\sin(\pi)$ b. $\cos\left(\dfrac{3\pi}{2}\right)$ c. $\sin\left(\dfrac{5\pi}{2}\right)$

Applying Your Skills

In Exercises 13–20, match the given function with its graph among those shown in Figs. 8.1.22–8.1.29. Try to do this without turning on your graphing calculator or computer.

13. $f(x) = 2^x - 1$
14. $f(x) = 2 - 3^{-x}$
15. $f(x) = 1 + \cos x$
16. $f(x) = 2 - 2\sin x$

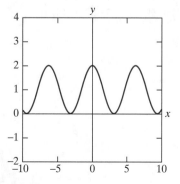

FIGURE 8.1.22

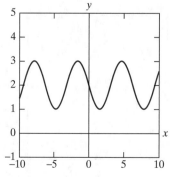

FIGURE 8.1.23

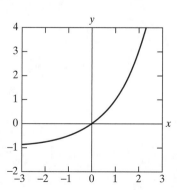

FIGURE 8.1.24

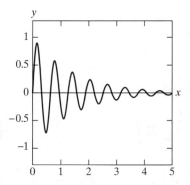

FIGURE 8.1.25

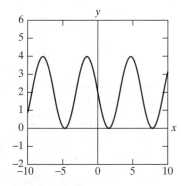

FIGURE 8.1.26

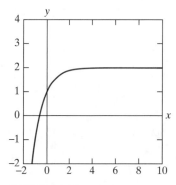

FIGURE 8.1.27

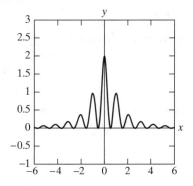

FIGURE 8.1.28

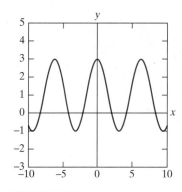

FIGURE 8.1.29

17. $f(x) = 1 + 2\cos x$

18. $f(x) = 2 - \sin x$

19. $f(x) = 2^{-x}\sin(10x)$

20. $f(x) = \dfrac{1 + \cos 6x}{1 + x^2}$

In Exercises 21–26, find both $f(g(x))$ and $g(f(x))$. Then compare the graphs of these two compositions.

21. $f(x) = 1 - x^2$ and $g(x) = 2x + 3$
22. $f(x) = -17$ and $g(x) = |x|$
23. $f(x) = \sqrt{x^2 - 3}$ and $g(x) = x^2 + 3$
24. $f(x) = x^3 - 4$ and $g(x) = \sqrt[3]{x + 4}$
25. $f(x) = \sqrt{x}$ and $g(x) = \cos x$
26. $f(x) = \sin x$ and $g(x) = x^3$

Use a graphing calculator to find numerically (accurate to four decimal places) each solution of the equations in Exercises 27–33.

27. $2x = \cos x$
28. $x + 1 = 3\cos x$
29. $x - 1 = 3\cos x$
30. $x = 5\cos x$
32. $x^2 = 100\sin x$
33. $\cos 3x = x^3 - 3x^2 + 1$

8.2 TRIGONOMETRIC MODELS AND PERIODIC DATA

When we look at the Atlanta monthly average temperature function

$$f(t) = 61.33 + 18.54 \sin\left(\frac{\pi t}{6} - \frac{\pi}{2}\right) \tag{1}$$

discussed in Section 8.1, we see that

- the constant term 61.33 is the year-round average temperature, while
- the coefficient 18.54 is the maximal variation up-and-down from this average.

More generally, the function

$$f(t) = A + B \sin\left(\frac{\pi t}{6} - \frac{\pi}{2}\right) \tag{2}$$

describes a 12-month variation with a year-round average of A and a maximal variation of B up and down from this average.

EXAMPLE 1 — Interpreting a Trigonometric Model

Suppose that the monthly average temperature function of a more temperate locale is given by the trigonometric model

$$f(t) = 49 + 17 \sin\left(\frac{\pi t}{6} - \frac{\pi}{2}\right). \tag{3}$$

a. Find the minimal average temperature of this locale.
b. Find the maximal average temperature of this locale.
c. Find $f(4)$ and explain its meaning in terms of the situation.

SOLUTION

a. Since the year-round average temperature is 49°F and the maximal variation away from this average is 17°F the minimal average temperature of this locale occurs in January ($t = 0$), and is $49° - 17° = 32°F$.
b. Similarly, the maximal average temperature occurs in July ($t = 6$), and is $49° + 17° = 66°F$.
c. Substituting $t = 4$, we get

$$f(4) = 49 + 17 \sin\left(\frac{4\pi}{6} - \frac{\pi}{2}\right) = 49 + 17 \sin\left(\frac{\pi}{6}\right) = 49 + 17 \cdot \frac{1}{2} = 57.5.$$

This means that the average temperature in May in this locale is 57.5°F.

SECTION 8.2 Trigonometric Models and Periodic Data

The annual average temperature function in (3) can be rewritten in the form

$$f(t) = 49 + 17 \sin\left(\frac{2\pi(t-3)}{12}\right). \tag{4}$$

The number 12 in the denominator is the **period** in months—12 months in a year. The number 3 subtracted from t in the numerator is the **delay** (in months) between time $t = 0$ (January) and the time $t = 3$ (April) when the temperature reaches its yearly average of 49, since $f(3) = 49 + 17\sin(0) = 49 + 17 \cdot 0 = 49$. The formula in (4) illustrates the way a typical periodic function is constructed.

> **DEFINITION: The General Periodic Function**
>
> The formula
>
> $$f(t) = A + B \sin\left(\frac{2\pi(t-D)}{P}\right) \tag{5}$$
>
> defines a periodic function where
>
> - A = the **average value** of the function;
> - B = the **amplitude** of its oscillation, that is, the amount by which it oscillates above and below its average;
> - P = the **period** during which its value makes a complete cycle and returns to its original value; and
> - D = the delay between time $t = 0$ and the next time when the function reaches its average value (so if the average value occurs at time $t = 0$, then $D = 0$).

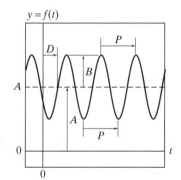

FIGURE 8.2.1 The graph of the periodic function $f(t) = A + B\sin\left(\frac{2\pi(t-D)}{P}\right)$.

The meanings of the average A, amplitude B, period P, and delay D of a periodic function are illustrated graphically in Fig. 8.2.1.

We can use (5) to construct a mathematical model for a periodically varying quantity if we know the values of these four parameters. In particular, observe that the period P can be described either as the time between successive maximum values or as the time between successive minimum values.

EXAMPLE 2 Constructing a Trigonometric Model

The temperature during a typical late July day in Atlanta varies periodically between a minimum of 70°F at 4 AM and a maximum of 90°F at 4 PM.

a. Find a trigonometric model giving the temperature as a function of t, the number of hours past midnight.

b. Use this model to predict the temperatures at midnight and at noon during the day.

c. Find when during the afternoon the temperature is exactly 87.5°F.

SOLUTION

a. The *period* during which the temperature repeats itself daily is $P = 24$ hours. Given that the temperature varies up and down between 70 and

90, we see that the *average* temperature during a day is $A = 80$, with an *amplitude* of variation of $B = 10$. Finally, we reason that the average temperature of 80 occurs halfway between the minimum at 4 AM and the maximum at 4 PM, and hence at 10 AM. Hence the *delay* between time $t = 0$ (midnight) and the time 10 AM of average temperature is $D = 10$ hours. When we substitute these four values in (5), we get the mathematical model

$$f(t) = 80 + 10 \sin\left(\frac{2\pi(t - 10)}{24}\right) \tag{6}$$

giving the temperature at time t during the 24-hour day.

b. Thus

- the temperature at midnight is

$$f(0) = 80 + 10 \sin\left(\frac{2\pi(0 - 10)}{24}\right) = 80 + 10 \sin\left(\frac{-20\pi}{24}\right) = 75°F, \text{ and}$$

- the temperature at noon is

$$f(12) = 80 + 10 \sin\left(\frac{2\pi(12 - 10)}{24}\right) = 80 + 10 \sin\left(\frac{4\pi}{24}\right) = 85°F.$$

c. To find when the temperature is 87.5°F, we need to solve the equation

$$80 + 10 \sin\left(\frac{2\pi(t - 10)}{24}\right) = 87.5.$$

Since the maximum temperature of 90°F occurs at 4 PM, we expect the temperature to be 87.5°F somewhat before and somewhat after that time. Figure 8.2.2 shows a graph of the trigonometric model in the window $0 \leq X \leq 24$, $65 \leq Y \leq 95$, with the intersection point (13.2394, 87.5) labeled.

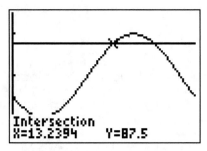

FIGURE 8.2.2 When is the temperature 87.5°F?

Since 0.2394 hours is about 14.36 minutes, the temperature reaches exactly 87.5°F at a bit after 1:14 PM, that is, approximately $2\frac{3}{4}$ hours *before* 4 PM.

You can see from the graph that, as we expect, the temperature falls to 87.5°F about the same length of time *after* 4 PM. You should use **CALC intersect** to verify that the temperature is again 87.5°F when $t = 18.7606$, or a bit before 6:46 PM.

Fitting Periodic Data

We now discuss the use of a graphing calculator's sine regression function to fit given data with a periodic function. For this purpose, let's rewrite the formula for a general periodic function in the form

$$y = a \sin(bx + c) + d \tag{7}$$

(with x instead of t as the independent variable) that a graphing calculator ordinarily uses. The coefficients in (7) are given in terms of those in (5) by

$$a = B, \quad b = \frac{2\pi}{P}, \quad c = -\frac{2\pi D}{P}, \quad d = A. \tag{8}$$

You should check that, when these relations are turned inside out to express the coefficients in (5) in terms of those in (7), the result is

$$A = d, \quad B = a, \quad P = \frac{2\pi}{b}, \quad D = -\frac{Pc}{2\pi} = -\frac{c}{b}. \tag{9}$$

EXAMPLE 3 Identifying the Parameters in a Trigonometric Model

Find the average value A, the amplitude B of oscillation, the period P, and the delay D of the periodic function

$$y = 3\sin(4x - 2) + 5. \tag{10}$$

SOLUTION

Comparing (10) and (7), we have $a = 3$, $b = 4$, $c = -2$, $d = 5$. Then the relations in (9) give

$$A = 5, \quad B = 3, \quad P = \frac{2\pi}{4} \approx 1.57, \quad D = -\frac{-2}{4} = 0.5.$$

The graph of (10) is shown in Fig. 8.2.3.

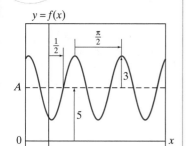

FIGURE 8.2.3 The graph $y = 3\sin(4x - 2) + 5$ in Example 3.

The calculator's **SinReg** (**Sin**e **Reg**ression) function (on the **STAT CALC** menu) fits a periodic function $y = a\sin(bx + c) + d$ to given x- and y-lists **L1** and **L2** of numerical data. We must decide in advance what the period P of the desired function is to be. Then the command

SinReg L1, L2, P, Y1

calculates the coefficients a, b, c, d and stores the resulting periodic function as **Y1**. The period P to enter may be determined either

- by inspection (perhaps graphical) of the given data, or
- by the situation being modeled.

The next two examples illustrate these two possibilities.

EXAMPLE 4 Using Data to Determine the Period

Fit a periodic function to the data in the following table and determine its average value, amplitude, period, and delay:

x	0	$\pi/12$	$\pi/4$	$5\pi/6$	$\pi/2$
y	3	5	7	5	3

SOLUTION

We first enter the x data into **L1** and the y data into **L2**. The graph (Fig. 8.2.4) of this data in the window $-\pi \leq x \leq \pi, 0 \leq y \leq 10$ appear to correspond to *one* of the two arches of a sine function.

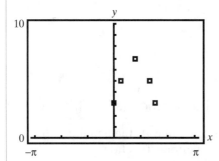

FIGURE 8.2.4 Graph of the Example 4 data.

Hence the x-range $\pi/2$ of these data is *half* of the period P of the corresponding periodic function. We therefore choose $P = \pi$ and enter the command

SinReg L1, L2, π, Y1

The results shown in Fig. 8.2.5—$a = 4$, $b = 2$, $c = 0$, $d = 3$—yield the periodic function

$$y = 4\sin(2x) + 3 \qquad (11)$$

whose graph fitting the original data points is shown in Fig. 8.2.6.

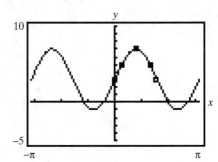

FIGURE 8.2.5 The resulting periodic function $y = 4\sin(2x) + 3$.

FIGURE 8.2.6 The graph $y = 4\sin(2x) + 3$.

Indeed, you can check that this function fits the given data *exactly*, not just approximately. Finally, the relations in (9) give us the average value $A = 3$, amplitude $B = 4$, period $P = 2\pi/2 = \pi$, and the delay $D = 0$.

EXAMPLE 5 Using the Situation to Determine the Period

Find a periodic function fitting the Atlanta monthly average temperature data given at the beginning of this section.

SOLUTION

With the 12 months January through December numbered $t = 0$ through $t = 11$, we again enter the given data in **L1** and **L2**. The command

SinReg L1, L2, 12, Y1

(generating a function of period $P = 12$) yields the results shown in Fig. 8.2.7—$a = 19.42$, $b = 0.48$, $c = -1.35$, $d = 59.79$—and thus gives the periodic function

$$f(t) = 19.42 \sin(0.48t - 1.35) + 59.79, \qquad (12)$$

whose plot is shown in Fig. 8.2.8.

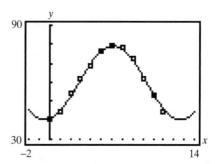

FIGURE 8.2.7 Fitting the monthly temperature data.

FIGURE 8.2.8 The periodic approximation $f(t) = 19.42 \sin(0.48t - 1.35) + 59.79$.

We see that this function approximates the given data quite well. However, do you see that something about this picture is not quite right? Although we asked for a function of period 12, it appears that the period of $f(t)$—which is the distance between successive minima—is larger than 12. Indeed, the P-relation in (9) gives

$$P = \frac{2\pi}{b} = \frac{2\pi}{0.48} \approx 13.09$$

(months) instead of 12. The reason is this: The periodic function of the form in (7)—the one that best approximates the data—simply does not have period 12. Sometimes you've got to take what you get.

It's sometimes said that trigonometric data modeling is "more art than science." We really would like to have an approximating function that really does have period 12. A known trick by which we can attempt to force this to happen is to enter the temperatures for 2 years rather than for a single year. We enter the months $0, 1, 2, \ldots, 23$ in **L3** and store 2 years of temperatures in **L4**, as indicated in Fig. 8.2.9.

Then the command

SinReg L3, L4, 12, Y1

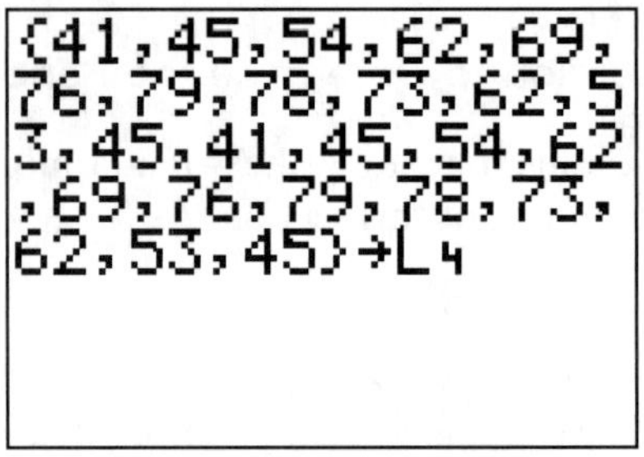

FIGURE 8.2.9 Storing 2 years of temperatures.

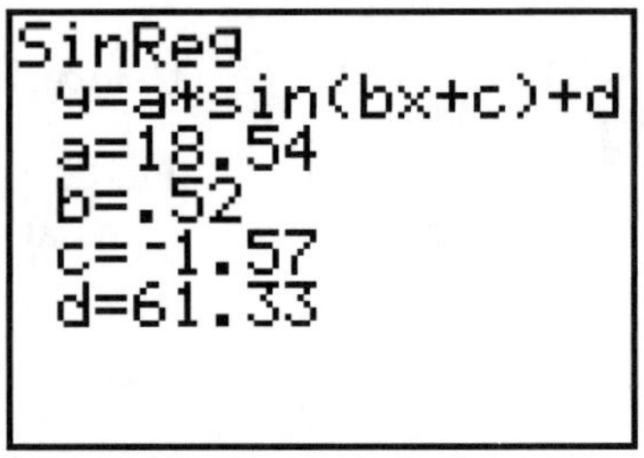

FIGURE 8.2.10 The new periodic approximation $f(t) = 18.54 \sin(0.52t - 1.57) + 61.33$.

yields the results shown in Fig. 8.2.10—$a = 18.54$, $b = 0.52$, $c = -1.57$, $d = 61.33$—and thus gives the periodic function

$$f(t) = 18.54 \sin(0.52t - 1.57) + 61.33 \qquad (13)$$

The plot shown in Fig. 8.2.11 indicates that this function approximates the given data quite well. Moreover, the P-relation in (9) now gives

$$P = \frac{2\pi}{b} = \frac{2\pi}{0.52} \approx 12.08,$$

much closer to the ideal 12-month period we sought. Finally, (13) is the periodic approximation with which we began in Example 1 of Section 8.1.

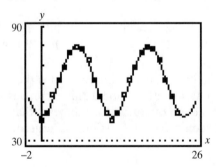

FIGURE 8.2.11 The new periodic approximation $f(t) = 18.54 \sin(0.52t - 1.57) + 61.33$.

8.2 Exercises | Building Your Skills

In Exercises 1–4, the average value A, the amplitude B, the period P, and the delay D of a periodic function are given. Sketch the graph of this function, and label A, B, P, and D on this graph. Use your graphing calculator to check your graph.

1. $A = 5$, $B = 2$, $P = 1$, $D = 0$
2. $A = 3$, $B = 5$, $P = 2$, $D = 0$
3. $A = 4$, $B = 3$, $P = 2$, $D = 1$
4. $A = 10$, $B = 5$, $P = 4$, $D = 3$

Find the average value A, the amplitude B, the period P, and the delay D of each of the periodic functions given in Exercises 5–10.

5. $f(x) = 3\sin(x) + 7$

6. $f(x) = 7\sin(\pi x) + 3$

7. $f(x) = 3\sin(x - 2) + 5$

8. $f(x) = 5\sin(2\pi x - \pi) + 3$

9. $f(x) = 10\sin(4x - 2) + 5$

10. $f(x) = 15\sin\left(\dfrac{\pi x}{3} - \dfrac{2\pi}{3}\right) + 10$

Applying Your Skills

11. The average temperature during a year in a certain location is given by a function of the form

$$f(t) = A + B\sin(kt).$$

a. Determine the values of the coefficients A and B so that the average temperature here varies during a year between a minimum of 36°F and a maximum of 72°F. *Hint:* What are the largest and smallest possible values of $\sin(kt)$?

b. Determine the value of the coefficient k so that the period of this temperature is 12 months. *Hint:* One oscillation of $\sin(kt)$ is complete when $kt = 2\pi$.

12. The temperature during a single July day in a certain location is given by a function of the form

$$f(t) = A + B\cos(kt).$$

a. Determine the values of the coefficients A and B so that the temperature varies during a day between a minimum of 74°F and a maximum of 79°F. *Hint:* What are the largest and smallest possible values of $\cos(kt)$?

b. Determine the value of the coefficient k so that the period of this temperature is 24 hours. *Hint:* One oscillation of $\cos(kt)$ is complete when $kt = 2\pi$.

In each of Exercises 13–18, use the method of Example 4 to fit a periodic function $y = a\sin(bx + c) + d$ to the data given in the table. Determine the average value, amplitude, period, and delay of this periodic function.

13.

x	0	$\pi/2$	π	$3\pi/2$	2π
y	4	7	4	1	4

14.

x	0	$\pi/6$	$\pi/2$	$5\pi/6$	π
y	5	6	7	6	5

15.

x	0	1	2	3	4
y	3	7	3	−1	3

16.

x	0	4	12	20	24
y	7	12	17	12	7

17.

x	0	1	2	3	4
y	5	11	17	11	5

18.

x	0	$\pi/2$	π	2π	3π
y	13	17	21	25	21

In each of Exercises 19–22, the monthly average temperatures of a U.S. city are given in °F. Use 2 years' worth of data to find a periodic function with period $P \approx 12$ that models these temperatures.

19. Boston, MA

Jan	Feb	Mar	Apr	May	June	July	Aug	Sept	Oct	Nov	Dec
29°	30°	39°	48°	58°	68°	74°	72°	65°	55°	45°	34°

20. Minneapolis, MN

Jan	Feb	Mar	Apr	May	June	July	Aug	Sept	Oct	Nov	Dec
12°	18°	31°	46°	59°	68°	74°	71°	61°	49°	33°	18°

21. Houston, TX

Jan	Feb	Mar	Apr	May	June	July	Aug	Sept	Oct	Nov	Dec
50°	54°	61°	68°	75°	80°	83°	82°	78°	70°	61°	54°

22. Seattle, WA

Jan	Feb	Mar	Apr	May	June	July	Aug	Sept	Oct	Nov	Dec
41°	44°	47°	50°	56°	61°	65°	66°	61°	54°	46°	42°

At the equator (0° latitude), the length of each day (measuring from sunrise to sunset) is always the same (a little bit more than 12 hours per day). In each of Exercises 23–26, the number of hours between sunrise and sunset on the 15th of each month at a certain latitude are given. Use 2 years' worth of data to find a periodic function with period $P \approx 12$ that models these data.

23. 10° latitude

Jan	Feb	Mar	Apr	May	June	July	Aug	Sept	Oct	Nov	Dec
11.58	11.82	12.07	12.35	12.57	12.70	12.67	12.47	12.30	11.92	11.67	11.53

24. 20° latitude

Jan	Feb	Mar	Apr	May	June	July	Aug	Sept	Oct	Nov	Dec
11.03	11.35	12.00	12.60	13.07	13.33	13.27	12.83	12.28	11.70	11.20	10.93

25. 40° latitude

Jan	Feb	Mar	Apr	May	June	July	Aug	Sept	Oct	Nov	Dec
9.62	10.70	11.88	13.23	14.37	15.00	14.82	13.80	12.52	11.17	10.02	9.33

26. 60° latitude

Jan	Feb	Mar	Apr	May	June	July	Aug	Sept	Oct	Nov	Dec
6.63	9.18	11.68	14.52	17.07	18.82	17.52	15.77	13.00	10.18	7.62	5.90

27. Refer to Exercises 23–26. What happens to the pattern of the length of days as you move farther from the equator? Which parameter in your mathematical model describes this phenomenon?

Chapter 8 | Review

In this chapter, you learned about trigonometric functions and models. After completing the chapter, you should be able to

- determine whether a relation described numerically, graphically, or symbolically represents a trigonometric function;
- find the output value of a trigonometric function for a given input value;
- find the input value(s) of a trigonometric function for a given output value;
- solve an equation or inequality involving a trigonometric function;
- determine the amplitude, the period, and the delay of a trigonometric function;
- find the sine function model that best fits given data.

Review Exercises

In Exercises 1–6, the given information (table, graph, or formula) gives y as a function of x. In each case, the function is quadratic, logistic, or trigonometric. Determine the type of function.

1.

x	−5	−3	−2	−1	0	1	2
y	−2.5	2.5	0	−2.5	0	2.5	0

2.

x	−2	−1	0	1	2	3
y	4	2.5	2	2.5	4	6.5

3.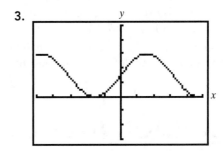

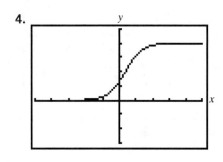

5. $y = 0.4(x - 1)^2 + 3$

6. $y = 0.4\cos(x - 1) + 3$

7. Find $\sin(\frac{4\pi}{3})$.

8. Find two values of x for which $\cos x = \frac{1}{2}$.

9. Find numerically the solution to the equation $3\cos 2x = 2x^2$.

10. The accompanying table gives the monthly average minimum temperature in °F for Columbia, Missouri. Find a trigonometric function with period $P \approx 12$ that models these temperatures.

Jan	Feb	Mar	Apr	May	June	July	Aug	Sept	Oct	Nov	Dec
21°	24°	32°	44°	54°	64°	68°	66°	58°	47°	33°	25°

11. (Chapter Opener Revisited) In the discussion that introduced this chapter, we looked at a plot that displayed the average maximum temperature in Topeka, Kansas. The plot consists of two annual cycles of the same average monthly high temperatures (25 months, beginning and ending in January). Twelve months of the data are shown in the following table, with $t = 0$ represents January 1 in the first year considered:

Month t	0	1	2	3	4	5	6	7	8	9	10	11
Average Maximum Temperature (°F)	35	44	53	66	76	86	92	90	82	71	54	43

a. Enter the months (0–25) in **L1** and the average maximum temperatures in **L2**, beginning and ending with January's temperature. Then use sine regression to find the trigonometric model that best fits these data.

b. Our goal is to create a function in which the period is approximately 12 months. Find the period of the trigonometric model, and compare it to the desired value of 12 months.

c. Find the average value of the temperature function. What does this average value mean in terms of the temperature of Topeka?

d. Find the amplitude of the function's oscillation, and use the amplitude to determine the maximal average maximum temperature and the minimal average maximum temperature. (The wording may sound strange, but here we are talking about the largest average maximum temperature and the smallest average maximum temperature.)

e. Find the delay for this function, and use the delay to determine the month(s) in which the temperature function attains its average value.

f. What is Topeka's average maximum temperature in March?

g. In what month(s) is the average maximum temperature 75°F?

INVESTIGATION Are U.S. Cities Getting Warmer?

In this activity, you will create function models based on temperature data for one of the 100 largest cities in the United States.

From the latest census data available, select the xyth largest city, where x and y are the last two digits of your student identification number. This is your city. (For example, if your student ID number ends in 37, you should find the data for the 37th largest city in the U.S. If your ID number ends in 00, choose the 100th largest city.)

Find the historical temperature data for your city for the years 1950, 1970, and 1990. Record the maximum and minimum temperatures for each month during those years.

1. Use 24 months of data to find the best-fitting trigonometric model for the monthly maximum temperatures for each of the years 1950, 1970, and 1990. Compare your three models with respect to amplitude and average value.

2. Find the best-fitting trigonometric model for the monthly minimum temperatures for each year. Compare your three models with respect to amplitude and average value.

3. Do your models provide evidence to support the theory of global warming? Why or why not?

Answers to Selected Problems

CHAPTER 1

Section 1.1

1. B is a function of A; domain = $\{10, 10.5, 13, 15, 16, 20\}$; range = $\{3, 4, 9, 13, 15, 23\}$.
3. B is not a function of A.
5. $f(2) = 4$; if $f(x) = 2$, then $x = 6$.
7. $f(2) = 6$; there is no value of x for which $f(x) = 2$.
9. (a) Calories are not a function of fat grams.
   (b) Fat grams are a function of calories.
11. (a) Average math score is a function of average verbal score.
    (b) Average verbal score is a function of average math score.
13. (a) Year is a function of Dow Jones Average because for each milestone Dow Jones Average, there is only one year in which that value was first attained.
    (b) Dow Jones Average is not a function of year because in 1995 the Dow Jones Average first reached two different milestone levels, 4000 and 5000.
    (c) 1991
    (d) 4000

Section 1.2

1. B is a function of A because each input value has only one output value; no two points lie on the same vertical line.

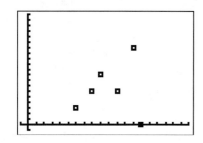

3. The graph represents a function; domain = $\{1, 2, 5, 7, 8, 10\}$; range = $\{5\}$.
5. The graph does not represent a function.

7. $f(2) = 4$; if $f(x) = 2$, then $x = 1$
9. $f(2) = 1$; if $f(x) = 2$, then $x = -3$ or $x = 3$
11. Weight is not a function of height because one height (e.g. 74 inches) corresponds to several different weights.
13. (a) The number of deaths is a function of the magnitude because for each of the magnitudes given, there is only one number of deaths.
    (b) The earthquakes of magnitudes 5.7 and 6.7 have the same number of deaths, as do the earthquakes with magnitudes 6.2 and 7.7.
    (c) The earthquake with magnitude 6.4 had about 58 deaths.
15. (a) [1979, 2003]
    (b) [624585, 1527858]
    (c) 1.0 million
    (d) 1989

Section 1.3

1. independent s; dependent R; $R(s) = 8\sqrt{s}$
3. independent r; dependent C; $C(r) = 2\pi r$
5. y is a function of x because each month has only one record high temperature.
7. y is not a function of x because a Christmas tree of a given height has many different prices.
9. y is not a function of x because each positive value of x has two different y-values.
11. $f(-1) = 1; f(0.5) = 4; f(\sqrt{2}) = 2\sqrt{2} + 3$
13. $f(-1) = -1; f(0.5) = 0.5; f(\sqrt{2}) = \dfrac{1}{2\sqrt{2} + 1}$
15. $a = 3$
17. $a = \pm 7$
19. domain = $\left[\dfrac{3}{2}, \infty\right]$; range = $[0, \infty]$
21. (a) BMI is a function of height because for each height there is only one BMI.
    (b) 66 inches
    (c) 20.3 (correct to one decimal place)
23. (a) $V(n) = 3000 - 560n$
    (b) $1320

(c) 2 years old
(d) domain = [0, 5.357]; range = [0, 3000]

Section 1.4
1. increasing [0, 3]; decreasing [−3, 0]
3. increasing [0, 4]; decreasing [8, 10]; constant [4, 8]
5.

x	f(x)	$\frac{\Delta f}{\Delta x}$
−6	−128	22
−3	−62	22
0	4	22
2	48	22
5	114	—

The average rate of change is constant on [−6, 5].

7.

x	f(x)	$\frac{\Delta f}{\Delta x}$
1	18	−6
3	6	−4.5
5	−3	−3
7	−9	−1.5
9	−12	—

The average rate of change is increasing on [1, 9].
9. decreasing [1999, 2003]; constant [1996, 1999]
11. (a) [0, 1] and [3, 4]   (b) [1, 3]
    (c) [2, 4]   (d) [0, 2]

Chapter 1 Review
1. y is not a function of x.
2. y is a function of xs.
3. y is a function of x.
4. y is not a function of x.
5. y is a function of x.
6. y is not a function of x.
7. domain = {300, 320, 430, 530, 550, 680, 710, 750}; range = {5, 9, 20, 26, 29, 33, 43}
8. domain = [0, 4]; range = [1, 3]
9. domain = (−∞, ∞); range = [−3, ∞)
10. increasing [−1, 1] and [3, 4]; decreasing [−2, −1] and [1, 3]

11. (a)

Year	0	1	2	3	4	5	6
Percentage	65.3	64.5	64.8	65.8	64.7	68.6	69.4

(b)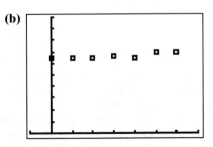

12. (a) $W(h) = 100 + 5(h − 60), h \geq 60$
    (b) 130 pounds
13. 13.35 hours
14. increasing [1984, 1985], [1993, 1994], [1996, 1998], and [2000, 2001]; constant [1983, 1984] and [1990, 1991]

CHAPTER 2

Section 2.1
1. (a) B is a function of A.
   (b) B is a linear function of A.
3. (a) B is not a function of A.
5. Initial population = 123 thousand; constant rate of change = 6 thousand.
7. Initial population = 487.139 thousand; constant rate of change = 20.558 thousand.
9. $P(t) = 42 + 5t$
11. $P(t) = 324.175 + 15.383t$
13. $P(t) = 375 + 12t, t = 0$ in 1987
15. $P(t) = 227.625 + 17.234t, t = 0$ in 1991
17. $P(t) = 35500 + 1700t, t = 0$ in 1991.
    (a) 61,000
    (b) February 2014
19. $P(t) = 45325 + 1092t, t = 0$ in 1985
    (a) 61,705 thousand
    (b) March 2012
21. February, 2001
23. (a) $C(t) = 358 + 37.667t, t = 0$ in 1995
    (b) January 2012
    (c) $772.337 million

25. **(a)** $P(t) = 397 - 4.9t, t = 0$ in 1990
    (b) 333.3 thousand
    (c) Yes
27. March, 2010
29. **(a)** $A(t) = 3.95 + 0.07t$
    (b) $8.15
    (c) $Q(t) = 0.99 + 0.15t$
    (d) $9.99
    (e) $A(t)$ is cheaper if the customer uses more than 37 minutes of long distance a month; $Q(t)$ is cheaper if the customer uses less than 37 minutes of long distance a month; the plans have the same cost if exactly 37 minutes of long distance are used.
31. **(a)** $L(a) = 74.4 - 0.94a$
    (b) 57.48 years
    (c) 45 years old

Section 2.2

1. $f(x) = 2x + 3$
3. $f(x) = -7x + 5$
5. $y = \dfrac{3}{2}x$
7. $y = -5$
9. $y = \dfrac{1}{2}x - \dfrac{7}{2}$
11. The slopes are the same; the points lie on a line.
13. The slopes are different; the points do not lie on a line.
15. $P(t) = 375 + 12(t - 1987)$
17. $P(t) = 227.625 + 17.234(t - 1991)$
19. **(a)** $K(F) = 273.16 + (5/9)(F - 32)$
    (b) -459.688 K
    (c) $255.382°F$
21. **(a)** $L(C) = 0.00213C + 124.899$
    (b) 125.004 cm
    (c) $83.286°C$
23. **(a)** $F(t) = 705 - 6(t - 1987)$
    (b) 2007
25. **(a)** $T(c) = 61 + 1.5(c - 14)$
    (b) $76°F$
27. **(a)** $P(t) = 628 - 3.1(t - 1990)$
    (b) 587.7 thousand
    (c) -3.1 thousand people per year
    (d) July 2063

Section 2.3

1.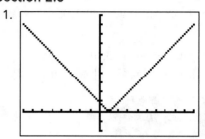

3.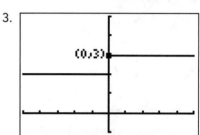

5. $f(x) = \begin{cases} -x - 1 & \text{if } -2 \le x \le -1 \\ x + 1 & \text{if } 1 < x \le 2 \end{cases}$

7. $f(x) = \begin{cases} (1/2)x & \text{if } -2 \le x \le 2 \\ 2x - 3 & \text{if } 2 < x \le 4 \end{cases}$

9.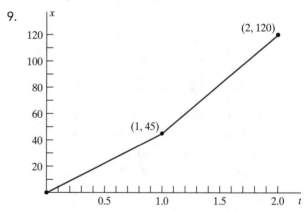

$d(t) = \begin{cases} 45t & \text{if } 0 \le t \le 1 \\ 75t & \text{if } < t \le 2 \end{cases}$

11.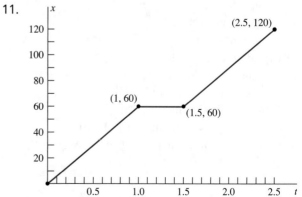

$$d(t) = \begin{cases} 60t & \text{if } 0 \le t \le t \\ 60 & \text{if } 1 \le t \le 1.5 \\ 60t & \text{if } 1.5 \le t \le 2.5 \end{cases}$$

13. **(a)** $P(t) = \begin{cases} 45 + 10.9t & \text{if } 0 \le t \le 20 \\ 263 + 7.5(t - 20) & \text{if } t > 20 \end{cases}$ where, P is measured in thousands and t is years after 1950.
    (b) 1955
    (c) 563 thousand

15. **(a)** $P(t) = \begin{cases} 163 + 3.38t & \text{if } 0 \le t \le 50 \\ 332 - 2.24(t - 50) & \text{if } t > 50 \end{cases}$ where, P is measured in thousands and t is years after 1900.
    (b) 271.160 thousand
    (c) 2025

17.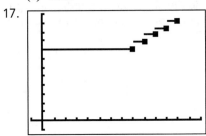

19. **(a)** $P(x) = \begin{cases} 0.39 & \text{if } 0 < x \le 1 \\ 0.63 & \text{if } 1 < x \le 2 \\ 0.87 & \text{if } 2 < x \le 3 \\ 1.11 & \text{if } 3 < x \le 4 \\ 1.35 & \text{if } 4 < x \le 5 \end{cases}$
    (b)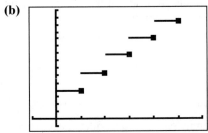

21. **(a)** $C(t) = \begin{cases} 195 + 30.5t & \text{if } 0 \le t \le 8 \\ 439 - 53(t - 8) & \text{if } t > 15 \end{cases}$ where, C is measured in millions and t is years after 1985.
    (b) 1988 **(c)** 1994 **(d)** 1075 million

23. **(a)** $171 **(b)** $480 **(c)** $1830
    (d) $T(I) = \begin{cases} 0.03I & \text{if } 0 < I \le 16000 \\ 480 + 0.05(I - 16000) & \text{if } x > 16000 \end{cases}$

25. **(a)** $88
    (b) $430
    (c) $2410
    (d) $T(I) = \begin{cases} 0.02I & \text{if } 0 < I \le 12500 \\ 50 + 0.04(I - 12500) & \text{if } 12500 < I \le 25000 \\ 550 + 0.06(I - 25000) & \text{if } I > 25000 \end{cases}$

27. **(a)** $222.30 is owed in Michigan; $171 is owed in Connecticut on the same income.
    (b) $1677 is owed in Michigan; $1830 is owed in Connecticut on the same income.
    (c) $T(I) = 0.039I$
    (d) A taxpayer in Michigan pays more on income less than $29,090.91; a taxpayer in Michigan pays less on income more than $29,090.91.

Section 2.4

(Values calculated with calculator in three-decimal-place mode.)

1. **(a)** 5.774 (thousand) **(b)** 11.547 (thousand)
3. **(a)** 17.321 (thousand) **(b)** 34.641 (thousand)
5. **(a)** $P(t) = -3.6t + 354$
   (b) SSE = 720 and average error = 13.416 (both in thousands)
7. **(a)** $P(t) = 15t - 28775$
   (b) SSE = 2500 and average error = 25 (both in thousands)
9. **(a)** $P(t) = 18.570t + 346.800$, where P is in thouands and t is years after 1950
   (b)

Actual Population (thousands)	Predicted Population (thousands)	Error (thousands)
334	346.8	−12.8
573	532.5	40.5
697	718.2	−21.2
876	903.9	−27.9
1111	1089.6	21.4

   (c) 1275.3 thousand or 1,275,300

11. **(a)** $P(t) = -4.040t + 446.800$, where P is in thousands and t is years after 1950
    (b)

Actual Population (thousands)	Predicted Population (thousands)	Error (thousands)
439	446.8	−7.8
405	406.4	−1.4
382	366	16
329	325.6	3.4
275	285.2	−10.2

    (c) 244.8 thousand or 244,800

13. (a) The plot is not approximately linear; a linear function is not a suitable model.
    (b) not applicable
15. (a) The plot is approximately linear; a linear function is a suitable model.
    (b) $L(A) = -0.903A + 76.508$
17. (a) $S(t) = 81.700t + 130.140$, where S is in millions and t is years after 1988.
    (b) The prediction of 702.040 million is approximately 20 million less than the actual value.
    (c) The prediction of 1273.940 million is nearly 500 million less than the actual value.
    (d) The discrepancy is the result of CDs sales growing at a greater rate after 1996.
19. (a) $A(C) = 34.435C - 327.111$
    (b) For each additional country represented at the Winter Olympics, approximately 34 more athletes participate.
    (c) $2427.689 \approx 2428$ athletes
21. (a) $P(Y) = 88.271Y + 3176.197$
    (b) $4500.26 (rounded to the nearest cent)
    (c) The model underestimates the pay for 2, 4, 16, and 20 years of service.
23. (a) $W(h) = 7.333h - 348.333$
    (b) $179.643 \approx 180$ pounds
    (c) No, the average American male is not a professional athlete.

Chapter 2 Review

1. y is neither a linear function of x nor a piecewise-linear function of x.
2. y is a linear function of x.
3. y is a piecewise-linear function of x.
4. y is neither a linear function of x nor a piecewise-linear function of x.
5. y is a linear function of x.
6. y is neither a linear function of x nor a piecewise-linear function of x.
7. 1.4
8. (a) $P(t) = 227 - 2.3t$, where P is in thousands and t is years after 2000.
   (b) 215.5 thousand
   (c) 2011
9. (a) $G(t) = 51 + 1.7t$, $t = 0$ in 1989
   (b) For each additional year after 1989, the percentage of students graduating in 6 years increases by 1.7 percent.
   (c) 79.9%
10. (a) $p(x) = 60 - 1.25(x - 20) = -1.25x + 85$
    (b) 24 sprinklers
    (c) The price can be no more than $47.50.
    (d) The p-intercept is the price at which the demand x is zero; at a price of $85, she will sell no sprinklers. The x-intercept is the demand when the price p is zero; if the sprinklers were free, she could give away 68 sprinklers a month.
11. (a) $C(d) = \begin{cases} 95 & \text{if } 0 < x \leq 10 \\ 95 + 1(x - 10) & \text{if } x > 10 \end{cases}$
    (b) $C(16) = \$101$; this is the cost for grinding a tree stump of diameter 16 inches.
    (c) 23 inches
12. (a) $P(t) = 2.443t - 5.409$
    (b) $124.51
    (c) For each additional minute of grinding and polishing time, the price of the tableware increases by about $2.44.

CHAPTER 3

Section 3.1

1. 10
3. 0.405
5. $124.30
7. $72.21
9. 15% = 10% + 5%. You get 10% of A when you move the decimal point one place to the left. You get 5% of A when you divide the result by 2. The sum of these two values is the 15% tip; the total amount paid is the sum of the three amounts you have written down.
11. $P(t) = 130 \times (1.063)^t$, with P in thousands, and $t = 0$ in 2000.
13. 12 years (to the nearest year)
15. (a) $34,0
    (b) $50,787
    (c) The family pays less tax under thtax proposal.
17. (a) $A(t) = 2000 \times (1.0345)^t$
    (b) $2623.46
    (c) 14 years (to the nearest year)
19. (a) and (b)

Column % Interest Rate	Column B Number of Years to Double	Column C Column A × Column B
4	18	72
6	12	72
8	9	72
9	8	72

(c) The product of the percent interest rate and the number of years for the investment to double is 72.
21. (a) $P(t) = 635 \times (1.0149)^t$, P in thousands and $t = 0$ in 1990.
(b) 2005
(c) 769.619 thousand or 769,619 people; this overestimates the actual 2003 population by approximately 36,000 people.
23. (a) $P(t) = 257 \times (0.0075)^t$, P in thousands and $t = 0$ in 1990.
(b) 283.217 thousand or 283,217 people; this overestimates the actual 2003 population by approximately 4000 people.
(c) 2049
25. (a) 6 million
(b) 12%
(c) The predicted value of 81.314 million is approximately equal to the actual number.
(d) The predicted value of 91.072 is 10 million larger than the actual number.
(e) The model predicts that the number of people enrolled is increasing every year. The number actually decreased from 1999 to 2000.
27. (a) $E(t) = 46.4 \times (1.0851)^t$, E in thousands and $t = 0$ in 1994
(b) $105.006 billion or $105,006,000,000
(c) 2002
29. A negative initial guess leads to the negative square root of the number.

Section 3.2
1. $76.26
3. $68.46
5. $532.86
7. It is less than 120, because we are adding back 25% of a smaller number.
9. $P(t) = 35 \times (0.97)^t$, P in thousands and $t = 0$ in 2000.
11. $19,019.80
13. (a) $V(t) = 1000 \times (0.98)^t$
(b) $885.84
(c) to the nearest month, 11 months
15. (a) $P(t) = 333 \times (0.9941)^t$, P in thousands and $t = 0$ in 1990.
(b) 308.344 thousand or 308,344 people; this value is less than 700 people smaller than the actual population.
(c) 2019

17. December, 2010
19. (a) $D(t) = 586.8 \times (0.9826)^t$, $t = 0$ in 1950.
(b) 435.4 rounded to the nearest tenth
(c) 1988
21. (a) $R(t) = 30 \times (0.834)^t$
(b) to the nearest day, 10 day
(c) 2.363 grams
23. (a) $I(t) = 400 \times (0.71)^t$
(b) 2.024 hours
(c) to the nearest milligram, 143 mg

Section 3.3
1. $y = 10 \times 2^x$
3. $y = 100 \times (0.2)^x$
5. $P(t) = 100 \times (1.75)^t$
7. $P(t) = 120 \times (1/3)^{t/4}$
9. $P(t) = 75 \times (2)^t$
11. $P(t) = 100 \times (2)^{t/2.5}$
13. 0.731 year for t given in years
15. -0.240, which is an annual decrease of 24.0%
17. (a) -0.046, which is a decay rate of 4.6%
(b) 12.991 thousand or 12,991 people
(c) 2014
19. (a) 0.052, which is an annual increase of 5.2%
(b) 3.333 million or 3,333,000 people (rounded to the nearest)
(c) 2007
21. (a) 2051
(b) 1995
23. (a) 11.0%
(b) $121.944 billion or $121,944,000
(c) 2008
25. (a) $P(t) = 49 \times (6)^t$
(b) 1127.096 ≈ 1127 bacteria
(c) 3:21 PM
27. 3 hours, 52 minutes
29. (a) $A(t) = 15 \left(\dfrac{10}{15}\right)^{t/5}$
(b) 7.841 SU
(c) 33.394 months
31. (a) $A(t) = 557.271 \left(\dfrac{556.902}{557.271}\right)^t$, A in thousands and $t = 0$ in 2002
(b) 0.1% decrease
(c) to the nearest year, 164 years
33. June 29

Section 3.4

1. (a) $y = 4.843 \times (1.464)^x$
   (b) 0.390
3. (a) $y = 15.263 \times (1.396)^x$
   (b) 1.601
5. (a) $y = 40.818 \times (0.645)^x$
   (b) 0.475
7. (a) $y = 111.438 \times (0.496)^x$
   (b) 1.112
9. (a) $P(t) = 566.381 \times (1.022)^t$, P in thousands, $t = 0$ in 1960
   (b) 2.2%
   (c) 1352.523 thousand or 1,352,523 people
11. (a) $P(t) = 881.191 \times (0.981)^t$, P in thousands, $t = 0$ in 1960
    (b) -1.9%
    (c) 409.101 thousand or 409,101 people
13. (a) $P(t) = 575.672 \times (1.016)^t$, P in thousands, $t = 0$ in 1960
    (b) 1.6%
    (c) 1086.234 thousand or 1,086,234 people
15. $42.922 \approx 43$ feet
17. (a) 2003
    (b) to the nearest dollar, $1,194,216
19. to the nearest game, 15 games
21. (a) 94.4%
    (b) 62651.530 thousand or 62,651,530 subscribers
23. to the nearest minute, 1:54 PM

Chapter 3 Review

1. neither
2. exponential
3. linear
4. linear
5. exponential
6. neither
7. exponential
8. $F(t) = 0.082 \times (1.041)^t$; $t = 0$ in 1938; to the nearest cent, $2.50
9. 646 days
10. 2023
11. 2.5% annual decrease
12. $R(t) = 203.265 \times (1.071)^t$, R in millions and $t = 0$ in 1984.

CHAPTER 4

Section 4.1

1. (a) $1040.60
   (b) $1040.74
   (c) $1040.79
   (d) $1040.81
   (e) $1040.81
3. (a) $13,557.59
   (b) $13,580.10
   (c) $13,588.80
   (d) $13,591.04
   (e) $13,591.41
5. 23.105 years
7. 11.552
9. 3.56%
11. 3.05%
13. $x = 0.693$
15. $2.079 = 3 \times (0.693)$; the solution to $e^x = 2^3$ is three times the solution to $e^x = 2$.
17. 4.5% compounded semiannually
19. (a) $A(t) = 3000\left(1 + \dfrac{.0425}{52}\right)^{52t}$; $t = 0$ in 2004
    (b) May 2032
    (c) $4587.97
21. (a) $A(t) = 4800e^{0.052t}$; $t = 0$ in 2000
    (b) $17,612.62
    (c) April, 2013
23. 6.3% compounded monthly, 6.25% compounded continuously, 6.2% compounded daily
25. (a) $A(t) = 300 \times e^{0.10t}$
    (b) $447.547 \approx 448$ bacteria
    (c) 5:49 PM
27. 3590 years old
29. November 1, 2005; $121.29 interest

Section 4.2

1. (a) 64
   (b) 1
   (c) 125
3. $2e^{3x-4}$
5. (a) $2(\ln 3)$
   (b) $\ln 2 + \ln 5$
   (c) $3(\ln 2) - 3(\ln 3)$
7. $2^{(3^4)} = 2^{81} > 2^{12} = (2^3)^4$

9. -0.693

11. $g(x) = 4x + 3$

13. $x = -\dfrac{\ln 17}{\ln 5}$

15. $x = \dfrac{1}{3}\left(5 + \ln \dfrac{1}{7}\right)$

17. $x = \dfrac{1}{10}e^{-2}$

19. (a) $P(x) = 26.6\left(\dfrac{34.5}{26.6}\right)^{x/25}$, P is in millions and $x = 0$ in 1975.
    (b) 38.282 million or 38,282,000 people
    (c) $x = 25\dfrac{\ln 2}{\ln(34.5/26.6)}$, so the year is
    $1975 + 25\dfrac{\ln 2}{\ln(34.5/26.6)}$
    (d) $x = 25\dfrac{\ln 2}{\ln(34.5/26.6)} \approx 66.636$, so the year is 2041.

21. (a) $P(x) = 906\left(\dfrac{629}{906}\right)^{x/33}$, P is thousands and $x = 0$ in 1970.
    (b) 521.208 thousand or 521,208 people
    (c) $x = 33\dfrac{\ln(400/906)}{\ln(629/906)}$, so the year is
    $1970 + 33\dfrac{\ln(400/906)}{\ln(629/906)}$
    (d) $x = 33\dfrac{\ln(400/906)}{\ln(629/906)} \approx 73.936$, so the year is 2043.

23. $\dfrac{2.5(\ln 2)}{\ln 3} \approx 1.5773$ hours

25. (a) 300 flies
    (b) 2.5%
    (c) 426 flies
    (d) about 48.159 day

27. (a) 300   (b) 2.44%
    (c) $t = -\dfrac{\ln 0.5}{0.0244} \approx 28.4077$ years
    (d) 265.5445 grams

29. $A(t) = 100e^{-0.000121t}$
    (b) 46%
    (c) $t = -\dfrac{\ln 0.31}{0.000121} \approx 9679.1982$ years

Section 4.3

1. $y = 4.843e^{0.381t}$

3. $y = 19.8196e^{0.0986t}$

5. $y = 410.169e^{-0.792t}$

7. $y = 4.961 + 9.988\ln x$

9. $y = 3.064 + 6.974\ln x$

11. $y = 11.225 + 2.911\ln x$

13. (a) exponential model: $P(t) = 287.394e^{-0.008t}$
    (b) 7.665 thousand
    (c) 199.349 thousand or 199,349 people

15. (a) logarithmic model: $P(t) = -90.420 + 176.591\ln x$
    (b) 7.916 thousand
    (c) 600.408 thousand or 600,408 people

17. (a) exponential model: $P(t) = 68.119e^{0.027t}$
    (b) 1.2977 thousand
    (c) 262.764 thousand or 262,764 people

19. (a) $A(t) = 9.992e^{0.157t}$, $t = 0$ in 1990.
    (b) September 2004

21. $T(t) = 350 - 299.621e^{-0.004t}$; 1:53 PM

23. (a) $C = 43.598 - 19.988\ln W$
    (b) $-32.490°F$

Chapter 4 Review

1. logarithmic
2. linear
3. exponential
4. logarithmic
5. exponential
6. neither
7. exponential
8. 0.4%
9. $16467.77; 7.296 years, or about 7 years, 4 months
10. (a) $V(t) = 35.749e^{0.142t}$, V in billions and $t = 0$ in 1980.
    (b) 2008
11. (a) $872.731 + 401.961(\ln x)$
    (b) 2166.594 thousands or 2,166,594 dollars

CHAPTER 5

Section 5.1

1. The average rate of change is not constant, so the function is not linear. The function is quadratic.
3. The average rate of change is constant, so the function is linear
5. $a = 2$
7. $a = -4$
9. Figure 5.1.16

11. Figure 5.1.15
13. Figure 5.1.17
15. (a) There are two solutions if $c < 25$.
    (b) There is one solution if $c = 25$.
    (c) There are no solutions if $c > 25$.
17. 10 years, 298 days
19. 6 years, 184 days
21. (a) 344.86 thousand, or 344,860 people
    (b) March, 2024
    (c) 485.26 thousand, or 485,260 people
23. (a) 5.982% ≈ 6%; this value is close to the actual value of 6.4%
    (b) 1995 and 2003
    (c) in August, 2007
25. (a) 240 feet per second
    (b) 10 feet
    (c) 15.053 seconds
27. (a) 7.746 seconds
    (b) 6.340 seconds
    (c) 9.465 seconds
29. 144 feet deep

Section 5.2

1. There is no maximum value; the minimum value is -1.
3. The maximum value is 8; the minimum value is -1.
5. The maximum value is 6, there is no minimum value.
7. The maximum value is 5, the minimum value is -10.
9. The maximum value is 9, the minimum value is -3.5.
11. (a) The minimum population is 185.617 thousand, or 185,617 people.
    (b) January 2010
    (c) November 2025
13. (a) 1.563 seconds
    (b) 139.063 feet
    (c) 4.511 seconds
15. (a) 1610 feet
    (b) 1210 feet
17. (a) March 1981
    (b) 319.660 thousand or 319,660 people
    (c) June 1992
19. (a) May 1999
    (b) 6.102%
    (c) -0.75%; this is not reliable because the percentage of households leasing a car for personal use cannot be negative.

21. (a) Imports were decreasing on $[0, 1.929]$, so from January 2001 until December 2002.
    (b) December 2002
    (c) 2006
23. (a) 976.563 feet
    (b) 3906.25 feet

Section 5.3

1. average error $= 1$
3. average error $= \sqrt{18/4} \approx 2.121$
5. average error $= \sqrt{209/5} \approx 6.465$
7. $q(x) = 1.667x^2 + 0.333x - 4.000$; average error $= 0.001$
9. $q(x) = -0.083x^2 + 1.506x + 2.288$; average error $= 0.208$
11. $q(x) = 4.233x^2 - 4.580x + 24.996$; average error $= 3.107$
13. (a) $P(t) = 21.371 - 1.914t + 0.213t^2$
    (b)

t	Actual Population (Thousands)	$P(t)$ (Thousands)	Error (Thousands)
0	17	21.371	-4.371
10	34	23.531	10.469
20	63	68.291	-5.291
30	152	155.651	-3.651
40	288	285.611	2.389

(c) 458.171 thousand, or 458,171 people

15. (a) $P(t) = 10.543 + 1.801t + 0.111t^2$
    (b)

t	Actual Population (Thousands)	$P(t)$ (Thousands)	Error (Thousands)
0	8	10.543	-2.543
10	45	39.653	5.347
20	90	90.963	-0.963
30	160	164.473	-4.473
40	262	260.183	1.817

(c) 378.093 thousand, or 378,093 people

17. (a) $P(t) = 1387.829 + 138.121t - 15.696t^2$, P in millions and $t = 0$ in 1988.
    (b) 39.793   (c) 1996
19. (a) $P(t) = 770.727 + 45.431t + 2.532t^2$, $t = 0$ in 1985.
    (b) 1981   (c) 1989

(d) The 1989 date because it is within the range of the given data values, whereas the 1981 date is before the first data point.

21. (a) $R(t) = 0.539t^2 + 17.417t - 4.371$, $t = 0$ in 1956.
    (b) 2508
    (c) 2002
23. (a) $M(t) = -0.005t^2 - 0.160t + 27.624$, $t = 0$ in 1980.
    (b) 0.124
    (c) 19.224 gallons

Chapter 5 Review
1. y is a quadratic function of x.
2. y is not a quadratic function of x.
3. y is not a quadratic function of x.
4. y is a quadratic function of x.
5. y is not a quadratic function of x.
6. y is a quadratic function of x.
7. y is a quadratic function of x.
8. Based strictly on the table, the domain is {0, 0.2, 0.4, 0.6, 0.8, 1.0} and the range is {0. 0.76, 10.24, 13.44, 15.36, 16}. However, since the situation involves the height of a ball for times between 0 and 1 second, we can reasonably conclude that the domain is [0, 1] and the range is [0, 16].
9. domain is $[-3, 1]$; range is $[-2.5, 2]$
10. domain is $(-\infty, \infty)$; range is $(-\infty, 4]$
11. (a) $R(x) = -1.25x^2 + 85x$
    (b) 34 sprinklers
    (c) $1445
12. (a) The ball passes the building on its way down after 3 seconds.
    (b) The ball remains in the air for 5 seconds.
13. (a) $P(t) = 72.514 + 0.277t + 0.079t^2$, P in thousands and $t = 0$ in 1950.
    (b) 1995
14. (a) $P(t) = 11.765 - 0.356t + 0.005t^2$, $t = 0$ in 1930.
    (b) 11.345%
    (c) 5.428%
    (d) 1965

CHAPTER 6

Section 6.1
1. $x = -2, 1, 3$
3. $x = -3, -1, 2, 4$
5. $x = -5/2, 2/3, 3$
7. $x = -15/4, -2, 1, 11/3$
9. $x = -15/8, 7/6, 13/3$
11. 2.4941 in $\times$ 4.4941 in $\times$ 7.4941 in
13. $2\frac{15}{16} = 2.9375$ inches or $13\frac{5}{16} = 13.3125$ inches
15. $x = 2$ ft $5\frac{5}{8}$ in or $x = 10$ ft $10\frac{5}{8}$ in
17. $x = 14$ ft 10 in or $x = 31$ ft 1 in
19. (a) $x^3 - 3x^2 + 1 = 0$
    (b) $x = 7.832$ inches
21. (a) $3x^3 - 9x^2 + 8 = 0$
    (b) $x = 1$ ft 2.713 inches

Section 6.2
1. $x = 2, y = 3$
3. $x = 10, y = -3$
5. $x = -1/5, y = 4/5$
7. $x = 27, y = 37$
9. $x = 57, y = 38$
11. 54 nickels and 23 dimes
13. 15 dimes and 34 quarters
15. 15 hamburgers and 9 cheeseburgers
17. 480 bottles of wine and 216 bottles of liquor
19. 10 gallons of 6% solution and 20 gallons of 12% solution
21. 36 grams of pure gold and 84 grams of 14-karat gold
23. 30 pounds of 75% brass and 120 pounds of 90% brass

Section 6.3
1. 93
3. 27
5. $x = 3, y = 1, z = 2$
7. $x = 4, y = 2, z = 5$
9. $x = 5, y = -3, z = 7$
11. $x = 5, y = -3, z = 7$
13. $x = 39, y = 27, z = 44$
17. 77 one-ounce coins, 108 half-ounce coins, 180 quarter-ounce coins
19. 162 gal of reddish paint, 32 gal of bluish paint, 50 gal of greenish paint
21. $1.25 for a hot dog, $0.85 for an order of French fries, $0.65 for a soft drink
23. $90 per TI-82, $95 per TI-83, $100 per TI-85, $120 per TI-86
25. $1.49 per hamburger, $1.79 per cheeseburger, $2.29 per roast beef, $2.00 per chicken sandwich, $2.19 per ham sandwich

Section 6.4
1. $P(t) = 2284.03 - 2.3789\,t + 0.00062\,t^2$, $P(2000) = 6.228$ (billion)

3. $a = 277/14212$, $b = -172/14212$, $c = 523/14212$
5. center $(1, -2)$ and radius 5
7. center $(-5, -10)$ and radius 17
9. $P(t) = -281.919 + 0.1672t$
11. $P(t) = -1920.27 + 1.0079t$
13. $P(t) = 31160.877 - 31.5134t + 0.0080t^2$
15. $P(t) = -50680.25 + 50.13665t - 0.012375t^2$
17. $P(t) = -7605540 + 11539.4t - 5.836t^2 + 0.000984t^3$
19. $P(t) = 9249722 - 14041.6t + 7.105t^2 - 0.00120t^3$
21. $P(t) = 327409 - 507.5751t + 0.262115t^2 - 4.50833 \times 10^{-5}t^3$; average error $= 1.091$
23. $P(t) = 319035 - 476.262t + 0.235858t^2 - 3.87083 \times 10^{-5}t^3$; average error $= 3.054$
25. $P(t) = -2.057864 \times 10^7 + 42410.4t - 32.7750t^2 + 0.0112567t^3 - 1.44974 \times 10^{-6}t^4$; average error $= 0.770$
27. $P(t) = -1.665644 \times 10^7 + 34077.8t - 26.1377t^2 + 0.00890736t^3 - 1.3794 \times 10^{-6}t^4$, average error $= 0.538$
29. center $(1, 2, 3)$ and radius 13

Chapter 6 Review
1. $f(x)$ is exponential (E-II)
2. $f(x)$ is a quartic polynomial (D-I)
3. $f(x)$ is linear (A-I)
4. $f(x)$ is logarithmic (F-III)
5. $f(x)$ is a quadratic polynomial (B-I)
6. $f(x)$ is a cubic polynomial (C-I)
7. $x = -12/5, 5/3, 4$
8. $x = -25/8, -5/4, 5/6, 10/3$
9. A hot dog costs $1.50, a bag of chips $0.50, and a soft drink $1.00. Hence the student was charged $$4.50.
10. **(a)** The interpolating cubic polynomial is $P(t) = -3963905593 + 5963852.85t - 2990.95t^2 + 0.5t^3$.
    (b) This gives $P(1999) \approx 63\%$, which seems unreasonably large, so the cubic model does not seem a good one for predicting rap music sales.
11. **(a)** $E_3(x) = 7296.79 - 380.943x - 27.9586x^2 + 4.74553x^3$
    (b) $E_4(x) = 7262.41 - 237.656x - 108.358x^2 + 19.0742 x^3 - 0.796037x^4$
    (c) Taking $t = 12$ for 1998, we calculate $E_3(12) \approx 6900$ and $E_4(12) \approx 5261$. The latter fits in much better with the given data, so the quartic model seems more reliable for short-term prediction. However, the cubic model predicts greater future pollution, so you might use it to argue for stronger pollution control laws. Unfortunately, this kind of deliberate obfuscation is not uncommon in the political use (or misuse) of mathematics.

CHAPTER 7

Section 7.1
1. value of $f(x)$ is 150 when $x \approx 17.9176$
3. value of $f(x)$ is 3 when $x \approx 0.2027$
5. $P(t) = \dfrac{12500}{50 + 200e^{-0.1792t}}$
7. $P(t) = \dfrac{7}{1 + 6e^{-2.7081t}}$
9. $P(t) = \dfrac{30000}{300 - 200e^{-0.02877t}}$
11. $t = 0.6616$
13. $t = 24.0942$
15. $P(t) = \dfrac{25000}{50 + 450e^{-0.08109t}}$; $P = 350$ when $t = 37.54$, in August 2003.
17. $P(t) = \dfrac{6000}{25 + 215e^{-0.09039t}}$, P measured in thousands; $P = 120$ when $t = 23.81$, in June 2002.
19. $P(t) = \dfrac{0.04}{0.002 + 19.998e^{-1.56645t}}$, P measured in thousands; $P = 10$ after $t \approx 5.88$ weeks.
21. $P(t) = \dfrac{500}{5 + 95e^{-0.3025t}}$, P given in percentage points; $P = 35$ after $t \approx 7.69$ weeks.

Section 7.2
1. $f(x) = \dfrac{6.335}{1 + 6.100e^{-0.782x}}$, average error $= 0.015$
3. $f(x) = \dfrac{15.850}{1 + 1.819e^{-0.870x}}$, average error $= 0.013$
5. $f(x) = \dfrac{36.633}{1 + 5.111e^{-0.361x}}$, average error $= 0.064$
7. **(a)** $P(t) = \dfrac{673.865}{1 + 1.075e^{-0.024t}}$
   (b) $P \approx 549.7$ thousand in 2025.
   (c) $P = 90\%$ of limiting population of 673.9 in the year 2054.
9. **(a)** $P(t) = \dfrac{255.598}{1 + 13.95e^{-0.161t}}$
   (b) $P = 255.5$ thousand in 2025.
   (c) $P = 90\%$ of limiting population of 255.6 million in the year 1990.
11. **(a)** $P(t) = \dfrac{324.348}{1 + 6.955e^{-0.091t}}$
    (b) $P = 318.4$ thousand in 2025.
    (c) $P = 90\%$ of limiting population of 324.3 million in the year 2005.

13. $P(t) = \dfrac{64.173}{1 + 4.333\,e^{-0.318\,t}}$. The limiting percentage is 64%, so it is possible for this candidate to win. He has majority approval after eight and a half months of campaigning.

15. $P(t) = \dfrac{110.937}{1 + 2.705\,e^{-0.128\,t}}$. The limiting population is about 111 fish. It takes about 18.6 months to reach 80% of this number.

17. $P(t) = \dfrac{35.645}{1 + 8.106\,e^{-0.704\,t}}$. About 35.65% of the population will eventually suffer this disease. It will take about 7 weeks 1 day for 95% of this limiting population to contract it.

19. (a) $P(t) = \dfrac{865.082}{1 + 10.22\,e^{-0.016\,t}}$ (with $t = 0$ in 1900).
    (b) Limiting population 865 million
    (c) $P = 300$ million in the year 2005.

21. $P(t) = \dfrac{2751.239}{1 + 4.005\,e^{-0.026\,t}}$ (with $t = 0$ in 1950). This model predicts a limiting population of about 2.75 billion for China, and that it will hit 2 billion in the year 2041.

23. $P(t) = \dfrac{13.323}{1 + 2.264\,e^{-0.026\,t}}$ (with $t = 0$ in 1975). This model predicts a limiting world population of only about 13.3 billion, as compared with 26 billion in Exercise 22. Evidently world population growth slowed during the last quarter of the 20th century. It predicts that the world population will not hit 10 billion until the year 2048.

Section 7.3

1. If we first store $50 \to P$, $0.005 \to k$, $200 \to M$ then the iteration $P + k*P*(M - P) \to P$ gives increasing values of P that rapidly level off at $P = 200$.

3. P increases and eventually levels off at $P = 500$.

5. P oscillates above and below 300 as it eventually levels off at $P = 300$.

7. P eventually bounces back and forth between $P = 149.25$ and $P = 232.57$.

9. P immediately starts bouncing back and forth between $P = 300$ and $P = 600$.

11. Verification of results in text.

13. Answers vary, depending on the choice of r. As r gets closer to 2, fewer iterations are necessary to get close to the limiting value.

15. Verification of results in text.

17. Verification of results in text.

19. Verification of results in text.

Chapter 7 Review

1. $f(x)$ is a logistic function with formula of type C.
2. $f(x)$ is a polynomial function with formula of type A.
3. $f(x)$ is an exponential function with formula of type D.
4. $f(x)$ is a logarithmic function with formula of type B.
5. (a) $f(10) \approx 16$
   (b) $f(0.2291) \approx 10$
6. The number of fish in the lake after t months is given by $P(t) = 80000/(100 + 700\,e^{-0.11819\,t})$. There are 600 fish after about 25.76 months, during August 2002.
7. (a) $P(t) = 70.179/(1 + 94.687\,e^{-0.362\,t})$
   (b) At age 30 weeks, Ripley will weigh just over 70 pounds.
   (c) Her limiting weight is about 70 pounds 3 ounces, so she will be a bit lighter than the average female.
8. Using Equation (6) in Section 7.3, the discrete logistic solution with limiting population $M = 180$ and growth rate parameter $k = 0.0062$ is described by the iterative model $P_{n+1} = P_n + 0.0062\,P_n(180 - P_n)$. Starting with $P_0 = 60$, the population P_n approaches the limiting population of 180 rapidly, but briefly oscillates up and down about it:

n	0	1	2	3	4	5	6	7	8
P_n	60	104.640	153.531	178.727	180.138	179.984	180.002	180.000	180.000

CHAPTER 8

Section 8.1

1. (a) $\pi/6$
   (b) $5\pi/6$
   (c) $7\pi/6$

3. (a) $3\pi/4$
   (b) $5\pi/4$
   (c) $7\pi/4$

5. (a) 18°
   (b) 20°
   (c) 36°
7. (a) 54°
   (b) 80°
   (c) 252°
9. (a) 1/2
   (b) −1/2
   (c) −1/2
11. (a) −1/2
    (b) −1/2
    (c) 1/2
13. Fig. 8.1.24
15. Fig. 8.1.22
17. Fig. 8.1.29
19. Fig. 8.1.25
21. $f(g(x)) = -4x^2 - 12x - 8$, $g(f(x)) = 5 - 2x^2$
23. $f(g(x)) = \sqrt{x^4 + 6x^2 + 6}$, $g(f(x)) = x^2$
25. $f(g(x)) = \sqrt{\cos x}$, $g(f(x)) = \cos(\sqrt{x})$
27. $x = 0.4502$
29. $x = 1.4277$
31. $x = -5.9245$, $x = -3.2472$, $x = 0$, $x = 3.0485$, $x = 6.7574$, $x = 8.5939$

Section 8.2

In Exercises 1 and 3 you can check your sketch by graphing $y(x)$ in the window $-P < x < 2P$, $A - 2B < y < A + 2B$.

1. $y(x) = 5 + 2\sin 2\pi x$
3. $y(x) = 4 + 3\sin(\pi(x - 1))$
5. $A = 7, B = 3, P = 2\pi, D = 0$
7. $A = 5, B = 3, P = 2\pi, D = 2$
9. $A = 5, B = 10, P = \pi/2, D = 1/2$
11. (a) $A = 54, B = 18$
    (b) $k = 1/6$ if t in months
    It probably is more instructive to solve Exercises 13–17 by inspection of the data to determine the function's average, period, and amplitude than to use the calculator's sine regression facility.
13. $y = 3\sin x + 4$, $A = 4, B = 3, P = 2\pi, D = 0$
15. $y = 4\sin(\pi x/2) + 3$, $A = 3, B = 4, P = 4, D = 0$
17. $y = 12\sin(\pi x/6) + 5$, $A = 5, B = 12, P = 12, D = 0$
19. $f(t) = 22.05\sin(0.52t - 1.73) + 51.38$
21. $f(t) = 15.95\sin(0.52t - 1.59) + 67.90$
23. $f(t) = 0.58\sin(0.52t - 1.18) + 12.14$
25. $f(t) = 2.77\sin(0.52t - 1.16) + 12.20$
27. The constant terms in the formulas of Exercises 23–26 are all a bit more than 12, the annual average length of days in hours at any latitude. However, we see that the coefficient of the sine term increases significantly as the latitude increases. This means that there is more *seasonal variation* in the length of days at higher latitudes. The following table exhibits this fact dramatically. (Do you see where we got these figures from the answers to Exercises 21–26?)

Latitude	Shortest Day (Hrs)	Longest Day (Hrs)
10°	11.56	12.72
20°	10.93	13.33
40°	9.43	14.97
60°	6.27	18.35

At a location on the equator (0° latitude) the length of a day should be 12 hours year round. But at 60° latitude (near Anchorage, Alaska or St. Petersburg, Russia, for instance), we have winter days only 6½ hours long—sunrise at 8:45 AM and sundown at 3:15 PM—and summer days 18½ hours long—sunrise at 2:45 AM and sundown at 9:15 PM.

Chapter 8 Review

1. Trigonometric
2. Quadratic
3. Trigonometric
4. Neither
5. Quadratic
6. Trigonometric
7. $\sin(4\pi/3) = \sqrt{3}/2$
8. $x = \pm \pi/3$ are two values;
9. $x \approx \pm 0.6449$
10. $f(t) = 23.49\sin(0.52t - 1.63) + 44.66$

Index

A

Advertising campaign problems, 316
AIDS, growth of, 304, 345
Ajax City example, 43 ff.
Akron population, 192
Albuquerque population, 317
American Eagle gold coins, 280
American Indian languages problem, 135
Amplitude of a trigonometric function, 367
Annual growth rate, factor, 104
Annual interest rate, 101
Annual percentage rate (APR), 162
Annual yield, effective, 160
Apparent temperature, 2, 36
Area:
    Circle, 21
    Rectangle,
Arrow example, 209
AT&T "one rate" plan, 54, 57
Atlanta temperatures, 349–351, 367
Atmospheric pollution problem, 193
Atmospheric pressure problem, 146
Austin population, 223
Australia, cigarette consumption, 88
Average error, 82
Average rate of change, 28, 41

B

Babylonian square root algorithm, 108
Ball tossed upwards, 198, 237
Bank failures problem, 234
Base constant, 127
Baseball example, 217
Bell South long distance plans, 54
Benford's Law, 180
Bent tree problems, 254
Best fit, 79, 83, 137, 183, 225
Blackberry device problem, 147
Body Mass Index (BMI), 8, 18, 25
Botox problem, 53
Bounded population, 305
Brazil exports problem, 112
Buffalo population, 90, 145
Bureau of the Public Debt, 5
Buttermilk problem, 147

C

Caffeine half-life example and problem, 119, 122
Cake example and problem, 142, 147
California earthquakes, 16
Canada, household spending, 121
Cancer deaths problem, 135
Car stopping distances, 13
Carbon-14 problem, 180
Carrying capacity, 311
Cassette tape sales problem, 92
CD sales problem, 92
Celsius temperature, 40, 56
Center of population of U. S., 150, 195
Central conic equation example, 294
Change, constant rate of, 40
Chaos, 339
    And period doubling, 335
Chaos game, 347
Charlotte population, 74, 77, 80–81, 85, 192
Chesapeake (VA) population, 192
China:
    GDP problem, 146
    Trade volume, 139, 195
Cholesterol-fat example, 9
Cigarette smoking data, 35
    Consumption example, 87
Circle, area of, 21
Circle definitions of sine and cosine, 353
Circle equation example, 293
Cleveland population, 145
Climate (versus weather), 348
Climate changes in U.S. cities, 376
Coefficient determinant, 261
Coefficient matrix, 264, 274, 277
Coin example and problems, 267, 269
College Board, 8, 26
Colorado Rockies, 238
Common logarithm, 167
Composition of functions, 360
Compounding of interest:
    Continuously, 157, 159
    n times annually, 154
    Quarterly, 153
    Semiannually, 152
Concave downward curve, 31
Conception, odds of, 11, 30
Constant percentage growth, 101
Constant rate of change, 40
Consumer credit problem, 53
Consumer Price Index (CPI) problem, 93
Continuous growth and decay, 162
Continuous growth model, 181
Continuous growth rate, 162, 175
Continuously compounded interest, 157, 159
Cookbook example, 22, 49
Cookie sales investigation, 36
Cooling, Newton's law of, 142
Corvette prices, 16
Cosine and sine graphs, 356
Cosine function, 352
Cost curves investigation, 302
Cotton production data, 223
Crude oil imports data, 222
Cubic equation, 241
    Solving, 242
Curve-fitting and regression:
    Exponential, 137, 181
    Linear, 84
    Logarithmic, 188
    Logistic, 317
    Polynomial, 289 ff.
    Quadratic, 226
    Sine, 369

D

Danbury-Hartford distance example, 68
Data Modeling:
    Exponential, 137, 181
    Linear, 78, 286
    Logarithmic, 187

Data Modeling (*continued*)
  Logistic, 317
  Polynomial, 289 ff.
  Quadratic, 223, 286
  Trigonometric, 367
Decreasing function, 26
Degree-radian conversion, 353
Demand-price function, 95
Dependent variable, 23
Depreciation, straight line, 25
Determinant method (solution of equations), 262, 273
Determinant formulas, 274
Determinants, 261
Diabetes (diagnosed in U.S.), 33
Discrete population model, 331
  Fractional population model, 337
Domain of a function, 6
Dow Jones Average, 9

E

e (the famous number), 156
Effective annual yield, 160
El Paso population, 192
Elimination, method of, 258, 271
English language, evolution of, 134
Equation:
  Cubic, 241
  Exponential, 174
  Graph of, 61
  Logarithmic, 175
  Quadratic, 202
  Quartic, 241
  Trigonometric, 358
Error:
  Actual and predicted, 80
  Average, 82
  Sum of squares of, 80
Euros (and dollars), 25
Exponential and logarithmic equations, 174–175
Exponential function, 127
  As an inverse function, 172
  Natural, 157
Exponential growth and decay, 162, 184
Exponential model, 126
Exponential regression, 137, 181
Exponents, laws of, 165–166
Extrapolation, 86, 292

F

Factor theorem, 204
Fahrenheit temperature, 40, 56

Falling and rising lines, 61
Fat-cholesterol example, 9
Fertility odds, 11, 30
First marriage, 33
First-class postal rates, 25, 74
Fish stocking example and problems, 311, 316
Fitting:
  Exponential models, 137, 179
  Linear models, 84
  Logarithmic models, 185
  Logistic models, 317
  Polynomial models, 289 ff.
  Quadratic, 226
  Trigonometric models, 369
Flat tax problem, 110
Floating ball problems, 255
Ford Explorer, value of, 195
Foreign-born population, 240
Fractal, 343, 347
Function, 4
  Composition, 360
  Cosine, 352
  Decreasing, 26
  Demand-price, 95
  Domain, 6
  Exponential, 127, 157, 184
  Graph of, 63
  Increasing, 26
  Input, 4, 23
  Inverse, 174
  Linear, 41
  Logarithm, 172
  Logistic, 306
  Natural decay, 115
  Natural exponential, 157
  Natural growth, 104, 127, 129
  Output, 4, 23
  Periodic, 367
  Piecewise-linear, 66
  Price-demand, 95
  Quadratic, 199
  Piecewise-linear, 66
  Range, 6
  Sine, 352
  Square root, 24
  Squaring, 23
  Step, 71
  Trigonometric, 350, 352, 366
  Value of, 4
Function machine, 23

G

Garland population, 91
Gasoline, family income spent on, 91
Gasoline cost problem, 112

Georgia:
  Cotton production data, 223
  State revenue, 91, 106
Grant, President Ulysses S., 38
Graph of:
  Equation, 61
  Function, 63
  Linear function, 47, 60
  Step function, 71
Graphic, numeric, and symbolic viewpoints, 68, 124
Graphic solution of equations, principle of, 242
  Of pairs of linear equations, 256
Graphical maximum-minimum method, 219
Greeley, Horace, 150
Growing season versus latitude, 15
Growth and decay, natural, 162, 184
Growth of Internet, 98, 149
Growth rate, continuous, 162, 172

H

Half-life, 117
  Of caffeine, 119
Hartford-Danbury distance example, 68
Heart disease death rates, 121
Heat index, 2, 36
High and low points on parabolas, 214–215
Higher-degree polynomial models, 289 ff.
Hispanic unemployment rates, 212
HIV virus and deaths, 304, 345
Horizontal and vertical lines, 62
Hospital occupancy, 12

I

Ibuprofen problem, 122
Income tax example, 71
Income tax problem:
  Connecticut, 76
  District of Columbia, 77
  Louisiana, 76
  Michigan, 77
  Mississippi, 76
Increasing function, 26
Independent variable, 23
Input, 23
Input-output process, 109
Interest rate, nominal, 161
Internet, growth of, 98, 149
Interpolation, 86, 292

Inverse functions, 174
Investigations (projects):
    Changing climates of U.S. cities, 376
    Cost curves and minimum average cost, 302
    Exploring rate of change, 36
    Quadratic models in baseball, 238
    Modeling used car prices, 96
    Population projections, 149
    Sierpinski triangle, 346
    Used car loan, 196
Iteration, iterative formula, 108
    Babylonian, 108

K

Kaiser Permanente, 8
Kelvin temperature, 65
Kentucky Derby earnings problem, 146
Knight-Ridder Newspapers, 95

L

Ladder examples and problems, 250
Languages, American Indian, 135
Latitude and length of days, 374–375
Laws of exponents, 165–166
Laws of logarithms, 171
Lazarus, Emma, 240
Least squares polynomials, 294
Lexington (KY) population, 54
Life expectancy problem, 55, 91, 194
Light intensity problem, 145
Limiting population, 305, 311
Line:
    Point-slope equation, 57
    Rising and falling, 61
    Slope-intercept equation, 47
Linear function, 41
    Distinction among equation, formula, and graph, 60
    Graph of, 47, 60
Linear modeling applications, 86 ff.
Linear equations, pairs of:
    Applications, 266
    Determinant solution, 261
    Graphical solution, 256
    Method of elimination, 258
    Possible solutions of, 256
    Symbolic solutions, 260
    Using matrices, 264
Linear population model, 42
Linear regression, 84

Linear systems of equations, 270
    Applications, 280
    Method of elimination, 258
    Solution using determinants, 274
    Solution using matrices, 277
Logarithm,
    As an inverse function, 172
    Common, 167
    Natural, 168
Logarithmic and exponential equations, 174
Logarithmic model, 189
Logarithmic regression, 188
Logarithms, laws of, 171
Logistic function, 306
Logistic model, 310
Logistic regression, 317
Los Angeles Dodgers, 94
Louisville population, 193
Lung cancer example, 87

M

Mandelbrot, Benoit, 340
Mandelbrot iteration, 342
Mandelbrot set, 342
Marine species in aquaria, 91
Marriage, first, 33
Matrices, 263
Maximum and minimum values, 214
Maximum-minimum method, graphical, 219
Medical equipment spending problem, 134
Method of elimination, 258, 271
Miami population, 192
Microcomputer transistors problem, 134
Migration, unauthorized, 55
Milk consumption problem, 236
Milwaukee population, 66
Missouri farms, 32
Mixture problems, 267, 270, 281, 284
Model:
    Continuous growth, 181
    Exponential, 126
    Linear, 42, 286
    Logarithmic, 189
    Logistic, 311
    Natural growth, 104, 129, 162, 181
    Piecewise-linear, 67
    Polynomial, 289 ff.
    Quadratic, 205, 286
    Trigonometric, 366
Money in circulation problem, 235
Mountain Dew, 75

N

Natural decline, 114, 131
Natural exponential function, 157
Natural growth applications, 175
Natural growth, function, 104, 127, 129, 131
Natural growth and decline, 104, 131, 162
Natural growth model, 104, 129, 162, 181
Natural logarithm, 168
Newark population, 90
Newton's law of cooling or heating, 142, 184
Nominal interest rate, 161
Nuclear power plant problems, 135, 145, 193, 194

O

Ohio State University graduation rate, 95
Oil field problem, 2
Old Faithful geyser, 38, 95
Olympics, 92–93
Optimal model:
    Exponential, 183
    Linear, 83
    Quadratic, 225
    Natural growth, 137
Oscillating population, 335
Output, 23

P

Parabolas, translated, 200
    Finding vertex of, 216
    High and low point on, 214–215
Pascal triangle, 346
Percent, symbol, 100
Percentage increase, 100
    Decrease, 113
Percentage rate, annual (APR), 162
Period, of population oscillation, 336
Period doubling, 339
    And chaos, 335
Periodicity of sine and cosine functions, 356
Phoenix heat, 2, 36
Phoenix population, 145
Piecewise-linear function, 66
    Cost function, 68
    Income tax, 71
    Model, 67
Point-slope equation, 57
Political modeling, 320

Polynomial, least squares, 294
Pollution, atmospheric, 193
Popcorn tray examples and problems, 246, 254
Population model:
  Continuous growth, 162
  Discrete, 331
  Linear, 42, 286
  Logistic, 311, 324
  Natural growth, 175, 323
  Quadratic, 205, 286
  Polynomial, 289 ff.
Population of:
  Akron, 192
  Albuquerque, 317
  Austin, 223
  Buffalo, 90, 145
  Charlotte, 74, 77, 80–81, 85, 192
  Chesapeake (VA), 192
  Cleveland, 145
  El Paso, 192
  Garland (TX), 91
  Lexington, 54
  Louisville, 193
  Miami, 192
  Milwaukee, 66
  Newark, 90
  Oklahoma City,
  Phoenix, 145
  Providence, 59
  Raleigh, 145
  Rochester, 74
  St. Louis, 10, 53
  San Antonio, 145
  San Diego, 90, 145
  San Francisco, 74
  San Antonio, 145
  Tucson, 74
  United States, 28, 75, 104, 106, 136, 181, 227, 321 ff.
  Virginia Beach, 66
  World, 286 ff., 295 ff.
Population projection investigation, 149
Possible solutions of linear pairs of equations, 256
Price-demand function, 95
Private airports, lighted runways, 222
Projectile motion, 208
Providence population, 59
Public debt, 5

Q

Quadratic equation, three cases, 203
Quadratic formula, 202
Quadratic function, 199
  Factor theorem, 204
  Graph of, 200–201

Quadratic model, 205
  Baseball investigation, 238
  Population, 206
  Projectile motion, 208
Quadratic regression, 226
Quartic equation, 241
  Solving, 248

R

Radioactive decay, 130
Radian measure of angles, 353
Radian-degree conversion, 353
Raleigh population, 145
Range of a function, 6
Rate of change:
  And slope, 47
  Average, 28
  Constant, 41
  Constant percentage, 101
  Investigation, 36
Regression and curve-fitting:
  Exponential, 137, 181
  Linear, 84
  Logarithmic, 188
  Logistic, 317
  Quadratic, 226
  Sine, 369
Relative humidity, 2, 36
Rising and falling lines, 61
Roast example and problem, 185, 193
Rochester population, 74
Rodents in a forest, discrete model, 330
Rule of 72, 111
Rule of three, 124
Rumor problems, 316

S

St. Louis population, 10, 53
San Antonio population, 145
San Diego population, 90, 145
San Francisco population, 74
SAT scores, 8
Scandinavian countries, cigarette consumption, 87
Seven-Up sales, 17
Sierpinski triangle investigation, 346
Sine and cosine graphs, 356
Sine function, 352
Sine regression, 369
Slope as rise/run, 47
Slope-intercept equation, 47
Solution matrix, 264

Solution matrix, 278
Solving equations:
  Cubic equations, 242
  Exponential and logarithmic equations, 174–175
  Quadratic, 202–203
  Quartic equations, 248
  Trigonometric equations, 358–359
Sphere equation problem, 299
SSE (sum of squares of errors), 80
Statue of Liberty, 240
Step function, 71
Straight line:
  Depreciation, 25
  Point-slope equation, 57
  Slope-intercept equation, 46
  Types of, 62–63
Subway sandwiches, 9
Super Bowl commercial, 4
Symbolic viewpoint, 21

T

T-Mobile cell phone service, 19
Temperature:
  Apparent, 2, 36
  Celsius and Fahrenheit, 39–40, 56
Texas Rangers pitchers, 15
Tobacco production problem, 234
Tooth Fairy problem, 148
Topeka temperatures, 348, 376
Translated parabola, 200
Trigonometric function, 350, 352, 366
  Amplitude and period of, 367
Trigonometric model, 366
Trigonometric graphs and periodicity, 356
Tucson population, 74
Tuition costs, 26

U

Unauthorized migration, 55
United Auto Workers (UAW) membership, 17, 36
United States:
  Born abroad, 28, 240, 302
  Center of Population, 150, 195
  Life expectancy, 91, 194
  Money in circulation, 195
  Oil production, 148
  Population, 28, 75, 104, 106, 136, 181, 227, 321 ff.
  Public debt, 5
UPS next day delivery rates, 70

Used car prices, 96
Used car loan investigation, 196

V

Value problems, 266, 269, 280, 284
Variables, dependent and independent, 23
Vehicle leasing data, 212, 222
Vertex, of a parabola, 200, 216
Vertical and horizontal lines, 62
Viewpoints: graphic, numeric, symbolic, 68, 124

W

Waffle House example, 3
Warrant Officer pay scale, 93
Word problems, 266
World population, 286 ff., 295 ff.

Y

Yellowstone National Park, 38
Yucca City example, 48